GSM and Personal
Communications Handbook

For a complete listing of the *Artech House Mobile Communications Library,* turn to the back of this book.

GSM and Personal Communications Handbook

Siegmund M. Redl
Matthias K.Weber
Malcolm W. Oliphant

Artech House
Boston • London

Library of Congress Cataloging-in-Publication Data

Redl, Siegmund M.
  GSM and personal communications handbook / Siegmund Redl,
Matthias Weber, Malcolm Oliphant
    p.   cm. — (Artech House mobile communications library)
  Includes bibliographical references and index.
  ISBN 0-89006-957-3 (alk. paper)
  1. Global system for mobile communications. 2. Personal
communication service systems.   I. Weber, Matthias K.
II. Oliphant, Malcolm W.   III. Title.   IV. Series
TK5103.483.R44   1998
621.3845'6—dc21   98-4710
                                         CIP

British Library Cataloguing in Publication Data

Redl, Siegmund M.
  GSM and personal communications handbook—(Artech House mobile
  communications library)
  1. Global system for mobile communications
  I. Title   II. Weber, Matthias K.   III. Oliphant, Malcolm W.
  621.3'8456

  ISBN   0-89006-957-3

Cover and text design by Darrell Judd.

© 1998 ARTECH HOUSE, INC.
685 Canton Street
Norwood, MA 02062

International Standard Book Number: 0-89006-957-3
Library of Congress Catalog Card Number: 98-4710

10 9 8 7 6 5 4 3 2 1

Contents

Preface

Use of the *global system for mobile communications* (GSM) continues to spread throughout the world. It works, it is efficient, and it is well liked. As is true of any mature but vital and growing system, the services and equipment based on the GSM specifications are still evolving to accommodate its new users and operating environments. The new services, improvements, applications, and products are the new flavors offered in the GSM ice cream stand of wireless telecommunications networks. New terminals feature increased standby and talk times while their sizes shrink and their prices fall. The combination of competitive pricing and access to a growing menu of services, which is attractive to a wider variety of users, marks the transition of GSM from a high-end offering to a *consumer* product orientation.

Interesting features that go far beyond the point-to-point voice conversation link typical of traditional wireless services are a reality in GSM networks. Sophisticated data services with access to the Internet, video connections, ISDN links, and supplementary services, which are expected in wireline digital networks, are becoming a reality in GSM-based networks.

Why have the authors of the previous work *An Introduction to GSM* (Artech House, 1995) chosen to write again on the same subject? Considering the metamorphosis just described, one answer is clear: GSM remains an evolving standard. Its use and applications are no longer restricted to the initial aims of the early experimenters, the mothers and fathers of GSM. Another reason appears when we compare the coverage of our first book with this volume. *An Introduction to GSM* was meant to be—and was widely accepted as—a first confrontation with the subject. It focused on *what* GSM is rather than on *how* it works and *how* it can bring a growing catalog of services and applications to its users. The subject was treated so as to explain the system architecture and radio techniques used to convey information from one point to another. Some chapters on testing—still an important subject, especially because GSM is supposed to be an open standard—were added to enhance understanding, and further illustrate the techniques and processes. The first book, then, can be considered an illustrative brochure describing GSM to those considering its purchase and use. This volume is a logical extension of the earlier book; it is a user's manual for those who wish to exact efficiency and new features from GSM.

When *An Introduction to GSM* came to life, GSM services, as well as those of its *digital cellular system* (DCS) cousin, were struggling in their infancy. Networks were just starting to offer their services based on the status of the standards termed, at that time, *GSM Phase 1*. The situation has completely changed today. GSM has moved from the showroom to the customer's garage. Accessories have been added and copies have been made. Today we see three different standards based on the core GSM technology: GSM 900, DCS 1800 (GSM 1800), and PCS 1900 (now called GSM-NA). Today, network operators are introducing *GSM Phase II* services and products. Noting that GSM has found favor far beyond its original European roots, researchers and developers in the industry and standardization bodies are working on *GSM Phase 2+* services, features, and products. What are these new services, what do they offer, and how do they work?

The world has adopted GSM as the most widely deployed digital cellular standard with expanding interworking and roaming capabilities. With wider deployment comes greater variety. Since the publication of our first book, we have seen other systems employing different wireless access techniques adopt key GSM properties. The number of new applications, varieties, and flavors of GSM grows with its acceptance. As the world's dominant wireless protocol, the industry has accepted the responsibility of exploring ways in which GSM can interwork with other wireless access systems. Success in these efforts will yield a platform capable of even greater capacity for the improved services and bandwidth allotted to carry them.

Still, GSM is not the only technology setting out to win the hearts and cash of those who want to use or offer digital wireless communications services. Competitive technologies such as *code division multiple access* (CDMA) are poised to take their share of the market as they offer their own set of applications. Whatever these systems may be—cellular radio, *personal communication systems* (PCS), *specialized mobile radio* (SMR), *wireless in the local loop* (WLL), cordless phones, or even satellite-based systems—GSM will thrive. The better these systems work together for the benefit of their users, the more all the providers, whatever technology they select, will win.

What are the important issues that drive the variations and added features? What are the trade-offs and compromises? What are the limitations? Where are the solutions? These matters are treated in the widely accepted style of the authors' first GSM volume. The answers are in the details. Subjects like digital baseband technology, new radio techniques and implementation schemes, and intelligent networks are shrouded in specialized language and mathematics. This increased specialization, which is not unique to mobile radio, frustrates managers, marketing specialists, and others taxed with the responsibility of financing and deploying GSM networks, and building the devices and the equipment on which these networks depend. Technical specialization tends to stifle effective communication among people. Just as with our first book, this one is written for those who must manage GSM projects and the growing variety of technical specialists needed to run them. Jargon and specialized mathematics are *avoided,* and new terms are explained as they are introduced. Your authors have worked diligently to explain obscure but important concepts, processes, and devices in clear language without the aid of sophisticated mathematics shorthand. Moreover, the explanations are animated with some of the excitement and passion of the scientists and engineers. The actual work must, however, be left to the technologists and the specialists, for it is only with the appropriate tools (mathematics, software, jargon, and experience) that the devices and features in the GSM networks can be manipulated efficiently and designed at a price such that many people can afford them. The treatments in this book, therefore, include appropriate references for each of the subjects covered. Some of the matters, particularly the radio techniques covered in the last chapter, are so vast in their scope that small tutorials to carefully selected references are included. Just like our first book, this one is an initial confrontation, a guide, or an orientation for further reading. Students and engineers new to GSM and digital mobile radio will, therefore, find this book helpful.

Even as we finished the chapters, we knew there was still lots to tell about GSM. This book illuminates just another episode in its continuing story. The reader is invited to view this latest volume as a complement to *An Introduction to GSM*. Though some of the chapters make direct reference to the earlier book, anyone familiar with the fundamentals of GSM will be rewarded here.

This book has 11 chapters sorted into three parts. Part I has three chapters. The first chapter is a status report on where GSM is deployed in the world today and where it is likely to be accepted in the future. Some market sizes and other figures are offered. The confusion over the PCS and PCN designations is explored in the light of competing wireless systems. Chapter 2 traces the phased deployment of GSM. A means toward understanding current and future enhancements and system variations is possible when we understand how GSM adapted to conflicting national and regional requirements from its original, narrow European goals. Chapter 3 looks at the influence of new technologies, such as CDMA, other wireless services, such as TETRA, and certain social and economic realities, such as the North American market, on GSM as well as how GSM tempers those influences.

Part II consists of Chapters 4 through 10, which explain in detail the huge number of features and services of GSM. GSM introduces user services and improvements in phases. Chapter 4 starts by describing these services as it traces their phased introduction into the networks. Chapter 5 describes how teleservices and bearer services are handled in networks. Chapter 6 covers *short message service* (SMS), a service that is not found in wireline networks and a popular feature that removes so much of the intrusive nature of basic cellular service. Chapter 7 explores *supplementary services* (SS), which are those features common in many digital wireline networks that users find so helpful in their busy lives. Caller ID is an example. The subscriber identity module, also referred to as SIM card, which is viewed with growing interest and envy by proponents of some newer competing mobile radio technologies, is thoroughly explained in Chapter 8. Chapter 9 covers the latest features introduced into GSM, such as the SIM application toolkit, CAMEL, and features introduced for railway applications. Chapter 10 brings us back to the network side, describing parameters stored in various switches or registers. This discussion reveals the mysteries of when and how the charging clock ticks and how calls are actually routed through the network.

Part III concludes the book with only one chapter, Chapter 11, which covers some salient technology issues. The emphasis is on the handsets. GSM handsets are the most visible part of the network, and their variety and high quality have played a major role in the acceptance of GSM in so many markets. The handset's ability to carry all the features described in Part II is seen as a miracle by many of the people who purchase these wireless wonders in all their colors and shapes.

Acknowledgments

The authors wish to acknowledge the help of Dr. William R. Gardner of LSI Logic whose advice on the details in Chapter 11 was critical to our work. We also thank Vinay Patel of Hughes Network Systems and Mark Vonarx of Omnipoint for their generous help with Chapter 3. The authors also acknowledge the enduring patience and support of their employers.

A sincere thank you also goes to the personnel at Artech House for their help in bringing this book into existence. Many thanks go to the *unknown* reviewer(s) for their excellent work and many important inputs, which improved the quality of all the chapters.

Siegmund Redl offers his special thanks to his wife Johanna and son Christoph for their patience and support. Matthias Weber likewise extends special thanks to his wife Ilse and daughter Laura, who endured his sporadic hours, days, and weeks of absences while completing this work.

PART

I

GSM in the light of today

Contents

The changing scene—again

To the delight of its supporters and the surprise of its detractors, the *global system for mobile communications* (GSM) has, after a few false starts and sputters, found its place in the communications world—and what a place it is. GSM has brought low-cost and reliable mobile communications to most of the countries of the world. Its features and options are rich enough to satisfy the peculiar and disparate needs of the users in all of the GSM countries. A system definition that came to life through a Pan-European initiative for a single open cellular standard has spilled over the borders of Europe and continues to conquer new territory around the world. Not only is new territory conquered geographically, but also includes new user groups within countries (target customers), new services, and new applications. What happened? Where did all these 50 million phones and users come from only five years after the introduction of the first GSM-based cellular services?

3

What were the initial intentions for establishing the standard, and what were the early experiences? How did the GSM standard succeed in Europe only to then spread into new areas, new frequency bands, and new applications, and how will GSM continue its march around the globe? What is GSM's future? What is GSM's place in *personal communications systems* (PCSs)? And what is a PCS anyway? In this and in subsequent chapters, we will find answers to these questions. GSM is not what it used to be only five years ago, and it will change into something else next year. In this chapter we pause to see what GSM means today. We review market figures and perspectives, and discuss new applications and marketing schemes. We also dig into some technology and standardization issues, and consider key definitions of services and features around GSM.

1.1 The digital cellular evolution

With the introduction of digital technology as the second generation of wireless communications, the world experienced a tremendous growth in cellular subscribers. The potent combination of added value service features and applications, increased capacity, expanded coverage, and improved quality of service continues to make mobile telephony an indispensable commodity. Once users try reliable private communications unconfined by wires, they do not want to give it up.

Deregulation, growing demand, marketing prowess, competition, and open standards are the pillars on which the success of digital cellular rests [1]. Today, the term *personal communications* usually applies to new systems and services (PCSs) that are offered to an increasing portion of our world's population. The underlying technologies for personal communications are digital cellular ones, with GSM being the most widely accepted form today and for many years to come.

But what is the difference between PCS and cellular? One way to clear the confusion is to regard PCS as a deployment scheme that occurred after cellular. Because the deployment added significant additional wireless capacity, something had to attract new subscribers to fill the additional capacity: if cellular was for the business users and the wealthy, then PCS was for the mass market. From a pure technology perspective, *personal communications* describes a set of services that a customer might expect. *Cellular* refers to a range of technological solutions that may be used to deliver such services. The vast majority of PCS subscribers are receiving these services by using a cellular technology. However, there are a minority of "noncellular" technologies that can also be a basis for personal communications services, for example, cordless systems. Some PCSs may use more than one technology to deliver a comprehensive service. For now, let's consider PCS as an approach that can bring mobile communications services to a broader consumer market. *PCS* originally stood for a North American initiative with new spectrum allocations in the 1900-MHz band, thus the term *PCS 1900* for the GSM 900 derivative in North America. The original *PCN* term was introduced in the United Kingdom before PCS, and referred to the *personal communications networks* licensed in the 1800-MHz band; whereas the PCS term originated in the United States and originally referred to spectrum licenses auctioned in the 1900-MHz band. As such, the United Kingdom hosted the world's first PCSs, which were—and still are—referred to as PCNs, personal communications networks.

One alternative for the awkward PCS 1900 term is *GSM-NA*, which stands for GSM North America. For the general user who neither knows nor cares what a hertz is, the new 1900-MHz allocations are simply extensions to the cellular network. But because the new frequency allocations were auctioned off to new operators, fierce competition in certain trading areas arose for new subscribers. Once we review the whole market, a general approach for defining personal communications services and related marketing efforts is discussed later in this chapter.

Eventually, innovative marketing concepts shake each other out, buzzwords settle in our minds, and the imaginative blurs into the familiar. What finally counts in the end is what service we can get and what it will cost.

Before we commence our discussion of the different flavors of personal communications through digital cellular or simply wireless access systems, we need to look at how the market developed its services and their prospects for the future. Then, we need to look at GSM's dominant place in the picture.

1.2   Basic market figures
and the system standards

The dynamic market for personal wireless communications shows dramatically increasing growth rates. At the end of 1995 more than 85 million users were subscribing to cellular telephone services. The year 1996 saw more than 133 million users. Predictions for the year 2000 and beyond are constantly continuing to rise, and they currently range between 300 and 500 million subscribers.

For comparison, in 1996 the world's population (more than 5 billion) had only about 800 million fixed-line telephone installations, which was a 16% penetration. The cellular penetration accounted for only around 2% of the world's population. Cellular subscribers, worldwide, exceeded the 100 million mark in 1996 with plenty of momentum to achieve multiples of this figure by the turn of the century. Penetration in some countries (Scandinavia and Australia) is already at 30% with a trend toward 40%.

Another trend is for revenues from mobile services to exceed those achieved by fixed-line services even though the amount of traffic generated through mobile phones is much less than in the fixed networks.

Private mobile radio (PMR)—which is increasingly dominated by so-called trunking systems, cordless telephony, paging, and messaging—*wireless local-area networks* (WLANs), and *wireless in the local loop* (WLL) are other wireless business sectors that either saw a proportional growth with cellular or still have a huge growth potential. The growth potential is particularly bright for WLL, because many developing regions are now provided with flexible, uniquely tailored, and cost-effective wireless access to telephony services. Deregulation, new spectrum allocations, and new network operators will install WLL systems to provide services in developed countries too. The fixed-wire plant is a very expensive structure that requires lots of maintenance and is sensitive to storms and vandalism. The "last mile" of the system, which provides the access to the home and office, as well as installations within offices and factories, is often more efficiently covered by radio transmission.

Whereas the majority of users still subscribe to analog cellular networks (67 million or 79% in 1995 [2]), digital systems are catching up. It is expected that in the year 1999 about 80% of the new mobile phones sold to customers will be based on some kind of digital technology.

Figure 1.1 depicts the growth of digital cellular subscribers—also at the expense of analog—from 1991 to 2000 (the figures for 1996 through 2000 were estimates) [2]. The analog systems comprise the well-established technologies: AMPS, TACS, NMT (450, 900), as well as some other minor systems. The digital systems include GSM and its derivatives (digital cellular system *DCS 1800* or *GSM 1800* and *PCS 1900* or *GSM-NA*), dual- and single-mode digital AMPS or TDMA (in the United States this is called *D-AMPS* or Interim Standard IS-54/136), personal digital cellular (abbreviated PDC in Japan) and code division multiple access (CDMA; also called *IS-95*). AMPS-based systems with a digital overlay (TDMA or CDMA) are widely deployed in North America and in some South American countries (Argentina, Brazil, Peru, etc.) to take advantage of the large number of AMPS phones already in use in these regions. Some Asian countries and a few other regions employ the same scheme on a diminished scale. The CDMA version is found very successful in South Korea and Hong Kong, and the TDMA version in Israel. We can expect to find such hybrid systems wherever a successful AMPS system is already deployed. IS-95-based CDMA will be seen in Japan, as this technology was chosen to be overlaid with existing analog (TACS-based) technology in order to increase network capacity (see discussion in Section 1.2.1).

GSM-based systems are not hybrid ones, and they are found almost everywhere: Europe, the Middle East, Africa, many Asian countries, and Australia. In some regions we find a mixture of cellular systems that

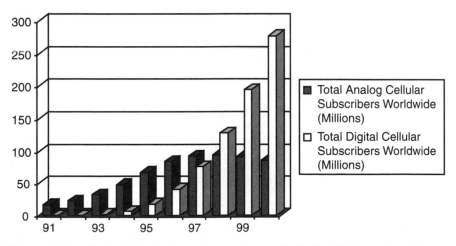

Figure 1.1 Subscribers in analog versus digital cellular networks worldwide (*From:* [2]).

simply coexist next to each other. Australia has both AMPS and GSM systems, and the AMPS system has been decreed to be taken out of service by 2000. No match can be found for the cellular salad in Hong Kong, where in 1997 we found the following system standards in operation: AMPS, CDMA, D-AMPS/TDMA, GSM, TACS, and CT2 (telepoint). Additional PCS licenses shall be awarded.

The Scandinavian countries (Sweden, Norway, Denmark, and Finland) have cellular penetration rates between 20% and 30%, and some are on their way to 40% in 1997. The Scandinavian trend is a predictor for the rest of the world. Worldwide, there will soon be more new subscribers signed up for services supplied by wireless systems than are being connected to pure wireline networks. Cellular service will replace certain fixed-line services for a variety of applications.

GSM networks will prevail in the year 2000, with a conservative estimate of 157 million subscribers representing more than 43% of the total cellular market [2]. GSM will grow with the general market. Figure 1.2 displays the worldwide trend with GSM's share in subscriber numbers. As was the case in Figure 1.1, the figures for 1996 through 2000 were estimates. The split between digital technologies (GSM, CDMA, IS-54/136, and PDC) predicted by one source [2] is typical of all those who analyze the market. The subscriber share will be about 57% for GSM, 22% for CDMA, 15% for TDMA (IS-54/136), and 7.5% for PDC [2].

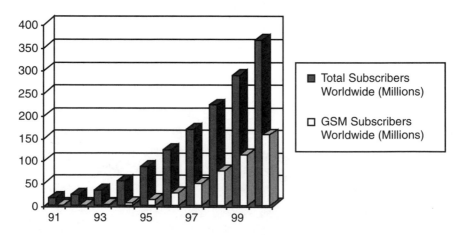

Figure 1.2 GSM subscribers versus total (analog and digital) cellular subscribers worldwide (*From:* [2]).

So, CDMA technology (U.S. Interim Standard IS-95, which is also applicable for North American PCS as SP-3384) is coming in second according to the current estimates of market penetration. Though there have been some stunning successes, particularly with PrimeCo (a large consortium of PCS operators in the United States), this technology is several years behind its digital rivals in terms of system deployment and product availability. Claims of CDMA's superiority and its potential to become a world standard have been constantly reduced to more realistic views in the past 3 years. With every year of delay in the broad introduction of CDMA-based services, and with more understanding of the cold technical issues and problems, the industry backers had to eventually realize that CDMA was not the magic solution for all wireless markets. The initial wild enthusiasm was replaced with a sober consideration of CDMA's real advantages, for example, a substantial reduction in frequency planning tedium and some new vocoder technology. As the industry turns its attention from exciting marketing promises to tedious engineering reality, it will discover how to take advantage of CDMA's benefits and improvements over today's TDMA technologies, and a great potential in many markets may be realized. When functional networks and attractive products finally become available at the right times, with adequate quality, and in the correct volumes, CDMA will have its day. Unlike the situation in South Korea where CDMA is the official cellular technology, and the European situation where GSM is the decreed protocol, North America was and still is a battlefield between GSM, CDMA, and IS-136.

More than 50% of the new PCS operators in the United States (covering an adequate potential subscriber base) have decided to and have already started to deploy IS-95 CDMA-based technology. The remainder is GSM and IS-136 territory. The mathematics behind how many "pops" are covered by which network operator and therefore by which technology is sometimes confusing. Let us stick with the simple statement that, for North American PCS, CDMA takes the lead, followed by GSM-NA and then IS-136.

The American regulators take no sides. The new PCS operators need to sort through all the proposed technologies for which sound arguments can be assessed. The religious has to eventually give way to the practical, claims have to be proved, and time has to be the arbiter. Decisions on which technology to support and deploy are also dependent on *intellectual property rights* (IPRs) and patents. GSM is typical of a system in which IPRs are shared among industry players who agree on certain conditions for licensing them. When agreement is reached among the participants, then everyone can participate in the system's development. Even though IS-95 is an open standard, most of the IPRs for CDMA are in the hands of just a few companies and individuals. When most of the IPRs are owned by a single player, innovation tends to be stifled because licensing restrictions cramp the resources of players who could otherwise make important contributions to the development of a system. A balance needs to be found between the valid interests of IPR holders who want to collect licensing fees, and the industry that needs to design, build, and market low-cost mass products. Worldwide, more than 40 major telecommunications equipment manufacturers have licenses for IS-95 CDMA. The balance is tested in the achievable volumes of products. Volumes, however, can only be achieved with clear industry commitment, functional products sporting attractive services, competitive pricing, and an early market presence. Claims that the sheer use of a certain technology will make things work out well do not count when it comes to investing large amounts of cash. Because CDMA is an innovative system worthy of serious consideration, it will be refined and a balance in recovering the costs of its development will be struck. GSM also has its IPRs whose holders are spread throughout the industry. Cross-licensing of patent rights is common practice among the holders. Eventually, CDMA's proponents and supporters may join forces to contrive a way in which it can coexist with the huge (and still growing) GSM deployments already in about 100 countries. If and when they do, CDMA will grow to be a major carrier of cellular traffic.

1.2.1 Cellular and personal communications services: market presence and potential

We just need to look at Figure 1.3 (the present) and Figure 1.4 (the future), taking into account subscribers on all cellular-based technologies including PCS and PCN, to see which regions have grown their subscriber bases. As you examine the figures, please note that there is a general growth of the whole cake in Figure 1.4 by a factor of approximately 4! Please also note that some regions are growing faster in population than others, and that the whole demographic behavior of such a wireless

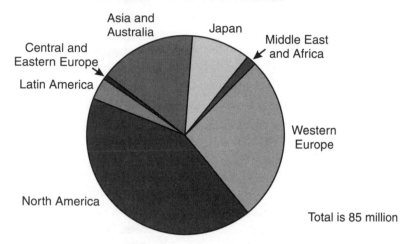

Figure 1.3 Cellular subscribers worldwide in 1995 (*From:* [2]).

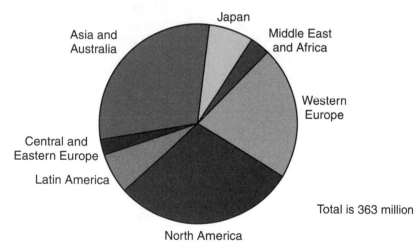

Figure 1.4 Cellular subscribers worldwide (including PCS) in 2000 (*From:* [2]).

subscriber base system is very dynamic in several planes, and difficult to model and describe in words. We want to look at two snapshot sample situations here, the present and the year 2000, with predictions of market evolution over the next 5 years.

Western Europe and North America   Today, Western Europe and North America account for the majority of cellular subscribers worldwide. With different well-accepted analog standards such as AMPS, NMT, TACS, Radiocomm 2000, and the C-Net system, and with the introduction of GSM in Europe with a strong foothold in countries such as the United Kingdom, Germany, Italy, and France, Western Europe had approximately 23 million subscribers in 1995. This was 27% of the worldwide subscriber base. The expected growth in Europe is predicted to be almost 80 million subscribers by the year 2000. As large as this number is, it accounts for only 22% of the increased worldwide subscriber base by 2000 [2]. This is due to the explosive growth in other regions.

In 1995, the United States and Canada had 35 million subscribers, which accounted for more than 40% of all the world's cellular users. Even as market penetration will grow substantially into 2000, the proportion of the world's cellular users represented by North America will shrink.

Asia and Japan   Asia, especially the "four tigers" (South Korea, Singapore, Hong Kong, and Taiwan), and Japan come in third and fourth, respectively, in terms of market size in 1995, but dominate the scene in 2000.

Japan, being a special case, accounted for enormous growth rates in the recent past and will fuel even more growth in the future. Since its deregulation in 1994 the Japanese market grew quickly from 8 million cellular subscribers in 1995 (already with 40% digital subscribers on *personal digital cellular* [PDC] systems) and will expand further to more than 25 million subscribers in 2000. There is a good potential for the future in Japan [2]. The traditional revenue per subscriber in Japan was, until recently, relatively high due to high tariffs, which often included leases of the terminal equipment. Deregulation and market liberalization in 1994, which allowed competition to Nippon Telephone and Telegraph (NTT DoCoMo) and the sale of handsets, sparked a tremendous run on wireless services. Besides the success of cellular (HCAP/J-TACS/N-TACS/N-MATS and PDC 800 and 1500), the *Personal Handy Phone System* (PHS), which provides two-way telepoint service, is chiefly responsible for the enormous success of personal communications in Japan. After the start of services in mid-1995, there was a steady growth in subscribers up to 3.5 million in mid-1996. Expectations for 1997 are as high as 8 million with a steady and sustained increase beyond the year 2000 [3]. This enormous acceptance by mobile users is due to good coverage (mainly in cities), low prices and cost of ownership, good service quality, and popular products. PHS competes very well with cellular in Japan. Cellular is doing very well in Japan too, so well that operators fear they will run into capacity problems. This is why they investigated the use of a CDMA overlay to the current J-TACS and PDC systems. Personal communication services are also deployed by four operators in Korea in the 1800-MHz band, based on IS-95 CDMA.

The biggest potential in the rest of Asia is found in highly populated countries where fixed-line penetration is low and wireless access (through cellular and WLL) is a decisive economic factor that governs growth.

Eastern Europe   There is a lot of market potential in many Eastern European countries. Since the fall of the Iron Curtain, we have seen a plethora of regulating efforts and commercial enterprises spring up that target the supply of wireless access technology for telecommunications services.

1.2.2   Meeting the demands

We sense that people naturally tire of being restricted to a wired infrastructure—and that once they try mobile radio they will not give it up. This is not the whole story, for you cannot try something to see if you like it unless you have the means and reasons to do so in the first place. We need a more disciplined examination of what people do with mobile phones and how the services appear. To gain an understanding of the growing demand for wireless communications services, we have to look at the way access to communications services is distributed, and then note the demand for those services in the light of telecommunications marketing practice. We also need to distinguish between two separate cases in our investigations: (1) the economically developed countries and (2) the developing countries.

1.2.2.1   Developed and developing countries

In economically developed countries, mainly North America, Western Europe, Japan, and a few other places, access to wireless services is a commodity that most business users today would term a *necessity* rather than a *luxury*. The operators in such countries can count on a subscriber base that generates a reliable revenue stream through business use and an increasing contribution through private luxury use of cellular services. The notion of using mobile phones in the first place originated within tiny segments of the industrialized economies that used inefficient precellular radio systems for routine mobile communications. These systems have almost completely passed from existence today, because they were extraordinarily inefficient and expensive. Cellular systems today are efficient enough to attract and carry mass traffic at a reasonable cost. Modern marketing schemes, in general, practice in mass markets and invade cellular markets as they respond to and create different user patterns, primarily through tariffing policies.

In many developing countries and regions (particularly the rapidly developing ones of South Korea, Taiwan, Hong Kong, Singapore, and, more recently, some Eastern European regions, China, and India) the conventional telecommunications infrastructure cannot keep pace with fast economic growth and the inherent requirement for reliable mobile access to basic telecommunications services that comes with economic development. The wired infrastructure is a very expensive machine that takes years to build and lots of money to maintain. Modern wireless technology is an attractive alternative that meets the demand for basic telephone services in developing economies. Cellular services and fixed WLL technologies kick in where time and money leave off. WLL will increasingly take over the role of copper access to residential and business customers, since wireless techniques provide low-cost access very quickly at a tiny fraction of the maintenance costs of fixed wire. The wired plant has a high fixed cost for its initial installation under streets and along poles. The cables need to be replaced over intervals of a few decades, and those who desire service demand a few meters of additional cable at the edges of the fixed network. This recurring expense is very high.

The adoption of wireless has also been encouraged by deregulation and introduction of competition, a relatively new and potent combination in developing economies, and many developed ones as well.

Wireless communications systems should be regarded as enablers of economic growth. This is why their development and installation are generally welcomed and supported in most of the potentially growing economies. Local telecommunications authorities and service providers join forces with experienced operators and investors (usually in the Western economies, though more are found in Asia today) in order to speed up the deployment of systems and services. Deregulation through the granting of multiple licenses to multiple operators is another catalyst to rapid growth.

The choice of equipment and standards in developing economies is devoid of philosophy and marketing hyperbole; it is based on availability, performance, and price. Because GSM is a mature technology, it is given preference in most regions. The system works, the equipment is available at relatively low cost, there is plenty of it, and the capability of offering international roaming with a multitude of other countries and operators is regarded as a political and social benefit.

1.2.2.2 Market implications for cellular and personal communications equipment and services

Whereas in Europe and Asia GSM-based systems are well represented, other markets, such as Japan and traditionally AMPS-based countries, are in favor of other competing digital cellular standards. Because the deployment of a TDMA overlay to AMPS in the United States was a disappointment in terms of market acceptance, and technical and commercial benefits to the operator were slow in coming, we can see some slowing in the acceptance of this technology exacerbated somewhat by CDMA claims. Still, the TDMA technology, particularly that represented by the IS-136 standard, is widely available and affordable, and will be more suitable than anything else in some cases. The PCS single-mode version for North America (IS-136) is one of the few low-risk choices for operators there.

Cellular CDMA can be called a revolutionary technology based on a well-established methodology [4], which has been late to the market when compared to its digital rivals. CDMA cellular technology was proposed, developed, and thoroughly described by Qualcomm Inc. of San Diego. Dual-mode cellular with CDMA as the digital half, as proposed for North America in IS-95 with AMPS interworking, was also discussed in some form in Japan (interworking with analog TACS). Single-mode CDMA versions are going into operation in North American PCS bands and are likely to appear in Japan too.

A number of factors and issues need to be considered by regulators (telecom ministries, commissions, etc.), operators, and investors when a choice for a particular network technology has to be made in light of the ystem quality required for the expected traffic mix and service requirements.

▶ *The cost of cellular infrastructure and terminals, operational costs, and typical performance:* Open standards allow multiple vendor possibilities, which reduces costs through competition enabled by easy compatibility. This is valid to some degree with terminals. The infrastructure makers always practice some form of protectionism, even with GSM. Buyers tend to end up with one supplier for the infrastructure, even though complete compatibility of network components through definition in the standards was a declared target. To a certain degree, it is possible to draw a line between the *base station subsystem* (BSS) and the *mobile services switching center* (MSC) among different vendors. Infrastructure cost mainly determines, in combination with prospective subscriber figures and revenues, the break-even point (investment recovered) and the share value for the investors.

▶ *Compatibility and ability to upgrade with existing equipment and dual-mode operation (e.g., AMPS plus TDMA or AMPS plus CDMA):* Operators who want to meet capacity needs in certain areas through the deployment of digital technology might want to reuse and upgrade existing (paid for) infrastructure for cost and compatibility reasons. The analog AMPS technology can serve as a fallback position when digital services are not available due to coverage or system loading.

▶ *Availability:* Even the best technology for a certain application is of no use when it is not available for deployment (network technology) and mass production (terminals) at a scale sufficient to generate revenue for the operator. The GSM case against CDMA is an excellent example. Due to GSM's 5-year head start and its wide support in the industry with a variety of IPR holders, many systems decisions were made in favor of this technology. Deployment of CDMA cellular service in the United States was postponed several times due to technical problems and the resulting lack of "ripe" network and subscriber equipment. Further delays also resulted from the fact that essential IPRs and first prototype products resided with only one industry player. Although there was wide support within the cellular industry for CDMA, some segments were made more cautious by the continued "hype" surrounding the technology even in the face of normal and expected technical delays. Once the implements for the technology are available, CDMA will get its share of the market. This share could have been larger had the technology been made to work a few years earlier.

▶ *Spectrum management:* Because there are only chunks of radio spectrum available to operators and this resource is seen as scarce and precious, each of the operators needs to exploit what is available to the greatest extent possible. Various standards and technology implementations from different manufacturers allow different degrees of optimum spectrum usage and smooth capacity adjustments. From a system technology point of view, CDMA is probably the better choice because minimal frequency planning is required. Even though initial claims of spectrum efficiency (over AMPS, TDMA, and GSM) have been constantly reduced from "more than 40 times AMPS" to more realistic figures, CDMA/IS-95 is in a very good position here. TDMA-based technology gains some ground in spectrum management efficiency through clever cell-splitting schemes, microcells, picocells, umbrella cells, frequency reuse patterns, intelligent overlays, *discontinuous transmission* (DTX), and frequency hopping.

1.3   Aspects on marketing the product

The transition from commercial or business use to a more consumer-oriented market can be seen in many countries and for all system standards today. Different needs and demands, different technical solutions and applications, and different marketing requirements will accompany the migration of cellular services to a consumer commodity.

The first users of cellular systems are those who specifically require mobile connections to the *public-switched telephone network* (PSTN). After a magic penetration threshold, say 10% (the number will be different for each country and market), is exceeded in a country, private users account for most of the additional growth. Relative cost of ownership (billing packages and terminal prices) as well as the perceived grade and value of the service are the key factors governing the transition beyond the threshold. Remember, these factors are to some degree operator driven (*subsidizing* phones, see Section 1.4.3) and can be a means for the operator to control usage. One example: Consumers use networks at different times than business users. This partitioning of use has an impact on system capacity and creates a new marketing playground for selling off-peak per-hour usage at lower prices. Operators have a certain freedom in controlling the factors of network usage against capacity by tariffing: high tariffs will not attract too many private users and may scare away some business users. This means that low usage and few subscribers can be a way to generate a sound revenue stream on a network that has not yet been built to full capacity. When more capacity is finally available, lower tariffs can attract the private users at the correct time. Competition in deregulated environments needs to be taken into account when making such rate calculations. Mixes of tariffs with different fixed monthly connection rates, and a variety of per-minute and per-second charges, allow a better match with customer expectations and behavior. The business user with typically hundreds of air-time minutes per month would rather choose a high fixed monthly rate with a lower per-minute charge. Some private users who would not use the phone much at all (tens of minutes a month) would rather pay a low monthly fee with a higher air-time charge.

One other basic figure that underlines the shift in the cellular customer base profile is the worldwide average annual service revenues per subscriber, which are expected to drop from approximately $1,000 in 1995 to less than about $700 during 2000. Still, the total service revenue market in 1996 was worth more than $68 billion, and this figure is expected to grow to more than $220 billion by 2000 [2].

At the time of this writing, less than 3% of the world's population had a mobile phone. In other words, 97% of the world's population does *not* have a mobile phone—what a potential market!

Technology becomes less of a marketing issue as rival standards deliver similar services at similar quality levels. Consequently, the religious battle between GSM and CDMA is better not fought in front of the end customer because she or he cares the least about the access scheme used on the radio interface or other technical babble.

1.3.1 Service providers

With the advent of deregulation and multiple network operators, cellular service providers appeared in abundance in many countries. This was particularly true for the introduction of GSM in Europe, when, for instance, in Germany 14 such organizations initially set out to collect a fortune. However, the advent of service providers also has a regulatory touch and impact.

Now, what is a service provider? A *service provider* is a reseller of air time. All other services offered through the service provider are arranged around the provisioning of air time. The network operator pays the service provider a commission for the air time sold to users. In other words, the service provider buys services from the network operators, which are then packaged for the user through adding some value. A service provider may market services from different networks at the same time. Differentiation among service providers is often made through different packages or "bundles" (which include the phone), convenient places (outlets) for people to look at phones, special service offerings, and unique features such as billing models and exchanges. Subsidizing of phones, which provides a low entrance threshold for a new subscriber, is made possible through long-term contracts (12 or 24 months).

For new private network operators having no or only narrow distribution channels, the service provider bridges the gap between the network operator and the cellular customer. In this context, distribution means:

▶ Selling air time;

▶ Handling (selling) phones and accessories;

▶ Dealing with SIM cards (in the case of GSM);

▶ Handling repairs and exchanges;

▶ Handling subscriptions and billing;

▶ Customizing service features and billing models.

The emergence and development of *value-adding service providers* have resulted in many changes and innovations in the cellular business. These changes will continue. Many reorganizations, mergers, and acquisitions have taken place. Such a shake-out, in general, leads to fewer organizations with broader service portfolios. Even though many network operators continue to build their own distribution channels and in-house service provider organizations, service providers play an important role in the development of subscriber bases and the efficient provisioning of customer services and packages. Service providers depend on the service features offered by the operators and the interoperability of their networks. The network operator's ability to offer constantly improving mobile telephone service of great quality enables the service providers to offer innovative and efficient customer services that attract users to the networks.

1.3.2   Fulfillment houses

The operator's primary job is to provide reliable mobile connections to the PSTN and collect fees for the service. Getting phones out to new customers is a difficult and distracting task for many operators, which is why so many of them take advantage of service providers. Some operators prefer much more control and bypass service providers, but hesitate to become involved with the phones themselves. A fulfillment house can relieve the operator of the drudgery of getting phones to customers and still allow some measure of control. Fulfillment houses understand that the phone is the customer's first and usually only vision of the network operator, and they are experts at getting phones out to customers quickly and correctly. Customers are invited, perhaps through TV ads, to call a toll-free number in order to arrange for delivery of a properly configured phone with which they can enjoy service. Most customers place their toll-free calls, make their billing arrangements, and receive their phones in the mail with a general feeling of efficient convenience. They think they are dealing directly with the operator and are seldom aware that an agent, a fulfillment house, has done all the work: taken the calls, warehoused the phones, tested and configured the phones (and SIM cards), packed the phones, shipped the phones, and collected the money.

1.4   Phones: shrink them, drop their price, and grow their features

To the user, the image of wireless communications services is conveyed primarily through the terminal. The little tool that allows the user to communicate while on the move *is* the network. It can also serve as a status symbol, a gimmick for the technology freak, or a plain and common accessory that one has to have, just like a wristwatch. The evolution of mobile terminals in their technology and functionality was a fast one, especially in the recent few years. It was directed by some old and basic requirements and some new ones too, but it was enabled by some important technological advances of the past 30 years.

1.4.1 What's your size?

Since the introduction of cellular services, cellular phones for all frequency bands and for all standards have decreased in size. In the past 13 years they have fallen in weight from 2 kg to way below 200g and 100g as they shrank from liters (2,000 cm^3) in volume to way below 0.2 liters (100 to 200 cm^3). This trend continues. Japan is considered the benchmark with below 100-cm^3 volume phones that have week-long standby times within the PDC and PHS systems. Heavy car-mounted units, a relic from the early days of cellular, have given way to pocket "Handies" and "Telefoninos." Many new cellular phones are not intended for car use at all. However, in order to change a car into a mobile phone booth, one merely has to buy a car mount kit. This kit, however, may be expensive when compared to the cost of the plain phone, which is subsidized (see Section 1.4.3). The car kit includes a power supply for charging, a microphone and speaker for hands-free operation, and an external antenna. Using the "handy" (phone) with or even without all these appliances is not advisable in a car. Driving a car demands our full attention. The attentive and responsible telephonist on the move stops the car in order to complete a call. In some regions, it's the law. Furthermore, in the case of the user who does not buy a car mount kit, radio transmission and reception with the built-in antenna are rather poor in the Faraday cages of cars. For operators who want to meet the anytime/anywhere expectation of their customers, the use of hand-portables from inside cars (without car kits) demands a more dense network with more base stations.

Today, cellular mobile stations are increasingly worn rather than carried along, thus underlining the "personal communicator" concept.

1.4.2 How long can you stand by?

Battery-operated phones offer longer standby and talk times than ever before. Systems standards can optimize battery operation by setting functional standby requirements such as active and passive monitoring periods for paging. In addition, current semiconductor technology, together with an infrastructure built up with small cells for lower transmit power, provides for several hours of talk time and standby times in excess of 1 week. Early mobile stations that were not connected to a car battery, but were carried around with bulky battery packs, had talk times of nearly 1 hour and several hours of standby.

Standby time greatly depends on the semiconductor technology used in the terminals. The standby current (measured in milliamps, mA) is a function of the time certain necessary entities within the terminal have to be active (duty cycles), the efficiency of the signal processing, and the silicon technology on which the products are implemented. Talk time depends, to a great degree, on the radio transmitter output power and transmit path power efficiency (power amplifier stage). The baseband sections that are active in conversation mode draw some considerable power from the battery and as such contribute to the discharge of the battery.

1.4.3   Ninety-nine cents?

Personal communication products have also become cheaper. Open standards, competitive markets, and other factors drive down the manufacturing cost of terminals by about 10% to 20% every year. But, watch out! Customer prices may not be what they appear. We all know the advertisements in which cellular phones are offered for an imaginative low price, say, "99 cents." Such subsidizing practices are common now worldwide, but they distort the pricing structures. The term *subscriber churn* describes one negative effect arising from this marketing tool. Subscribers disconnect and even switch service providers or network operators after a binding period (say, 12 months) is over. They do not let their original operator recover the investment for the phone and earn a profit, thus tying up an increasing portion of the operator's capital in phones wandering loose over the landscape. The subscriber is often attracted by a competing operator with lower tariffs and newer (and often better) phones that are subsidized again. The user's only hassle is that a new mobile phone number has to be assigned with the new subscription. With less customer loyalty, operators need to revisit this practice and either decrease their sponsoring share on the phones, increase binding periods, or add features that increase the inconvenience of changing networks (e.g., e-mail addresses). New consumer tariffs and prepaid models in some countries add services and features that allow the operator to tailor the subscriptions to the user in such ways as to prevent churn.

1.4.4 What can you do that I can't?

The increasing number of services offered by the mobile networks are supported in today's terminals. Such features include *short message services* (SMS), *supplementary services* (SS, known from the ISDN), receiving and transmitting faxes (supported by the relevant protocols and connection to appropriate *terminal equipment* [TE] supporting fax; e.g., a notebook computer), data services including Internet capability (supported by appropriate TE connected to or integrated with the phone), and a growing list of others. Thanks to open standards, a variety of hardware and software is available today. The hardware and software accessories do not always come from the original equipment manufacturers; instead they come from an increasing number of third-party vendors. Again GSM, as the world's most widely deployed digital standard, inherently offers an abundance of service features through its open definitions and standards.

A number of features are found in many modern phones. Some of these features that have increasingly little to do with the communications task include real-time clock with alarm, phone books, hands-free (for car use) function, memo functions (making use of memory and voice compression), answering machine (also found as a "Mailbox" in the network), and computer games (like Tetris). New flavors of personal communicators merging a *Personal Digital Assistant* (PDA), a fax machine, and a data terminal (for Internet browsing) with a digital cellular phone get lots of attention in the market. The Nokia Model 9000 Personal Communicator was the first product showing the way.

What can we expect for the future? Features like voice recognition ("call office" or "call mum")—as introduced by the Philips *Genie* model in early 1997—handwriting recognition (already seen in PDAs), and active echo cancellation (as opposed to echo suppression; see Chapter 11) are being discussed, developed, and introduced.

1.4.5 Multiple bands and multiple modes

Someone can use a single familiar terminal, in which dozens of phone numbers are stored, within any GSM network when traveling on the road and place calls on a private *Digital Enhanced Cordless Telephone* (DECT) System when in the office. There is a slight difference between multimode phones and multiband phones. The merging of different wireless radio standards into one terminal has been discussed for many years, and the first products have finally appeared.

Multiband refers to a terminal that can support a certain basic standard (say, GSM) but allows calls on different frequency bands employing the same protocol (GSM 900, GSM/DCS 1800, or PCS 1900). From a technology point of view, the implementation for GSM-based systems is *relatively* easy and cheap. The whole baseband section remains the same in multiband phones, while the radio section within the phone needs to be designed in such a way that more than one frequency band can be supported. This alone is still a challenging exercise. Multimode is a more general term that usually means a mix of a digital standard with a conventional analog capability, such as AMPS plus 800-MHz CDMA. The term also implies a multiband operation such as AMPS plus GSM in North America, where GSM can only appear in the 1900-MHz band in North America. Mixing modes in one phone is a bigger technical challenge. Some applications for multimode and multiband phones can be seen in a few cases.

First, a GSM operator may be granted GSM 1800 spectrum in order to meet some capacity needs, such as in hot spots or in corporate environments. With dual-band phones the operator can benefit from a larger flexibility in spectrum and user management. GSM 900 might be used for wide-area applications and GSM 1800 for small cells and in-building services.

Second, phones used for roaming between different networks (often on different frequencies) with the same number, such as between Europe (GSM 900) and in the United States (PCS 1900). Also, GSM 1800 network customers in the United Kingdom, for example, were lacking the possible roaming opportunities into neighboring European countries over the common GSM 900 networks until dual-band phones were available.

Another option for the three examples might be to use the SIM card as the roaming instrument rather then the whole phone. For example, a European takes his SIM card out of his GSM phone on his trip to the United States and rents a PCS 1900 phone when he gets there.

Fourth, pure North American applications use the AMPS system as the roaming vehicle. The currently specified AMPS-plus-TDMA and AMPS-plus-CDMA devices do the job in the 800-MHz band. In North America, and in some other countries as well, we already have dual-mode terminals and services: D-AMPS ("D" for dual mode or digital) appears in combinations of AMPS plus IS-54/136 (TDMA) or AMPS plus IS95 CDMA. Both of these dual-mode system standards offer digital services where they are available and support AMPS as a fallback position in all other places. The digital resources of the serving networks carry the additional capacity The objective to create more capacity in dual-mode networks can only be achieved if a certain high percentage of subscribers uses the dual-mode terminals. A more exotic single-band, triple-mode digital phone (TDMA/CDMA/GSM) would work in the American 1900-MHz band, but may not have a market.

Fifth, the exotic triple-mode device is not likely to appear in North America, but the triple-mode TDMA phone is already a reality: AMPS plus IS-136 in the 800-MHz band is added to digital-only IS-136 on the 1900-MHz band.

A 1996 initiative of the GSM MoU (Memorandum of Understanding) network operator association called for a *World GSM Phone* (also called simply the *World Phone*) that would operate in all three frequency bands in which GSM is deployed, namely, 900, 1800, and 1900 MHz. The World GSM Phone would allow "global roaming" among all three system standards. Given GSM's wide acceptance in so many countries, the "global" term applies.

Other flavors could be GSM plus DECT, or all North American PCS digital standards combined with AMPS. For Japan, a combination of the *personal handy phone* (PHP) and the *personal digital cellular* (PDC) systems is possible. Even GSM/PHP is an option being discussed.

Satellite communications technology will eventually appear, especially in areas where cellular or PCS coverage is not available. The dual-mode option (satellite plus cellular) is an ideal complement in isolated regions, and remains a solution for filling in some global roaming holes. In addition to multimode satellite communication, features such as geographic location functions, which make use of the *global positioning system* (GPS), can be added to the terminals.

The GSM (900/1800) plus DECT combination can be attractive for operators and users when both wide-area and local business locations need to be covered with one terminal. The subscriber's home would be included with appropriate tariffs. The GSM/DECT combination is a flexible solution for outside salespeople and telecommuters who need reliable communications through one number. Multiple numbers might be used for one subscriber with only one communicator; different private and business numbers can presort incoming calls depending on the alert tone assigned to each of them.

New licenses and spectrum allocations, deregulation, and the use of unlicensed bands (available to all operators and users) in combination with licensed bands (available only to particular operators and subscribers) may lead to different flavors of combinations and even hybrid systems in different areas worldwide. The success of such multimode technologies and their technical variations depends on market dynamics, competition, cost, and acceptance. The set of multiple-personality phones that will finally appear is much smaller than the number of different varieties we can imagine.

1.5   What is personal communications?

The PCS buzzword is not new in the worldwide wireless industry. The PCS term, in particular, is used and misused in North America. Since 1995, the notion of *personal communications* was coupled with the words *systems* and *services*. In fact, we should go back to the end of the 1980s, when the first personal communications concepts were discussed in the United Kingdom, still one of the most competitive cellular markets in the world. Can we define what personal communications stands for? What will it be? Is it a new standard or a new network? What does a personal communications system offer that other systems do not? When we gather together the different definitions, advertisements, explanations, and technology offerings of the mobile marketplace, we find some common problems, requirements, design targets, and technical solutions.

Changes in lifestyle and work habits have increased the mobility of people in recent years. People in industrialized countries are forced to increase their productivity by managing more information relative to *things*. Remote working and longer work hours have blurred the border between personal time and work time in the information society. The blending of work and home is enabled with supplements to plain old voice telephony. Now, many workers routinely add video, fax, and data communications to their voice communications tools. The process feeds on itself. Additional low-cost, non-voice communications services make it easier for employers to expect more performance out of their busy employees, and for customers to expect even faster service from their suppliers.

The communicating society finds itself using more and more value-added services that did not exist a few years ago. Point-to-multipoint information systems such as cell broadcast, short messaging, and supplementary (personalized) paging services seem ideal for alerting people to sports scores, business events, weather, and stock price changes. Packet data services, in addition to conventional point-to-point telecommunications, add interactive features. The convergence of communications equipment and computers continues to create new products and new services for mobile users. The challenge is to make these products reliable and simple to use, but affordable. The cycle is an enticing trap that guarantees further innovation, for the very tools and services that force people to carry their work around with them are the same things that seem to offer a way out of the trap. Look at how things have escalated. Fifteen years ago an airmail letter was fast enough for most needs. Today, a fax is seldom fast enough. The more features and gadgets people can use with their phones, the greater will be the number of people who feel they cannot function without at least one of them. The picture is not yet clear. We have listed some features and services people often associate with PCS and other personal communications terms, but we still don't know what it is. Let's try a more orderly approach

1.5.1 PCS: defining the requirements

Boiling the requirements down to a few visionary buzzwords would mean to say: *receive, revise, and originate calls or messages...*

- With one small terminal;
- Everywhere (in the world);
- Under one number;
- In any form;
- At any time.

Communications and related services become *personalized, customized,* and *location independent* in PCS. We connect to people, not places. We dial *one number* to reach *one person*, regardless of *where* this person is. The communicator, and all the things the user does with it, becomes part of the person. The person and the output of her labor become ubiquitous, in both time and space, to a growing number of people. What processes and services make this happen?

1.5.1.1   Cost of ownership

Subscriptions have to be affordable to a wide variety of mass users. We are talking about individual and personal services for the mass market that compete with, and to some degree replace, the wired phone services. To be successful, even in developing countries, such services must be priced at levels matching the wealth structures of the customers. Service revenues per subscriber are shrinking, because the majority of new entrants to wireless network services are private users and not business users. Western Europe will see a drop in annual revenues per subscriber of approximately 30% from 1995 to the year 2000 [2]. Similar decreases in the range of 20% to 40% are to be expected in all other regions worldwide.

1.5.1.2   Access, mobility, connectivity, and services

Access to both wireless and fixed networks will be supported *with mobility*. The user wants to be able to originate and receive calls that are routed through public switches, to and from practically anywhere in the world, without any restrictions. Furthermore, in order to be reachable all the time and at any location (with the option of some private imitations), some kind of sophisticated mobility management has to be employed. This should be transparent to the user without the need to call in with a roaming code for the area in which the user is currently located. Users do not want to participate directly in the mobility management processes. Services and applications have to be state of the art and scalable to the user. Bearer services should allow for high data rates, up to 144 Kbps (basic rate ISDN) and fast packet oriented delivery, say, for fast Internet access. The portfolio is completed with voice mailboxes, messaging, and alerting—each of which has to function well when users do not need to be reached or prefer not to be disturbed.

Combinations of business and private use, overlaid on even more combinations of office, residential, and wide-area use, challenge the system and its services. One terminal, one number (or explicitly two numbers: one for incoming private and another for business calls) will be possible. Different billing and even different radio access may be provided for different applications of the same communicator. The concept is to have a single terminal for each user. The user can be sitting still at home or walking in the street, at work or at leisure, alone in the office or with a client in a meeting. The user does not have the time or the skill to assist in mobility management tasks or to deal with the details of his *private branch exchange* (PBX), roaming, and handover.

1.5.1.3 Coverage and capacity

Network coverage and capacity must be able to cope with a high user density, a broad mix of traffic, and sharp spikes in peak usage. Peaks will occur during daytime business use and evening private use. In conventional cellular networks this peak business capacity was cleverly exploited by operators by marketing the off-peak periods to the private user at "moonshine" tariffs. Adequate coverage to the private user means that she can use her phone anywhere in a given area she may find herself most of the time—say, 90% of her time. The service should be available everywhere and anytime: at home, on the road, on the street, in the office, outside, or inside. Roaming agreements between operators, or the fact that a particular operator's service is nationwide, enhance the chances of being granted access to service all the time.

1.5.1.4 Voice quality

Voice quality has to be high. The trend is toward fixed-line toll quality without any compromises that are characteristic of first- and second-generation speech transcoder technologies. As a benchmark, the CCITT/ITU G721/726 *adaptive differential pulse code modulation* (ADPCM) algorithm delivers adequate quality, but at a relatively high rate of 32 Kbps. An abundance of speech coding algorithms is available that have been designed for use in communication systems with a requirement for low-bandwidth (baud rate) voice data, such as spectrally efficient wireless transmission systems, with few sacrifices in voice quality. Voice quality is determined by system design. Mobile systems dedicated to voice traffic can tolerate relatively high error rates on the channel relative to the lower error rates demanded of data services. Operators have to perform yet another balancing act as they configure their network densities in accordance with the mix of voice and data services their business plans require.

1.5.1.5   The "communicators"

Phones have to be conveniently small, lightweight, secure, and easy to use. They must guarantee long standby times in excess of 1 week and talk times of several hours on each battery charge. There are practical and technical limits to these requirements. If one has to move an ultrasmall phone from ear to mouth in order to listen and speak in a quasi-half-duplex mode, a tangible limit of size has been exceeded, which, in this case, is "undercut". Phones cannot be made as small as the technology allows. We still have to be able to use the phone with comfort.

Small, low-power phones also need additional support from the network architecture. The network's cell structure must be built densely enough to accommodate low RF power transmissions from the phone at the required grade of service. New data features that are supported by modem ports and larger displays need to be reliable and obvious in their use. Finally, security features, such as user authentication and encryption, must be supported without the hassle and confusion that usually accompany security measures.

1.5.2   PCS: the technical solutions to the requirements

Under the defined PCS circumstances, the basic demands are sufficiently met by a migration of current digital cellular technology into a technology that can handle higher capacity yet still allows for an upgrade path for services and applications. Existing digital technology has to be modified to accommodate more capacity while also being able to support additional services and applications.

Cordless telephone-based technologies might be enough for some applications. DECT or PHP technology may work in dense office environments. Because nothing solved all the problems at the time, PCSs were initially regarded as "next" or "third"-generation systems. Apparently, the feature-rich second-generation digital cellular systems such as GSM are well suited for use in systems characterized as PCSs, but there is a lot of room for improvement in compatibility, worldwide coverage, cost, effectiveness against user expectations, and service features. Such improvements are currently tackled in projects defining and designing the next generation.

1.5.2.1 The cost issue

The economies of scale must provide for lower cost network installation and operation for the provider of PCSs so that the lower cost of ownership can filter down to the user. Terminals, let's call them "personal communicators," must be affordable without hesitancy. All this is only possible with agreement on open standards and competition in a large marketplace. GSM is the most widely accepted digital cellular standard and has the advantages of (1) being defined through a feature-rich open standard, (2) having a large number of technology backers, (3) having established a multiple-vendor environment with lots of equipment available at competitive prices, and (4) having embraced mass production not only of terminals but of infrastructure equipment that can be tailored to local requirements. Other standards, such as the North American TDMA (IS-136) standard, can challenge GSM in many areas, and may be chosen for such special reasons as network upgrade and rollout considerations. Other systems such as CDMA (IS-95) may eventually be seen as an alternative to GSM once they are mature enough to reduce some of the risks associated with new technologies. Cordless technologies like PHP and DECT are an alternative for some applications. Infrastructure investment costs per subscriber are in the area of several hundred dollars and somewhat less for cordless technology. There is plenty of room for improvement.

1.5.2.2 Access, mobility, connectivity, and services

Unlike the messy situation of 10 years ago, mobility and connectivity today are handled well with current cellular technology. GSM, for instance, offers well-defined structures and procedures for subscriber mobility and security [5,6]. The only limitation for mobility is outside a local network's coverage area, and even that could be extended through roaming agreements with operators in other areas. These agreements can be made as invisible to the users as operators desire. The services offered through cellular and cordless technologies today are limited only by individual network and terminal capabilities. A large catalog of features is accessible to the personal communications user.

Concepts introduced by the *intelligent network* (IN) approach in modern communications systems allow easy and flexible introductions of service facilities. INs separate the network intelligence from the physical switching and transport entities through defined hierarchies, protocols, and interfaces. The distinctions on whether access, transport, and termination of communication links are of a wired or a wireless nature, or are stationary or mobile, are completely transparent to users and to many network entities. Personal communications becomes part of the *information superhighway* concept and benefits from its intelligent structures. From high-quality plain voice service and enhanced mailbox functions to packet- and circuit-switched data and fax services, from Internet access and video over radio to paging and (personalized) short messaging, everything is possible. Why not send a video postcard over your GSM phone—sorry, through your communicator—back home to your parents, or to a desk neighbor in the office, such as was proposed and advertised by Ericsson [7]?

Today's cellular system standards define a multitude of services and service platforms that fit the PCS requirements for applications around wireless personal communications. GSM, through its Phase 2 features, is a best-in-class example for that. Today, the freedom offered to tailor the service offering to the needs and requirements of individual customers or user groups enables the personal communications market. Combinations of services and applications offered by different servicing networks, including the mobile network and the Internet, become a reality with a single terminal.

1.5.2.3 Planning for coverage and capacity

By making use of modern cell planning methods, cell splitting, directional antennas and sectored cells, dynamic power control, frequency hopping, and DTX, second-generation digital cellular systems are well suited to supply high capacities and provide good coverage within an area to be served at whatever quality of service is desired. If not for the financial investment and frequency planning effort, coverage and capacities could be increased by merely installing more base stations.

One of the greatest challenges for all network operators is the acquisition and leasing of suitable sites, even for the tiny pico-sized base stations. Due to environmental, aesthetic, and health concerns, both imagined and real, more and more people refuse to allow radio transceivers to be installed in their neighborhoods. The human aspect is yet another parameter to be taken into account when planning a cellular radio network.

Cordless telephone systems, such as DECT, offer a *dynamic channel allocation* (DCA) technique that inherently makes best use of the available spectrum by only allocating channels that are tested and detected as unoccupied. Cordless technology, however, is best suited for low mobility and fixed wireless access applications. CDMA systems, through a very different approach, would work with far less frequency planning, but would still require thorough cell site and power planning [4]. It is also expected that CDMA systems may eventually provide a higher capacity for a given channel bandwidth, compared to TDMA techniques. A problem remains with increased loading of a CDMA channel; the transmission quality suffers and coverage decreases.

Higher system capacities were traditionally achieved in cellular networks by cell splitting and intelligent, dynamic frequency allocations. With the migration to higher frequencies such as those used in PCNs or in PCS networks (1800 and 1900 MHz, respectively), more capacity can be achieved through a combination of some physical effects. Higher frequencies have a lower range, as the path loss increases with frequency. With the assignment of even wider spectrum (more channels) and shrinking relative bandwidths (channel bandwidths and center frequencies), and with the reduction of peak power levels in TDMA microcell systems for PCNs and PCSs, there is more freedom for network designers. With smaller cells and tighter radio transmission power control, better reuse of spectrum and more capacity can be achieved. Compare the few hundred meters to a few kilometers separation of GSM 1800 cell sites to the somewhat larger spacing that can be achieved with GSM at 900 MHz. This means that a minimum capacity network built just for coverage will have more spectrum reuse and thus more capacity (measured in Erlangs per megahertz per square kilometer; see below and Chapter 3) at 1800 MHz than at 900 MHz. However, as traffic demand builds, the 900-MHz network can add extra cells to achieve exactly the same capacity.

The propagation of radio waves and spectral efficiency depends on a number of parameters and has to be modeled by taking into account a number of conditions and variables. More theoretical treatment can be found in [8,9], in which additional references are given.

There are a number of claims and scientific treatments from proponents of different technologies of what capacity can be achieved and at what cost. The comparisons are often expressed in Erlangs per square meter, or Erlangs per square kilometer, or (in the case of an in-building system) Erlangs per cubic meter. An Erlang is a dimensionless quantity used in statistical traffic studies. Traffic in Erlangs is equivalent to the number of calls carried in the circuits in 1 hour multiplied by the average duration of the calls in hours. One Erlang equals 3,600 call seconds per hour or 36 CCS (call century seconds) per hour. Expressed in this very general way, cordless telephone systems tend to have higher ratings than cellular systems. Cordless systems potentially have capacity ratings on the order of 600 to 10,000 Erlangs/km^2, while cellular systems, as deployed today, have ratings in the low hundreds. This kind of comparison is handy in fixed-wire systems, but quickly loses utility in the study of mobile systems. Cordless and cellular systems, for example, are not designed to do the same things. Because the capacity of mobile networks is dominated by a host of other more complicated factors such as network latency, traffic mix, and the availability of cell sites, such comparisons become oversimplifications that confuse rather than enlighten. The proper approach is to make the best of the chosen technology using the tools appropriate to the selected technology.

Full-service coverage may be achieved by dual-mode or dual-band systems and terminals, which are widely discussed in the industry and, in some cases, will eventually be introduced. A nationwide system, such as in North America where AMPS is the common interface, or a global system (GSM is the only example) would serve as the common denominator for a market. GSM 900 dual-band operation with GSM 1800 is part of ongoing specification work. A cordless standard supported in a dual-mode phone may be able to fulfill the low-range office and home functions. There are proposals to even include *low-earth-orbit* (LEO) satellite-based mobile systems to supply seamless coverage. This, however, should be seen as part of third-generation systems.

1.5.2.4 Voice quality

High voice quality is achieved through the digital transmission of speech codec (vocoder) data. Sampled voice signals are compressed and error protected. A variety of powerful algorithms have been developed for wireless applications and progress continues on the lower rate possibilities. Though there is plenty of redundancy in the human voice, there is limited freedom in voice coding. The aim is the highest quality at the lowest rate possible. The price we pay for low data rate voice with good quality is complexity, including lots of processor resources (and power consumption) and memory (cost). See Chapter 11 for a treatment of this subject.

Most of the North American GSM systems feel a threat from the competing vocoder technologies used in CDMA systems. The GSM full-rate vocoder is substituted by the so-called *"enhanced" full-rate* (EFR) vocoder, both of which use the same compression rate (13 Kbps) and channel coding. But the EFR offers much better voice quality. The EFR demands more complexity in signal processing and more program and data memory than the older full rate vocoder. In the newer CDMA and TDMA systems, voice compression technology has progressed beyond that initially achieved in the original GSM proposals. Half-rate voice codecs have also been defined and proposed for TDMA systems like GSM The half-rate process uses every other TDMA frame (one time slot) for a single voice communication link, thus doubling the network capacity if all the users were half-rate users. The penalty is even more complexity with some sacrifice in voice transmission quality that falls behind the already low rated—in perceived quality—GSM full-rate vocoder. The current requirements for the next generations of vocoder technology are simple: lower rates (like GSM half-rate, or below) or variable rate, with toll speech quality (as good as that found with *adaptive differential pulse code modulation* (ADPCM) or the *International Telecommunications Union* [ITU; formerly CCITT] standard G.721). Also, in order to ease the requirements on network density, improved immunity to high *bit error rates* (BERs) is preferred. Complexity and memory requirements should be within the range of feasible technology. Some thoughts are also given to increased protection of lower rate signals through more powerful channel coding in order to make voice codecs more robust in marginal reception conditions. Additional channel coding would allow a greater mix of data serves onto the networks.

1.5.2.5 Communicators

The evolution in cellular terminal technology paved the way for personal communicators to become attractive toys and tools that meet a wide variety of consumer tastes and requirements. Today, mobile phones offer a high degree of security with user authentication and encryption. GSM's *subscriber identity module* (SIM), which supports such features, is an excellent example. The advent of PCS can be seen as something that initiated a second evolution in product innovation as design cycles became shorter from generation to generation. The benefits of this second evolutionary phase will be fed back to traditional cellular applications, thus making the distinction between cellular and PCS harder to determine. The pager was, in fact, the first personal communicator. There are many differences between the old beeper, the alphanumeric pager, and the two-way messaging device. The trend follows cellular and personal communications systems: shrinking size, increased battery life, and a growing number of features. Modern paging services allow much more than plain point-to-point messaging. Much more customized information transfer supplements traditional point-to-multipoint service these days. GSM's SMS-CBCH adds some familiar paging services and features to mobile phones.

1.5.3   PCS and what system technology?

The general requirements for PCS can be met by today's second-generation digital cellular technology and its evolution, regardless of the frequency band in which it is deployed. From the service provision aspect, the technical details of which frequency band is used and which access technology and baseband algorithms are employed should be transparent and irrelevant to the PCS user.

The borders between digital cellular and PCS started to blur even before the first PCS networks went into operation. The cellular operators started to argue that what they had been providing for some time was already *personal communication services*. It remains to be seen how the new PCS operators force the cellular operators to offer and *market* personal communications services, which they will do only when they have to. The personal communication "marketing trick" was first successfully applied to the PCNs in the United Kingdom. The networks launched GSM 1800–based service shortly after GSM 900, and competed with cellular in many areas, especially in densely populated regions, which were the initial focus for PCN deployments.

PCS systems also reach out to replace fixed-line home access. This is possible even in industrialized countries, where fixed access is no commodity but a "given." With good coverage, sufficient capacity, careful tariffing, and innovative service offerings, PCS operators and their alliances can get their share as long as they can attract new users. Left alone, cellular operators would have eventually adopted the personal communicator and added PCS-like services to their networks. Competition from new operators licensed in new higher frequency bands forced the new operators to adopt something to distinguish themselves from cellular service: new phones, new features, and a new name—PCS.

GSM and its little brother and sister, GSM 1800 and PCS 1900 (GSM-NA), with their abundance of service features and platforms, are in a good position to be deployed in many areas where a new mobile network needs to survive in a highly competitive market. Other digital cellular technologies must try hard to compete with the GSM world standard.

1.5.4 Where is next?

As is the case with most business sectors, forecasting future development is not easy in wireless personal communications. We can look at what is available today, what is ready to be deployed very soon, and what is on the drawing boards of the people tasked with merging future technology with consumer and operator requirements into third-generation system definitions and products. PCS and current digital cellular already offer a great deal. What else can be improved and what can be added? More bandwidth is one: more bandwidth for higher baud rate data (*multimedia* services and applications), bandwidth on demand, improved circuit-switched and connectionless packet data services, and better speech quality. Better integration is another: even higher integration of residential, office, and cellular services into a single terminal can be achieved. A user should be able to *virtually* carry along his home service environment or profile into visited networks when roaming. This is possible, for instance, with a personalized subscriber module similar to the GSM SIM. High system capacities and lots of spectrum are required far beyond what is typical today. What is needed? System definition work has started all over again in order to accommodate expanded requirements. Careful consideration has to be given to service definitions and system concepts, integration of improved mobility management, radio management (including direct satellite access and private branch radio access), and network management.

The evolution of communications services and technology was not entirely self-driven. Some developments were put on track by regulatory bodies, concerted industry motivation, and standardization institutes. A lack of standardization leaves the industry awash in experiments that never end and high-priced products. Too much standardization squashes innovation. GSM has enjoyed a balance of influences. Europe, North America, and Asia reserve different research and development programs, which set the foundations for personal communications as we see it today, and as we are likely to see it tomorrow.

The international approach to the search for a logical extension of PCS into and beyond the third-generations systems is found within the *International Radio Consultative Committee* (CCIR) of the ITU-R, a United Nations body, called the *Future Public Land Mobile Telecommunications System* (FPLMTS), which is now referred to as *International Mobile Telecommunications 2000 System* (IMT-2000). A detailed treatment of FPLMTS/IMT-2000 can be found in [10].

ITU's standardization work on IMT-2000 aims to establish advanced third-generation global mobile communication services within the frequency bands identified by the *World Administrative Radio Convention of 1992* (WARC'92). Spectrum has been set aside in the 1885- to 2025-MHz and 2110- to 2200-MHz bands. With different commercial and technical requirements in different areas of the world, we will not see a single system standard being implemented worldwide for IMT-2000. Rather, we have to look at IMT-2000 as a family concept for third-generation wireless systems and services that offer somewhat more than today's second-generation systems.

1.5.4.1 Europe

In addition to spectrum allocations and operator licensing for the GSM 900 and GSM 1800 (PCN) bands, Europe is eagerly working on research and standardization for the communications scenario of the future. Programs in Europe include the RACE programs I and II (*Research and Development of Advanced Communications Technologies in Europe*) and the ACTS program (*Advanced Communications Technologies and Services*), with the final goal being to provide visions, definitions, and solutions for integrated broadband and radio telecommunication services [11]. The programs have also made contributions to third-generation mobile systems planned for the year 2000 and beyond. One such planned system is the European *Universal Mobile Telecommunications System* (UMTS). UMTS work is carried out in Europe by the *European Telecommunications Standards Institute* (ETSI). It is interesting to note that this initial work and responsibility were given to the *Special Mobile Group* (SMG), which is the group still in charge of the GSM specifications. Input for this work is also provided by the *3rd Generation Interest Group* (3GIG) within the GSM MoU.

ETSI SMG is defining UMTS as the European third-generation system based on the FPLMTS/IMT-2000 framework, operating within frequencies reserved within WARC'92. UMTS standards and final ITU recommendations are expected to be available before the turn of the century, with system introduction in following years [12]. Still, many issues need to be resolved before that can happen. The technology and the service features need to be agreed on, and they have to be capable of providing voice, data, Internet access, and images/video anytime and anywhere. We speak about flexible data bandwidth from 150 Kbps (high mobility) up to and beyond 2 Mbps (low mobility). Regulatory issues need to be settled: *who* will license *whom* to do *what* in *which* market? There are many questions, as you can see. Eventually answers will be found and agreed on, and then new "next-generation" systems and services can become a reality.

What is GSM's role in this picture? The GSM standards, which are still reviewed and revised within ETSI, have been and will continue to be adapted and expanded with features and technology requirements. Tricky matters such as satellite interconnection support—as a complement to, rather than a competitor of cellular—have all kinds of dark passages and traps for the industry. Can satellite interconnection really be standardized, or would it be better to accept what is available from the two or three satellite operators who get their birds up first? Which approach will serve personal communications applications better? GSM is seen in the vision of UMTS as a reference feature set and was proposed to provide the evolution platform on which to build the future. Although it is not a pure "third-generation system," GSM Phase 2+ standardization has a number of work items relating to UMTS. As such, UMTS is seen to provide somewhat more than was proposed by the more generic umbrella concept of IMT-2000. GSM and ISDN are also regarded as the core transport systems from which UMTS will have to evolve.

GSM Phase 2+ comes close to the objectives of UMTS/IMT-2000 by covering broadband services for personal communication. However, UMTS/IMT-2000 requires a better integration of mobile and fixed networks and fully integrated, spectrum-efficient, and cost-efficient cordless telephony. The deployment of UMTS networks and services will realistically have to occur gradually. The standardization work will be carried out in phases. The first core specifications, including a radio interface standard (UTRA, UMTS, terrestrial radio access), are expected in 1998, and ETSI UMTS Phase 1 standards in the following year. Preoperational trials followed by real commercial operation would be possible in 2001 and 2002, respectively. This is expected to be the right timing in order to start alleviating the congested GSM networks expected at the time and to offer more multimedia-type services, which are beyond the scope and capabilities of GSM today. This would include unrestricted and fast mobile access to the Internet and intranets.

Due to the required heavy investments and the complexity of yet another system rollout, a gradual transition employing dual-mode terminal equipment (GSM and UMTS)—ideally including seamless handover—is anticipated. In the terminal arena we are only limited by our own fantasies when trying to predict the flavors and combinations of technologies and applications that will emerge.

The WARC'92 spectrum in which UMTS services may be deployed in Europe appears in the following bands:

Cellular:	2 × 60 MHz paired (1920–1980 + 2110–2170 MHz)
Cordless asymmetric:	35 MHz unpaired (1900–1920 + 2010–2025 MHz)
Satellite:	2 × 30 MHz paired (1980–2010 + 2170–2200 MHz)

The total of 155 MHz of terrestrial spectrum has already been declared within the European wireless industry as far too little to support technically adequate and commercially viable services in mature (mass market) UMTS networks. Requests for as much as four times the spectrum were made. For comparison Europe has allocated 2 × 105 MHz of spectrum for GSM 900 and 1800 shared by up to four operators, and 20 MHz for DECT.

Apart from UMTS, the GSM platform is still the model whenever definitions or standardization for mobile networks is considered for new systems and applications. An example of a new application is the collection of LEO satellite systems. An easier convergence of systems and services between different networks is anticipated through LEOs that can work across equal and common network platforms.

The technical realization of the definitions and standards is left to the industry and their acceptance by the consumer. Timing the standardization work (avoiding delays) and the introduction of appropriate services is critical to success.

1.5.4.2 North America

The North American approach, which, due to the large size of its economy, is driven by the United States, is somewhat different from Europe. The term PCS originated in the United States, and under this buzzword the personal communication systems and services that will take the North American continent into the next century were discussed, refined, and marketed (see also Section 1.5.5). PCS evolved from second-generation system technology and can be seen as the next logical step toward the *universal personal telecommunications* (UPT) vision building on platforms emerging from and interconnecting to PCSs [13]. Such platforms are either proposed or are already available within the industry, in accordance with definitions and requirements set up by various industry committees.

The regulation of PCS and UPT is confined to spectrum assignments by the *Federal Communications Commission* (FCC) and the evaluation and recommendation of technology standards by the *Joint Technical Committee* (JTC) of the *Telecommunications Industry Association* (TIA) and the T1 committee. Eventually four physical layer radio interface standards were approved for PCS use in the newly assigned frequency band in North America. The four systems are PCS 1900 (GSM-NA), CDMA*One* based on Interim Standard number 95 (IS-95), TDMA based on IS-136, and a composite CDMA/TDMA/FDMA air interface proposed by Omnipoint (IS-661). A few other proposals appeared as the systems got ready to deploy. The GSM derivative is the PCS 1900 system. The remaining three systems are contrasted with GSM in Chapter 3.

With emerging PCS networks and with the amount of spectrum (part of FPLMTS frequencies from WARC'92) and licenses allocated, North America is likely to approach third-generation systems with a certain amount of attention paid to protecting the freedom of the self-regulating market and its own industry. This means that CDMA technology evolving from the narrowband IS-95 standard will be particularly favored. Backward compatibility with existing North American network and switching technology, including the installed cellular and PCS systems, will allow upgrades and operator flexibility for wideband services and new feature offerings. Adoption of a single North American wideband standard *outside* North America, for example, in parts of Asia and other continents, may be possible due to the strong foothold of North American wireless technology companies in other parts of the world.

1.5.4.3 Japan and Asia

In Japan the deregulation of cellular services and the introduction of PDC and PHP led to an extraordinary breakthrough of personal mobile communications. It also led to an extremely urgent need for spectrum and capacity. This need is driving today's activities in the Japanese wireless industry and operator community. The *Ministry of Posts and Telecommunications* (MPT), NTT, and the industry-based *Research and Development Center for Radio Systems* (RCR) are the main technological drivers behind the new cellular developments. Along with Japan and South Korea, many other Asian countries are engaged with ITU-R and make their own contributions to the definitions of third-generation mobile systems, such as IMT-2000. With European GSM becoming a world standard, the Asian wireless communications community—more than North America's—has lost out on catching the wave. A new chance is now seen in actively pushing and contributing to next-generation system standardization. It remains to be seen whether the results will match requirements and expectations, what compromises will be made, and what the regulatory and standardization policies will look like throughout the world.

As mentioned earlier, different interests in different areas and regions already demonstrate that agreements (e.g., with European standards) on basic system technology such as a *common air interface* (CAI) standard or a common network protocol are not likely to be reached. National and regional interests and schedules have already led to the establishment of different positions and specifications for inputs into IMT-2000 standards. For example, the likely proposals for the CAI are ranging from various wideband CDMA to wideband TDMA and other access methods. Also, a balance needs to be found between short-term necessity, long-term requirements, and technical and commercial feasibility.

1.5.5 GSM and PCS in the United States: an overview

PCSs in the United States have had an interesting past, and their development continues to be the basis of table conversations, both formal and informal, throughout land mobile industry. There were numerous clashes among the supporters of many different proposals for system implementations. The sale of frequency bands at extraordinary prices within a hands-off regulatory environment tested technical and nontechnical aspects of the proposals that were never seen before. There was plenty of confusion and suspense within the mobile radio industry, piles of money were at stake, and everyone learned many new lessons.

1.5.5.1 Personal communications in the United States: the potential market

The exploitation of the enormous potential of personal communications has become a major goal of today's telecommunication industry. Existing old analog cellular networks in the United States offer a limited number of telecommunication services and have generally tolerable voice quality with severe capacity constraints in only the largest markets. The AMPS network is, however, stable and too expensive to simply throw away; it took 12 years and $27 billion to build [14]. The cellular operators had already signed up all the available customers, and no number of clever marketing tricks and additional phone subsidies were going to maintain the subscriber growth rates on which the wireless industry had grown to depend. The American operators needed new customers. It is obvious that PCS systems had to offer more in order to be successful in the mass market, because the mass market expects wireline quality voice and no connection difficulties at all. Even if the expensive tariffs were reduced to attract new consumer-type users, the largest cellular networks could not handle all the new traffic. The demand for better quality, more attractive services, and more capacity meant more spectrum had to be allocated. Personal communication wireless networks are meant to cover that need, and they can fill the need as either complements or as a competition to cellular.

Cellular radio in the United States as of early 1997 claims about 40 million subscribers, which is a 13.5% penetration. Only a tiny fraction of these subscribers are on the new PCS networks, which are very new indeed. Predictions for the future cover a wide range of possibilities. It seems clear that the penetration will increase to about 35% by 2000 (which would be 100 million subscribers), and if the recent history of such predictions is any indication for the future, the 100 million figure is probably too low. It is, however, unlikely that most of the optimistic predictions for PCS's share of this growth will actually come true. The most optimistic portion for PCS the authors can justify is about one-quarter of the 100 million subscribers by 2000, and probably a bit less.

1.5.5.2 The PCS industry: the actions and the auctions

The "official" perspective for PCS in the United States was that the technology behind personal communications deployments could be anything. Whatever technology was selected, it needed to provide the required services at a certain quality and quantity level. Such a Wild West approach, it was argued, would leave lots of room for innovative equipment manufacturers and modern telecommunications service providers, thus optimizing the systems to the user's advantage. Another aim was also to provide a fair and common ground for competition, involving as many relevant market forces as possible. This includes all the manufacturers of mobile communication technology, operators, regulatory bodies, and the users. "Let the market decide" was the slogan.

A number of regulatory issues had to be addressed. The services offered by the PCSs had to be listed, spectrum allocations for licensed and unlicensed use had to be sorted out, spectrum licensing (for licensed operation) had to be organized, a map of geographic service areas for licensing had to be drawn, and standards needed to be reviewed. Actions for PCS started as early as 1989 with the FCC's "ruling over public frequency allocations" being the guide for regulatory and legislative initiatives [13].

One of the most important outcomes of these activities was the allocation of new spectrum for PCS in September 1993, and the decision to make large chunks of spectrum, confined to designated service areas, available to a limited number of operators through auctions. In addition, spectrum for low-powered unlicensed operation was allocated. All the allocations had taken account of prospective technical needs such as bandwidth and geographic coverage. The PCSs had to provide their services through a complicated matrix of frequency allocations and geographic areas, called blocks, within the 1850- to 1990-MHz band. The frequency bands have various sizes to accommodate different modulation techniques and to effect spectrum sharing with different kinds of incumbent fixed point-to-point microwave users. The frequency bands are further divided over the entire country into trading areas. The two types of trading areas were originally derived from the *1992 Rand McNally*

Commercial Atlas and Marketing Guide (123rd edition): (1) a *Basic Trading Area* (BTA) typically is an area around a city or other business area; (2) a *Major Trading Area* (MTA) roughly corresponds to the size of a state and consists of clusters of BTAs that are more or less economically independent. Only two operators are allowed in each MTA, and as many as five operators are allowed in each BTA. The largest blocks, the ones with the greatest population, are the A and B blocks, which are the MTAs. Smaller blocks are reserved for the BTAs, and have C, D, E, and F designations. The MTAs cover the BTAs. The blocks were sold to bidders during FCC *auctions* held in 1995 and 1996. We will not delve into the mountain of details except to say that they were far from hometown cattle auctions. Licenses for blocks A and B (the MTAs) raised more than $7.7 billion for the U.S. treasury. The winners of the bids for the 51 MTAs and the 493 BTAs are free to decide which technology and services they want to deploy. In a separate ruling, some narrowband PCS licenses were issued in the 900-MHz band during this time. Chapter 3 explains the details of the North America PCS frequency band allocations in the context of the different types of technologies selected by the operators.

1.5.5.3   Recovering the investments

Investment recovery, reasonable profit, and increased share value are only possible if service revenue is generated. This will not be easy for most of the American PCS operators. After major money is spent at auction for the blocks, radio spectrum used by microwave links may need to be cleared. Microwave users have a time frame of 3 to 5 years to migrate to new frequencies. Then, a radio system technology has to be chosen. The lowest risk selection is a technology that is proven to work, is available, and gives the users the required services at the expected quality. Operators who select low-risk, proven technologies expect the industry to be able to provide the required high-quality equipment in sufficient volume and on time. The prudent operator carries out all of these kinds of negotiations and plans before the licenses are awarded. The next step is to design the network and install the infrastructure. The network slowly comes to life as long as the bills are paid. The best efforts in network planning are compromised by site acquisition difficulties; insurmountable adversities drive the operator to second and third choice sites. Constant testing, tuning, and network design changes keep the engineering staff busy over long hours even when the equipment is installed "off the shelf."

Some operators supplement their own technical staffs with "friendly customers" or "early users" who are rather sophisticated, technically oriented customers tolerant of technical problems. These "customers" are heavy-duty users who, in return for a free phone and reduced connection charges, can be counted on to drive around and report all kinds of problems without the public relations overhead associated with typical paying customers.

Eventually the time comes to start advertising and recruit as many subscribers as possible. Where do the customers come from? They may come from other PCS operators in the area, existing 800-MHz cellular customers, or—of course—new customers from the general public. The challenge for the American PCS operators is that most of the American population that needs a mobile phone already has one. Taking customers away from other wireless operators without incurring high sales costs and subsidies requires sophisticated marketing skills. New customers come from the mass of those who want a phone but may not need one. Most of these kinds of new users will not be the high-volume users the cellular operators built their fortunes on during the past 15 years, and others may have credit problems. The PCS operators will need to distinguish themselves from established wireless services with their PCS features. They will need to adapt services and tariffs constantly to remain competitive. Once the customers are won, they need to take extraordinary measures to prevent churn. Operators have to establish a high level of service for their best users, maintain it, and make constant improvements.

The auditions are done, the bills are paid, the cast for the play is complete, some rehearsals (trials) have been performed, and the first scenes of the play have already begun. As services become available the dreaded shake-out begins; only a few of the operators will survive. The main characters have already armed themselves for the big fights; they have found allies and planned some conquests.

1.5.5.4 The standards—the challenge

The FCC did not mandate any standards to be used in the new allocated spectrum. Still, some basic rules about spectrum usage were stated, and recommendations on spectral etiquette were adopted [13]. The FCC presumed that spectrally efficient radio techniques could be accommodated in the smaller frequency allocations and reserved the larger ones for broadband services.

Due to adoption of wide open free market forces and deregulation, the industry (TIA/JTC) attempted to lend order to the chaos as it created a choice of seven standards (TAG-1 through TAG-7), which were named after the Technical Ad-hoc Groups that developed them. Among them we find wideband and narrowband CDMA systems, TDMA systems (both IS-136 and GSM), as well as DECT and PHP related approaches. They were all backed by their own supporters, and some even had potential operators. Seven standards also means seven different radio interface approaches. All of the proposed systems employ digital modulation and speech transmission techniques. There were no analog CAI proposals. Table 1.1 shows a list of the seven PCS standards proposals, together with some technical data for comparison.

Of the seven standards in the table, only three have lived to be deployed: IS-95 (CDMA), IS-136 (TDMA), and GSM (called PCS 1900 or GSM-NA in North America). Motorola recently proposed a 1900-MHz version of their analog IS-88 (N-AMPS) system, and the PACS and IS-661 systems may eventually find a place in specialized settings. The American PCS operators have taken on a daunting task. They have to build systems and attract customers without the time and other advantages the original

Table 1.1
List of American PCS Standards Proposals

	TAG-1	TAG-2	TAG-3	TAG-4	TAG-5	TAG-6	TAG-7
Standard derived from	new	IS-95	PACS	IS-136	DCS/GSM	DECT	(IS-661)
Access method	CDMA/ TDMA/ FDMA	DS-CDMA	TDMA	TDMA	TDMA	TDMA	DS-CDMA
Duplex method	TDD	FDD	TDD/ FDD	TDD/ FDD	TDD/ FDD	TDD	FDD
Modulation	QCPM	OQPSK / QPSK	π/4-DQPSK	π/4-DQPSK	GMSK	GFSK	OQPSK/ QPSK
Net bit rate (speech)	32 Kbps	8 and 13.3 Kbps	32 Kbps	7.95 Kbps	13 Kbps	32 Kbps	32 Kbps
Channel spacing	5 MHz	1.25 MHz	300 kHz	30 kHz	200 kHz	1728 kHz / 1250 kHz	5 MHz
Number and length of time slots	32 / 625 μs	—	8 / 312.5 μs	6 (3) / 6.7 ms	8 / 577 μs	32 / 417 μs	—

cellular operators enjoyed, and they have to accomplish this feat in a much more competitive market. Only a few can survive. There is a maximum of only two cellular operators in the 800-MHz bands, and some areas have only one. Both the A-band and B-band cellular operators have spent billions of dollars over more than a decade building their AMPS systems, which, today, are linked with mature roaming agreements and interworking network protocols (IS-41). The PCS operators, in some cases, have to build systems from scratch, and there can be as many as five additional competitors as they spend their money. Who will the survivors be?

The IS-136 proposal is championed by AT&T, which is putting much pressure on the wireless industry to supply triple-mode phones: (1) 800-MHz AMPS, (2) 800-MHz TDMA, and (3) 1900-MHz TDMA. AT&T has the resources, the marketing skill, and the brand recognition to make the IS-136 protocol succeed. With the triple-mode phone, AT&T will enjoy the advantage of not having to build an entire national network.

Two CDMA operators, notably PrimeCo and Sprint, need to build entire national networks unless dual-band, dual-mode (800-MHz AMPS plus 1900-MHz CDMA) phones appear at competitive prices. It is not clear when this will happen. Given the equipment shortages and continuing technical problems, both of these CDMA operators have done remarkable jobs building their networks in late 1996 and early 1997.

The first PCS operators to offer services to the American public were GSM operators. This was due to the relatively easy availability of GSM-based equipment. The success of the American GSM operators is studied in Chapter 3 in light of their IS-136 and CDMA competitors. Though it is too early to judge how the competing technologies will sort themselves out in the market, the struggles are an interesting study of the relationship of competing technologies with money, marketing skills, and luck.

References

[1]   Redl, S. M., M. K. Weber, and M. W. Oliphant, *An Introduction to GSM*, Norwood, MA: Artech House, 1995, Chaps. 1 and 2.

[2]   Market Trends—Cellular Services Worldwide, Dataquest 1996.

[3]   PHS International, www://phsi.com.

[4] Redl, S. M., M. K. Weber, and M. W. Oliphant, *An Introduction to GSM*, Norwood, MA: Artech House, 1995, Chap. 13.

[5] Redl, S. M., M. K. Weber, and M. W. Oliphant, *An Introduction to GSM*, Norwood, MA: Artech House, 1995, Chap. 7.

[6] Mouly, M., and B. Pautet, *The GSM System for Mobile Communications*, Palaiseau, 1992, Chap. 7.

[7] Ericsson GSM, http://www.ericsson.se.

[8] Balston, D. M., and R. C. V. Macario (eds.), *Cellular Radio Systems*, Norwood, MA: Artech House, 1993, Chap. 1.

[9] Turkmani, A., "The Mobile Radio Channel," Chap. 3 in *Personal Communications Systems and Technologies*, J. Gardiner, and B. West (eds.), Norwood, MA: Artech House, 1995.

[10] Fudge, R., and J. Gardiner, "The Future: Third-Generation Mobile Systems," Chap. 12 in *Personal Communications Systems and Technologies*, J. Gardiner, and B. West (eds.), Norwood, MA: Artech House, 1995.

[11] Gardiner, J., and B. West, "The Needs and Expectations of the Customer," Chap. 1 in *Personal Communications Systems and Technologies*, J. Gardiner, and B. West (eds.), Norwood, MA: Artech House, 1995.

[12] Rapeli, J. (Chairperson, ETSI Sub Technical Committee Universal Mobile Telecommunication System, SMG 5), *Standardization for Global Mobile Communications in the 21st Century*, http://www.etsi.fr.

[13] Russel, J., and A. Kripalani, "A North American Perspective," Chap. 10 in *Personal Communications Systems and Technologies*, J. Gardiner, and B. West (eds.), Norwood, MA: Artech House, 1995.

[14] Cena, A. M., and K. O. Nielsen, *Telecommunications Equipment*, New York: Bear, Stearnes & Co., December 1996.

Contents

From Pan-European mobile telephone to global system for mobile communications

If we dig up the roots of GSM, if we look back to the days when Europe's cellular land-scape was a patchwork of sundry incompatible analog mobile telephone systems, then we can understand the initial motivation and goals for a common Pan-European cellular standard [1]. Now, after we have passed a few standards and specifications milestones and tested and adopted the proposals in Europe and tried them out on the mobile communicating public in some countries beyond Europe's traditional borders, we see that something has changed. We see that the vision of the early proponents of GSM has been utterly transformed to a global reality that in some places is not even referred to as *cellular*. What were the initial goals and the early experiences with GSM? When and why were certain services and features introduced? What were the roles of ETSI and the GSM MoU in GSM's transformation, and how are they organized to meet their goals? What about *personal communications networks* (PCNs) and DCS 1800, and how did PCS 1900 appear in North America? As the GSM success story continues to be written, and as system standards continue to evolve to meet new requirements, we can revisit the original objectives and follow the changes to discover GSM's place in the new world of PCSs and PCNs.

2.1 GSM: what it was meant to be and what it became

2.1.1 The initial goals of GSM

Once upon a time…. The original humble objectives of the *Groupe Spécial Mobile*, the working party established by the *Conférence Européenne des Administrations des Postes et des Télécommunications* (CEPT) in 1982, and the European communications community was to define a new *common standard* for mobile phones in Europe. The common standard was to be implemented in each of the original 12 European signatory countries, and it would be a common *open* standard throughout Europe that would allow international roaming and compatibility among many equipment vendors so as to reduce prices and facilitate easy interworking. Moreover, GSM would offer a wide range of attractive new services, features, and applications unknown in the older analog systems, and those old systems would—by decree—eventually be replaced by GSM. GSM would create a common market of attractive and competitive dimensions new to the European markets. Some milestones in GSM's short history are listed in Table 2.1 [1].

2.1.2 The initial results

What was achieved through this common work and cooperative effort within the European wireless community? These were the initial results:

> ▶ A complete cellular system standard, not just the air interface, was written, describing wireless extensions to plain telephony that used

Table 2.1
Milestones in the History of GSM

Year	Milestone
1982	*Groupe Spécial Mobile* established by CEPT
1986	Permanent nucleus established
1987	GSM MoU group formed
1989 and ongoing	GSM becomes a *technical committee* (TC) within the recently founded *European Telecommunications Standards Institute* (ETSI); names change: Groupe Spécial Mobile becomes *Special Mobile Group* (SMG) within ETSI, and GSM stands for *Global System for Mobile Communications*
1990	Start of work on DCS 1800 based on GSM 900 Phase 1 specifications
1992	Commercial launch of GSM services in European networks
1993	Start of commercial DCS 1800 services (UK)
1994	First data services over GSM
1995	GSM Phase 2 core standards completed
	ETSI continues to work on Phase 2+ features
	PCS 1900 standardization within ANSI T1P1 in the United States
	117 GSM/DCS/PCS networks on the air in 69 areas worldwide
	Approximately 50,000 cell sites installed worldwide
1996	Start of commercial PCS 1900 services in North America, 6 networks/operators, 17 MTAs/18 BTAs, with about 200,000 subscribers by year's end
	More than 30 million subscribers in GSM networks; nearly 20 million new subscribers added in 1 year
	175 networks on the air in 92 areas
	215 network operators from 108 countries or territories are committed to GSM technology
1997	More than 15 new DCS 1800 networks and additional North American PCS networks go on air
	The first dual-band GSM/DCS handsets appear on the market
	GSM 1900 systems come up in smaller North American (C-block) markets
	More than 600,000 GSM subscribers in North American PCS
	World total GSM subscriber number by year's end is around 55 million in 200 networks/109 countries

digital representations and transmission of coded voice and user rate data with a TDMA radio interface, and a close compatibility to ISDN services, signaling, and network architectures.

▶ System specifications included a variety of services and application platforms as well as advanced security features such as encryption and user authentication.

▶ A phased introduction of the services was implemented that accommodated itself to realistic standardization and productivity schedules, and recognized the limited availability of system components [1].

▶ A tremendous level of support and interest from the whole manufacturing industry and prospective network operators was demonstrated even in countries not listed among the early signatories and supporters. The support was fueled by its own heat. The operators saw the chance to supply state-of-the art services with a relatively low initial investment and operating cost, and the manufacturers saw a huge market into which they could offer premium products at low cost.

Today we see a living system standard that has come alive in many countries and is still evolving to accommodate the services demanded by the diverse cultures that have adopted GSM.

2.1.3 First experiences

GSM 900 specifications for Phase 1 were frozen in 1990, and the first networks started commercial service in 1992. Initial difficulties with network coverage, terminal type approval, handset shortages, echo, and some control software problems were gradually overcome, and Phase 1 GSM technology became stable in a short time. The matrix of roaming agreements between GSM network operators quickly filled up, and the quality of service was accepted by the subscribers.

The early success of GSM cannot be wholly attributed to the standard and the products alone, because some countries and regions in which GSM was deployed were desperate for *any* kind of new cellular service. The availability of more bandwidth in the 900- and 1800-MHz bands, and the additional cellular radio capacity the new frequency allocations allowed, was a positive market driver in the United Kingdom, Germany, Italy, Portugal, and France. This was true even though, in some areas, GSM networks had to share spectrum with existing analog networks.

The introduction of Phase 1 services was initially restricted to plain voice telephony. Other features such as fax/data, voice mail, short messaging, and call forwarding/blocking were not available in most networks. Mobile data and fax only became a commodity to businesspeople on the move when GSM PCMCIA cards that allow GSM phones to work with portable computers finally became available.

The initial acceptance of data services, some of which had been available in many GSM networks since the start of 1995, was disappointing. According to the *Mobile Data Initiative* (MDI)—a cross-industry initiative aimed at promoting the use of data services [2]—data services in GSM networks accounted for only 0.5% of the traffic in 1995. This is very different from the situation in wired networks in which 46% of business traffic is data. As of 1995, data generated only 1% of the revenue in European GSM networks, and 1 out of 50 European GSM mobile phone users had a connecting device (PCMCIA card or special cable) for a mobile computer (notebook or laptop).

The GSM specifications describe a multitude of data bearer services from circuit switched to packet switched, both connectionless and connection oriented, transparent and nontransparent, asynchronous and synchronous—all of them at different rates. With the advent of more application features and tools such as professional mobile data equipment, Internet browsers, and e-mail through PDAs, data services will, it is hoped, command a more respectable share of air time and revenue. Not all of the specified services may be present in a particular GSM network, and some of the applications are not particularly easy to use; the marketplace will determine which ones are the most popular and which ones will attract the applications and accessories. The GSM standards furnish the platforms for the applications, and the evolving GSM standards (GSM Phase 2+; see later discussion) enhance the performance of the applications through higher data rates, new services, and even newer network features. It is up to the terminal manufacturers to design easier and more familiar user interfaces, and it is up to the operators to make new features and data services attractive and accessible.

2.1.4 PCNs and DCS 1800

PCN, the first conscious approach to personal communications, was a U.K. initiative, triggered by the *Department of Trade and Industry* (DTI), and was perceived as an essential complement and competitor to cellular. Spectrum was available at 1800 MHz, and GSM was selected over, for example, cordless techniques as the most appropriate technology to support the new services. One important reason for choosing an established standard (GSM 900) over the alternative of developing a completely new system for PCN was early and easy availability of low-cost network equipment and terminals. With all the air interface details worked out and tested, it was seen as a plug-and-play solution to establishing a PCN. A proposal for developing a completely new standard would have also led to less incentive for the industry to invest in the development of new equipment and new standards.

The main difference between GSM and PCN is in the amount of radio resources; the DCS 1800 system has 375 channels compared to GSM 900's 175 channels (including 50 channels in the extended frequency band). The signaling differs chiefly in its ability to accommodate the channel numbering and reduced power level assignments.

What impact did DCS 1800 have on the GSM specifications? When the GSM Phase 1 specifications were frozen, DCS was merely an add-on to GSM 900. Because the major difference was in the radio spectrum allocations, the specifications were edited as separate documents and the *DCS* prefix was substituted for the *GSM* designation. Specifications for *digital cellular system 1800* (DCS 1800) were born within ETSI in 1990 when adaptations of GSM Phase 1 documents (delta documents) were published. Almost every essential part of the GSM 900 standards was kept; changes were confined almost entirely to those required to accommodate the new RF band, and only individual pages were edited in the protocol definitions. With the advent of GSM Phase 2, DCS 1800 became an integral part of the GSM specifications. The *DCS* designation was replaced by *GSM*, thus *GSM 1800* instead of *DCS 1800*. Because the older DCS 1800 designation has been around for 6 years, and most of the literature refers to this form of GSM as such, we use the DCS 1800 reference in this book. The history of the PCNs and a more exhaustive treatment of the subject can be found in [3].

The first PCN services that used DCS 1800 technology started in late 1993 and early 1994. Networks opened up in the United Kingdom (*One-2-One* and *Orange*) and in Germany (*E-plus*); others followed in Europe (France, Sweden, Switzerland, Greece) and Asia (Singapore, Thailand, Malaysia, Hong Kong). Though there were many field trials and experiments, full-scale cellular service had never before been deployed in the PCN frequency bands. Different propagation characteristics were observed with the limited radio range and lower power class mobiles in DCS 1800 networks (1W as opposed to 2W GSM handhelds), and the much greater spectrum allocation (three times over GSM 900) led to new network planning techniques different from those already known in analog and digital cellular below 1 GHz; that is, much higher capacities were achieved.

New mobile PCNs often start out against existing, fully constructed, and thoroughly tested analog and digital cellular competition with nationwide coverage. How do the new operators attract and keep their customers? The customers are found in consumer mass markets, perhaps among those who have never owned a cellular phone, and in small regional enterprises, which may not need the wide roaming capability of GSM 900 and mature analog systems. Marketing efforts and the pricing of PCN services is tuned to these clients:

1. Lower cost service compared to conventional digital and analog services;

2. Quick activation—"walk out and call;"

3. Newer, state-of-the-art, and attractive terminals only available when advanced, second-generation PCNs appeared.

It was left to the U.K. markets to sort out the early prospects for the survival of the DCS 1800 operators. The first three U.K. licenses were initially awarded to Unitel, Microtel, and Mercury. After some reorganization, some commercial shake-outs, and some arrangements that split up coverage areas, only two operators, Mercury One-2-One (formerly Mercury and Unitel) and Hutchison Orange (formerly Microtel), now offer PCN services based on DCS 1800 technology. Investment logic and intensive market research led to this step in the direction of common sense after it was discovered that the PCN cake was not big enough for three consortia laying out three complete networks competing against two digital cellular operators (Cellnet and Vodafone) and two analog cellular operators (Cellnet and Vodafone). All of this occurred shortly after *telepoint* system's (using CT-2 cordless technology) disastrous failure in the United Kingdom. After 2 years of aggressive network rollout, and even more aggressive marketing, both PCN operators have together already earned a cellular market share of 17% in early 1997. This figure looks even better when we compare the PCN share of close to 42% among only the digital GSM-based standards (GSM and DCS), which says that most *new* customers are attracted by the PCNs. Most of the first subscribers were found in metropolitan areas (cities and major towns) where network rollouts began in order to satisfy the mandate of the licenses that required a certain degree of coverage within a relatively short time.

In other countries, some PCN licensees are also GSM operators, and certain coverage obligations imposed by regulations may be achieved through dual-band services, that is, GSM for wide-area coverage and DCS for busy spots, office campuses, and in-building service. This will be an attractive and low-cost alternative for solving the GSM 900 operators' coverage problems once dual-band GSM 900/DCS 1800 phones become available. GSM operators can use the spectrum in the DCS 1800 band to overcome capacity constraints in their GSM 900 networks. Sufficient quantities of dual-band GSM 900/DCS 1800 phones need to be available before this scheme can work; a significant number of users must be driven off the 900-MHz band, or most of the relatively small number of heavy users must be forced onto the 1800-MHz band. It is the European Union's policy that member countries must license at least one DCS 1800 operator by January 1, 1998, either to new network operators or to existing GSM operators.

Roaming is a major issue for PCN operators. Indeed, the DCS 1800 operators need the 900-MHz roaming resources to alleviate their capacity problems and fill their coverage holes more than the 900-MHz operators covet the DCS 1800 band. DCS 1800 customers can already enjoy SIM card roaming when they use their SIMs in rented GSM 900 phones abroad, but actual station roaming awaits dual-band terminals. For now, a GSM subscription is more competitive in international roaming situations.

One of the reasons the PCN operators are so strong in the United Kingdom is that they tried to distinguish their service from that of the GSM operators from the first day. The differentiation went beyond different marketing schemes, positioning of the service, and pricing; they offered somewhat different features and services from the 900-MHz operators. The PCN operators were very active in setting their own service standards and adding even more functionality to the GSM standard. The availability of a second line and number (alternate line service) for each subscriber (one for private use and another for business use) is just one example. Besides their own association (European PCN operators) the PCN operators are also represented in the GSM MoU Association through their own committee called the *Personal Communications Networks Interest Group* (PCNIG).

2.1.5 PCS 1900

The PCN experience in North America was, and remains, about as different from the European one as anyone can imagine. The reader should remember that GSM is a living standard that includes specifications for an entire network; it is not just an air interface definition. We are still modifying and enhancing the specifications to meet new expectations and markets. Turning his gaze from the orderly GSM world in Europe to the relative chaos of the American wireless landscape, the casual observer might throw up his arms in confusion over the collection of sundry incompatible air interfaces. The European and American approaches have their own peculiar logic with global consequences. Whereas the Americans allow different radio systems and technologies to fight each other in the market place, the Europeans confine the competition to the standards process. GSM was thoroughly tested in the early 1990s within the confines of the wealthy European community with the result that GSM was presented to the rest of the world as a fully functional, low-cost, low-risk way to deploy modern digital cellular service. The disadvantage of the orderly European approach is that it takes time; GSM does not necessarily represent the latest and greatest in digital wireless technology. The American way is to curtail sharply the lengthy standards process, and allow system proposals to prove themselves in the cold reality of the market. This fast prototype approach furnishes a stage on which the newest technologies can mature, but moves the expense of testing and proving those new technologies closer to the operators and their customers. So, since most of the world is not buried in mountains of excess cash that can fund elaborate experiments in radio technology, we see GSM deployed in more than 100 countries. The more high-tech American proposals, which are reviewed in Chapter 3, can only survive in more robust economies or under the auspices of a concerted national effort.

Why, then, did GSM appear in North America? Spectrum was finally reserved in North America for PCSs in the 1900-MHz band in 1994. The 1800-MHz band was too clogged with, for example, point-to-point government radio users to consider this more international frequency allocation. Chapter 1 explained the novel licensing approach for PCSs in the United States and the complex matrix of frequency allocations based on MTAs and BTAs. The frequency bands themselves are shown in Figure 2.1. We see that two 60-MHz bands—Base RX and Base TX, separated with an 80-MHz duplex offset—are dolled out in pairs (RX and TX) of 15- or 5-MHz allocations lettered A through F. Low-powered (in-building) PCS operation is also allowed in the unlicensed band, which is seen in Figure 2.1 as a 20-MHz resource between the licensed downlink and uplink allocations.

Operators are free to deploy any kind of radio system they can afford as long as the selected technology does not interfere with other licensees within the PCS bands or with radio services outside those bands. Each system has its own way of defining a physical channel within the American PCS spectrum (some of these schemes are disclosed in Chapter 3). The selected technology can even be a TDD technology that does not use separate uplink and downlink bands. The selected radio technology must, however, be disciplined enough to guarantee users a useful service. Unworkable and inefficient systems cannot continue to hold a license.

The original field of almost a dozen proposals was quickly reduced by economic reality to three major contenders: IS-95 (CDMA), IS-136 (TDMA), and PCS 1900 (GSM). The place of the IS-95 and IS-136 contenders, together with a few other interesting systems that may see new life in the future, is explored in Chapter 3, where we will see that the AMPS system retains an important influence on all PCS technologies and operators.

The PCS 1900 system is based on GSM Phase 2 specifications; even the channel numbering scheme was adopted from DCS 1800. In PCS

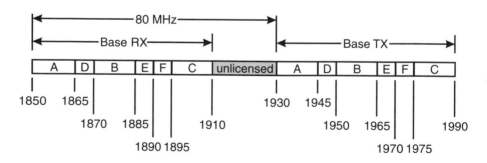

Figure 2.1 North American PCS frequency bands.

1900, the *absolute radio frequency channel number* (ARFCN), which is N in the following formulas, takes on the following values:

$$512 <= N <= 810$$

where the uplink (F_{ul}) and downlink (F_{dl}) frequencies are

$$F_{ul}(N) = 1850.2 + 0.2 (N - 512) \text{ MHz}$$

$$F_{dl}(N) = F_{ul} + 80 \text{ MHz}$$

Table 2.2 demonstrates how GSM channels are inserted into the North American PCS bands.

PCS 1900 specifications are not currently published by ETSI, but by an *American National Standards Institute* (ANSI) subgroup called T1P1. This U.S. specification body added features to the original ETSI standard that it felt were required to make the system more acceptable to U.S. needs. Most prominent among these North American adaptations was the *enhanced full-rate* (EFR) speech codec, which was eventually adopted by ETSI. The EFR was seen as a requirement because GSM arrived in North America late enough that it had to compete with the newer IS-95 and IS-136 systems, both of which had already started to upgrade to their own enhanced speech coding schemes. This is just one example of how T1P1 and ETSI work closely together. Another cooperative example is T1P1's push for the development and introduction of the 14.4-Kbps data

Table 2.2
Some North American PCS 1900 Channels

Band	ARFCN (N)	F_{ul}/MHz	F_{dl}/MHz
A	512	1850.2	1930.0
A	513	1850.4	1930.2
A	514	1850.6	1930.6
A	...	...	...
A	585	1894.8	1944.8
A	586	1865.0	1945.0
D	587	1865.2	1945.2
...	...	...	...

channel into the ETSI specifications. We can expect that GSM's future will be prolonged and made brighter by T1P1's influence, which is driven by the need to make GSM as viable as possible against IS-136 and IS-95 competitors.

Unlike all of its major competitors in North America, GSM is confined entirely to the 1900-MHz band. This is because the 800-MHz AMPS systems, which will remain the roaming vehicle for the North American continent, will not be taken out of service, and the two other major North American PCS systems are compatible with AMPS in the 800-MHz band. Chapter 3 fully explains the compatibility issues surrounding IS-136 and IS-95.

Nevertheless, GSM was selected by some North American PCS operators for exactly the same reasons so many other markets in the world have done so: it works. When the major technology decisions were made by the first PCS operators in 1995, the two most viable competing technologies, IS-136 and IS-95, were not ready for full deployment. Both IS-136 and IS-95 have dual-mode handsets as part of their specifications, but some of the operators choose to abandon the AMPS roaming option in favor of an early and assured presence in the PCS arena. They were not disappointed. It was an important event when the first PCS network in the Washington/Baltimore area went into commercial service at the end of 1995, well before any networks based on one of the competing standards. Some more GSM systems followed during 1996. Today, the IS-136 and IS-95 systems are up and running, and some of them compete directly with GSM systems in the PCS bands in the same markets. For the reasons explained in Chapter 3, the IS-95 and IS-136 systems will take the lion's share of MTAs and BTAs, leaving GSM in third place by 1999. But, the early GSM systems were deployed in the big-spender international-traveler markets such as New York and Washington, DC, and the GSM operators can secure an excess of efficient subscriber income by specifically catering to these high-volume users with lots of data services and international SIM card roaming. If the *world phone* (GSM 900/1800 and PCS 1900 multiband phone) ever appears, then the future will be even brighter. There was serious talk in the first half of 1997 about developing an AMPS/PCS 1900 dual-mode handset, but since it is not clear if there is a sufficiently large market to justify creating such a device we will have to wait to see what the future holds. For now, GSM subscribers in North

America enjoy more features (Caller ID, data services, SMS) than their IS-136 and IS-95 friends. Many GSM operators included some of the more advanced features in their basic packages from the very beginning, thus giving the customer more value compared to competing possibilities. This was possible because of the advanced stage of equipment and service packages already available from the GSM industry.

The American way of marketing is also evident when it comes to purchasing and activating a phone and a subscription. New customers are generally not subjected to a lengthy investigation beyond a credit check, and many clients need only show (or phone in) a credit number. Some American PCS operators have given up destructive and pointless handset subsidies, and some do not require clients to sign long-term contracts. When leaving the shop (or receiving his handset in the mail) the new subscriber's first call is directed to a customer service center, which activates the subscription over the air. If, after a few days, the subscriber begins to suffer from a case of buyer's remorse, he is generally allowed to cancel his contract and get his money back. To compete with existing cellular operators and future competitors, the PCS operators try to use every method they can to distinguish themselves from their established cellular competitors. They supplement the offerings with extra network features that make the service more personal, and, as any enterprise desiring to secure customers in a highly competitive environment would do, they are usually more accommodating and friendly in their routine dealings with customers than the established cellular operator. The phone and the service is sold as a *value* that is also represented in the initial cost of ownership for the subscriber. This approach reduces the risk of churn.

North American PCS 1900 operators are represented in the GSM MoU Association through their own group: the *North American Interest Group* (NAIG), but only a few operators are routinely active in international GSM meetings, shows, and conferences. As the IS-136 and IS-95 competitors get their networks in order, we will probably see American PCS 1900 operators traveling more within the global GSM community as they seek new ways to prosper.

2.1.6 UIC

When the members of the *Union Internationale de Chemin de Fer* (UIC, European Railway Operators Organization) started to investigate possibilities for a Pan-European system for railway communication in the 1990s, they naturally discovered GSM. From the UIC's point of view the advantage of GSM was that it was a proven and well-defined system for which infrastructure and terminal equipment were already available. In 1993 the UIC selected a system based on GSM and allotted some spectrum from 876 to 880 MHz for the uplink, and 921 to 925 MHz for the downlink. These frequency bands have been incorporated into the GSM Phase 2+ specifications with release 96 (see Section 2.4.3 and Chapter 4), and are sometimes referred to as the RGSM bands. The absolute channel numbers are just below the ones reserved for the extended GSM band (955 to 974 MHz). The UIC implementation uses the same power levels as specified in standard GSM, but there were some special requirements for their own services that have been incorporated into the GSM specifications. The special UIC features add some functions normally seen only in *professional mobile radio* (PMR) trunking systems. Some of these are *voice group call services* (VGCS), *voice broadcast services* (VBS), and *enhanced multilevel precedence and preemption* (eMLPP) service. These services, which are discussed in appropriate places in later chapters, are required so that information can be exchanged very quickly between a train and supervisory personnel. The VGCS and VBS features partition users into talk groups or fleets just as trunked radio systems do. The eMLPP service makes sure that high-priority messages or calls (train control and safety maters) get through even though system resources may already be tied up by other users with lower priorities.

The inclusion of the UIC requirements and frequencies into the GSM specifications should not be misinterpreted. The European railway organizations will run their own networks in the frequency bands allocated to them. They will also require dedicated mobile phones, which cover these frequency ranges and services, and which are more durable then the typical phones public GSM subscribers prefer. The system will be used exclusively for railway purposes, such as shunting where commands are exchanged among the driver in the engine and other personnel in the railroad cars. No roaming between GSM and UIC applications is planned. The UIC applications concentrate on services related to running railroads, and it is difficult to imagine what interest routine public and commercial communication services would have in them.

2.2 The role of the GSM MoU

The *Memorandum of Understanding* (MoU) group of GSM network operators was founded in September 1987. The initial membership comprised 13 operators from 12 different areas and countries. Although MoU was originally just European, now more than half the members are from outside Europe. The conditions for membership as an operator are that they must have been awarded a license and suitable spectrum, and must start commercial service using GSM-based technology (terrestrial or satellite) within 2 years.

The GSM MoU is an association with its legal incorporation based in Geneva, Switzerland, and a working office in Dublin, Ireland. MoU membership was initially limited to European members, but with the growing support of and interest in GSM worldwide the MoU was opened up to non-European members and is no longer limited to the operators of GSM 900 networks. By September 1997 there were 239 GSM MoU members in 109 countries or areas worldwide with 200 operational commercial networks. It is common within the MoU to talk about *areas* rather than *countries*, because in many countries licenses for network operation are granted regionally rather than for whole countries—the *area* term better describes the geographic dimension. Today, about half of the areas are found outside Western and Central Europe, in particular in Asia and the Asia-Pacific region (China, South East Asia, India, Australia, New Zealand), the Middle East, Africa, North America, and many areas within Eastern Europe and the *Commonwealth of Independent States* (CIS), including Russia. Members include traditional GSM 900 operators and operators (or soon-to-be operators) of DCS 1800 and PCS 1900 networks. Because this large membership comes from such diverse places as North America, Europe, and Asia, the different interests need a forum in which they can be adequately represented. The GSM MoU set up different interest groups to accommodate the diversity from different areas as well as a variety of technology issues. The regional interest groups cover the Arab States, Asia-Pacific, Europe, Central and Southern Africa, North America, and India. On the technology side there are, among others, the *PCN Interest Group* (PCNIG), the *3rd Generation Interest Group* (3GIG), the *International Roaming Expert Group* (IREG), the *Terminal Working Group* (TWG), the *Services Experts Rapporteur Group* (SERG), and the *Billing and Accounting Rapporteur Group* (BARG).

The charter of the GSM MoU is described in one of the GSM MoU's web pages [4]:

> The GSM MoU Association addresses issues which collectively face operators around the world. These include: roaming, customer-care/ marketing, technology, feature priorities, encryption, environmental health, smart cards, terminal approvals, and regulation.... There are also Regional and Technology Interest Groups which research and prioritise particular issues of relevance within each of the members' territories. These include: the Arab States, Asia Pacific, Europe, Central and Southern Africa, North America (PCS) East Central operators and India. There is also a PCN interest group. There are likely to be further developments to cater for other regions/countries e.g. CIS and China, given the significant number of regional/city licences being granted in these regions. The GSM MoU Association has programmes in place to investigate many topical subjects among members such as fraud, new markets, infrastructure choice and competition. Core aims of the signatories have been set out as follows: (1) The establishment of comprehensive systems for international roaming (2) Standardisation of cellular services throughout the world. This extends from Wireless Local Loop through pure cellular to satellite dual mode (3) The core of other systems employing different access methods (4) Terminal/SIM card/ service compatibility (5) Billing and accounting principles and methods (6) Common position on public issues (legal, regulatory, health, environment).

As a network operator organization, the GSM MoU persistently pursues the interests of its members. Without such an organization, neither the GSM systems and standards would be even close to the superior position they are in today, nor would GSM and its supporting industry be able to exploit today's worldwide potential for wireless and personal communication. The GSM MoU group, and the later association, provides a framework for commercial agreements between operators, especially to support roaming and billing. The GSM MoU Association continues its mission by lobbying for new spectrum, pushing for equipment and terminal availability and approval, and initiating and supporting standardization work and new applications. Expectations for GSM MoU membership in the year 2000 are predicted to be more than 230 members in about 120 to 140 areas with GSM-based networks.

2.3   ETSI and the Special Mobile Group

The *European Telecommunication Standards Institute* (ETSI) is located in Sophia Antipolis, which is a town near Cannes, France. On-line information may be obtained through [5]. ETSI was established in 1988. In the late 1980s the European Conference of Posts and Telecommunications Administrations organization CEPT (*Conférence Européenne des Administrations des Postes et des Télécommunications*) wanted to open up GSM standardization work beyond the telecommunications administrations and public operators. To include manufacturers and public operators in standardization work, GSM (then Groupe Spécial Mobile) was opened up by CEPT. The CEPT standardization work was finally moved to the newly founded ETSI, and GSM work continued under ETSI responsibility in 1989. ETSI is an independent organization supported by the telecommunications industry. Members and contributors consist of manufacturers (a majority of about 60%), public and private network operators, administrations, users, and research bodies.

As interest in GSM extended outside of Europe, the *GSM* term came to mean the *global system for mobile communications*, and the group working on the standards was named the *Special Mobile Group* (SMG). Because the standards comprise both GSM 900 and DCS 1800, the common specifications are now titled *digital cellular telecommunications system* from Phase 2+ onward. ETSI is a self-funded organization of more than 400 members from the telecommunications industry. Annual membership fees depend on a member's (company or organization) revenues. Funding is further supplemented though the sale of standards documents, including paper copies, CD-ROMs, and server access. More than 1,700 published documents and specifications were available by the end of 1996.

As one of the technical committees within ETSI, the SMG is responsible for the following [5]:

Developing and maintaining the specifications of the Pan-European digital cellular telecommunications system operating in the 900 MHz band known as GSM (global system for mobile communications), and of its variation in the 1800 MHz band, known as DCS 1800. Studying and defining all aspects of a third generation mobile system based on the UMTS concept as previously defined by TC RES, in support of ITU studies on a global system known as FPLMTS.

The mandate includes, but is not restricted to:

▶ Definition of services;

▶ Selection of efficient radio techniques;

▶ Elaboration of network architectures, signaling protocols, and interworking mechanisms;

▶ Establishment, maintenance, and release of standards.

The SMG works on three system standards: (1) GSM 900, (2) DCS 1800 (merged through Phase 2 specifications), and (3) UMTS/FPLMTS. The group is made up of 11 *sub technical committees* (STCs), which are named SMG1 through SMG11. The work is supported by a number of project teams (PTs, PTSMG). SMG's STCs are listed in Table 2.3.

Table 2.3
SMG Sub Technical Committees

Sub Technical Committee	Title of ETSI Technical Committee Special Mobile Group	Comments
SMG1	Services and Facilities	Specification of services for digital personal communications systems
SMG2	Radio and Speech Aspects	Responsible for the physical layer and for the study of all radio engineering aspects; work on speech codecs and acoustic performance has been moved to SMG11
SMG3	Network Aspects	Network architecture and signaling

Sub Technical Committee	Title of ETSI Technical Committee Special Mobile Group	Comments
SMG4	Data Services	Data and telematics services and interworking with other networks
SMG5	UMTS	Early work and coordinating definition of third-generation systems in cooperation with other bodies; now closed because work has been distributed over various other committees
SMG6	Operation and Maintenance	Network management functions
SMG7	Mobile Station Testing	Mobile station test specifications for type approval
SMG8	BSS Testing	Base station subsystem test specifications
SMG9	SIM Aspects	Specification of SIM card and ME interface
SMG10	Security Group	Responsible for all security aspects of GSM
SMG11	Speech Aspects	Responsible for evaluation, development, and standardization of GSM speech codecs and speech quality enhancements (e.g., tandem free operation)

2.4 Standards: the present and the future

Some background information on the establishment of GSM as a world standard is necessary in order to understand its implications and the scale of its dimensions. The technical details are discussed in Chapter 4.

As we have already made clear, GSM is no longer confined to Europe. Neither is it limited to the 900-MHz band, nor is it restricted to the limited cellular uses originally planned by its creators. Because there have been, and continue to be, plenty of other worthy system proposals and specifications, we wonder why we see so much evolution in the original GSM standard for mobile communications. Because there continue to be so many changes in the standard, it is fair to ask why GSM has been so widely accepted at the expense of all other systems. We can identify five answers.

First, many of the changes in the GSM standards responded to new frequency spectrum allocations that became available *after* the original specifications were created.

- Spectrum allocations grew from the two 25-MHz allotments (uplink and downlink) in the 890- to 960-MHz band initially reserved for the "Pan-European digital cellular radio telephone system." GSM Phase 2 extended the frequency range (EGSM) to include an additional 10 MHz below the original GSM frequency bands. The UIC's railway applications (see Section 2.1.6) tacked on another 4 MHz below the EGSM allocation. The UIC's operations are confined to 876.2 to 880.0 MHz in the uplink direction, and 921.2 to 925.0 MHz in the downlink direction.

- Today additional spectrum is available in the 1800-MHz range in Europe (2 × 75 MHz). There is also spectrum reserved in North America for American and Canadian PCSs in the 1900-MHz band (see Figure 2.1), which is also used by competing technologies.

Second, GSM matured early, just in time to respond to some opportunities and critical requirements in the markets.

- Deregulation of mobile communications occurred in many countries in the early 1990s. The wireless industry was ready to exploit a common and open standard in any new and attractive business sector.

- By the time they were demanded GSM was the only system that could demonstrate the improved coverage and quality of service operators were seeking.

- It takes a long time to build the ability to respond to the growing demand for new mobile personal communication services and applications and to offer them with the appropriate security, marketing prowess, and distribution channels. GSM offered the only forum with enough maturity with which the industry could work.

- GSM arrived in time to offer thoroughly tested, low-cost equipment developed and manufactured in enough variety for a huge global market.

Third, once a technology is adopted the *value* of staying with it increases. Brian Arthur [6] has identified five contributing factors that guarantee long life for a technology that is widely accepted.

1. *Learn by using:* The more we use something the more we know about it and the easier it is to make valuable changes; nothing is perfect.

2. *Network externalities:* The more people accept a particular technology, the more valuable that technology becomes to those who remain outside the technology and are hesitant in their decisions.

3. *Economies in production:* This applies to making GSM equipment just as much as in any other industry.

4. *Technological interrelatedness:* Subtechnologies (e.g., SIM cards and PC interfaces) born from the GSM infrastructure broaden its appeal.

5. *Increasing returns:* Those averse to risk, or confined to a strict deadline, adopt the leading technology because it is less likely to include unknown risks and surprises.

Fourth, the GSM specifications have always been *open,* and they define an entire mobile radio system rather than just an air interface. Also, many of the new adopters of GSM outside of Europe were looking for whole networks, not just RF extensions to existing fixed networks.

Open standards and the huge market potential partly enable the factors in the third reason just given and explain why there are so many manufacturers of GSM equipment. The need for a common and open standard is also due to the very high development costs of mobiles and infrastructure, so that the market is not fragmented. The original primary aim of the MoU was to agree that sufficient networks would start service at the same time, for this very reason. Competition, too, drives innovation and cost efficiency for attractive products.

Fifth, and not least, is that the main drivers for GSM over first-generation systems and other competing technologies were *roaming* and *security*. These have proven to be very important functions. Only manual roaming was supported in NMT, and none at all in the other standards. GSM roaming is totally automatic, user friendly, and secure. Roaming is important in Europe where there is much travel between countries—in fact, people are often surprised by how slow the United States has been to take up roaming, as opposed to just quoting a credit card number, resulting in fragmented bills and no incoming calls. Security was foreseen as being a major driver for GSM. It was recognized that the analog networks would soon be defrauded, and indeed they were—for many millions of dollars. Some networks even closed down until they could add security features, because fraudsters could make calls on anyone else's account and some customers could even deny having made expensive calls that they did make, since the operator could not be sure! GSM has not—at least so far—suffered in this way, unlike U.S. networks (except those based on GSM), despite varied attempts.

2.4.1 GSM Phase 1

To review, the first GSM services were introduced as a subset of service features. The reason for separating the features was that standardization and product development could not keep up with the demand to offer first services. Initial GSM 900 services (and subsequent GSM 1800 services) were restricted until 1996 to a subset of teleservices, bearer services, and supplementary services [1]. Not all the networks offered all the services to all their subscribers; many variations were possible. What was available? Bearer services supply a simple, low-level transmission path for data services and teleservices. Teleservices are higher level offerings (speech, Group 3 facsimile, and short message services) that call on communications protocols for data and speech transmission between two terminals. *Short message services* (SMS), until recently a distinctive feature of GSM, enable an abundance of applications and value-added services and features. Phase 1 networks, however, offered only limited point-to-point and point-to-multipoint SMS such as voice mail subscriber alerts. On the other hand, some supplementary services (call forwarding, call blocking) already allowed a number of call-related commodity features. Even though *calling line identity* (CLI) is a Phase 2 feature, many Phase 1 mobiles and networks do contain an "unofficial" version of it.

2.4.2 GSM Phase 2

With the completion of the Phase 2 specifications, and their official freezing in October 1995, the GSM standards comprise more than 150 documents. They offer a playground in which the industry can design a wide variety of services for their customers. Phase 2 work included the following main aspects:

▶ Merging of GSM 900 and DCS 1800 specifications into single documents: a single document lists and describes the standards, the requirements, and the parameters for both systems;

▶ Adding new radio-related definitions (extended GSM frequency band and channel number allocations, lower transmit power control levels for mobiles, etc.);

▶ Standardization of additional supplementary services;

▶ Standardization of the optional half-rate speech codec;

▶ Standardization of the optional enhanced full-rate speech codec (originally a Phase 2+ feature but moved into Phase 2);

▶ Improvements in SMS;

▶ Advanced SIM features.

Phase 2 specifications maintain a *compatibility* with Phase 1. This means that any Phase 1 terminal can still function, within its original Phase 1 specifications, in a network supporting Phase 2 features. Likewise, a Phase 2 terminal can work in a Phase 1 network, even though its enhanced Phase 2 capabilities will not be exploited. Compatibility is also related to the radio subsystem, which, even through all the evolution into Phase 2+, has remained unchanged in order to guarantee backward compatibility.

The introduction of GSM Phase 2 services in 1997 has already shown their first results because terminals that support Phase 2 features were available earlier than 1997. For example, 1996 saw the first Phase 2 mobile officially pass a type approval procedure for Phase 2. Type approval of terminals supporting Phase 2 (and initially Phase 1) was an issue with handset makers because of the lack of completed test suites and adequate test equipment. The details of the features are reserved for Chapter 4.

At the time of this writing, the GSM *half-rate* (HR) speech codec had not been deployed in a commercial network. Even though the industry claimed to be able to support the HR feature in the networks and terminals, operators have remained reluctant to offer it due to its inferior voice quality in nonideal radio environments or in the presence of background noise. *Half-rate* is probably an unfortunate—although technically correct—phrasing for the subject, as one perhaps logically would link it with *half-quality*. This conclusion is certainly not correct. Only a few networks, notably in Germany, had capacity problems in some areas that led to serious consideration about even introducing HR services. Discussions about how to market poorer quality—"Do we tell our customers?"—are still being pursued. Other suggestions were to introduce lower tariffs for HR mode, or to provide HR mobile stations with a preferred access to the network or to make no distinction at all. Another possibility would be to introduce HR in areas where voice quality is not yet an issue (where nobody has ever heard a better codec), and/or where the radio coverage is sufficient for adequate voice quality even when the HR codec is used. In any case, only a sufficient loading of the network with HR subscribers (something in the range of 30% to 50% of the network's traffic) will show reasonable network capacity benefits.

The *enhanced full-rate* (EFR) codec was initially specified by the North American market, which has the distinct advantage of having relatively few handsets already placed in subscriber's hands. This means that when the handsets equipped with the EFR finally appeared in late 1997, most users perceived GSM voice services in its EFR mode. This perception is critical to the survival of GSM networks that must compete with other networks sporting newer and more advanced technologies. The introduction of the EFR is more likely in networks outside North America than the introduction of the HR. The industry claims it is ready to deploy the EFR whenever it is asked to do so. We should note that equipment supporting the HR and the EFR should have the original GSM FR speech codec as a common denominator (fallback) to ensure roaming compatibility with other networks. Some technical detail on all the speech codecs is found in Chapter 11.

2.4.3 GSM Phase 2+

GSM is a system based on evolving standards, and the evolution continues beyond the current Phase 2 specifications. No sooner were GSM Phase 1 and Phase 2 services introduced than people asked for new features, enhancements, and services. New requirements grew out of GSM's success and the success of its competitors. Standardization bodies have already started to work on the specifications for even newer (and optional) features, which are listed under the heading of Phase 2+ enhancements. Examples are (1) the enhanced full-rate speech codec, which was adopted from the so-called US1 specification of ANSI T1P1 and then moved into GSM Phase 2; (2) *high-speed circuit-switched data* (HSCSD); (3) *general packet radio service* (GPRS); (4) *customized applications for mobile network enhanced logic* (CAMEL); (5) dual-band terminals; and (6) interworking with DECT. Some of the Phase 2+ features (e.g., CAMEL) will allow GSM to offer a number of services and service platforms that make use of or have been defined and created for intelligent networks.

One new standardization procedure has been introduced for Phase 2+. We have seen that the practice through Phase 2 was to eventually freeze the new specifications so that manufacturers could set about their designs; Phase 1 was clearly followed by Phase 2. Now we are already defining Phase 2+ features, and we learn that the demand for the new Phase 2+ features is so intense that we will never be able to freeze the specifications as a single effort in one set of documents. Because manufacturers still need some kind of design reference, a new naming convention was introduced: the *Release*. Now we will refer to the functionality of Release 96 or Release 97. There were more than 100 work items and proposals under consideration by ETSI SMG teams for GSM Phase 2+ in early 1997. We discuss some of these Phase 2+ features and proposals in more detail throughout this book. We discuss the particular features of Release 96 and Release 97 in Chapter 4.

2.5 GSM type approval issues

GSM type approval always was, and will likely remain, a source of irritation throughout the GSM industry. The original complaint from the industry remains that it still takes too long to make test tools available and then validate them. Type approval equipment was not initially available to approve GSM mobile phones for the commercial start of GSM services. When some form of type approval systems finally appeared, most of the phones could not pass the tests, and the costs associated with performing some of the tests were astronomical. The delays were the stuff of legends, and anyone could have become wealthy selling tickets to some of the arguments and meetings that resulted from the delays. This kind of chicken-before-the-egg situation was eventually sorted out between the industry and authorities when the *interim type approval* (ITA) was introduced in order to test at least a subset of the long list of test cases specified for declaring full compliance (*full type approval* [FTA]). Phase 1 specifications, especially those related to terminal testing (GSM 11.10), were constantly amended and changed, also to keep costs within reason. A similar situation occurred with Phase 2 in which we saw a continuation of efforts to control costs and allow for the more timely introduction of new features and products to support them. One of the reasons for the more than a year's delay between freezing the first Phase 2 specifications and the appearance of the first Phase 2 mobile phones on the market—with networks supporting Phase 2 features—was the delay in the development of Phase 2 type approval procedures the industry could use. However, lessons learned from Phase 1 were taken into account for improvements in this area.

Networks and their backbone functions are not subject to type approval testing. The first Phase 2 mobiles with FTA were anxiously awaited. Because FTA apparently does not guarantee a phone will function fully and properly in any network, extensive *field testing* has to be carried out beyond type approval in order to achieve compatibility. Some of the more experienced people in GSM practice say that field testing is the only workable way to clear software bugs or find ways to work around them.

The testing problem is not unique to GSM. Indeed, all digital radio systems share the testing burden with GSM, and the issue is *the* challenge within the test equipment industry today. What is the solution? Do we need more regulated testing to achieve and maintain quality and feature functionality? Does more testing mean we wait longer for more expensive products? Should we leave the testing procedures, at least partially, with the manufacturers as a kind of self-certification program? If so, then which parts should we leave with them? Perhaps we should increase the pressure on the test equipment industry, which increasingly views type approval testing as a looser business that is full of risks. There are many open questions. The debates and discussions will continue, and we are not likely to see quick solutions to the testing problems. For instance, the

GSM MoU initiated efforts to create a more independent and faster type approval and certification procedure for terminals supporting Phase 2+ features. A somewhat similar but more aggressive approach was taken by the advocates of the IS-136 system, a system plagued with its own unique testing challenges (see Chapter 3).

We cannot find a solution by simply discussing the subject, but we can put a handle on it with some background information. What is GSM type approval? What are the intentions and the issues? Who approves what?

2.5.1   The objectives

GSM is an open standard that allows many of manufacturers to offer their products to operators and subscribers. GSM is also an international standard, indeed, a global standard. This means that interworking is one of the key features of GSM, and it starts with the provision of a common ground for a competitive industry environment without any privileges or advantages for anyone. It ends with the user who wants to make use of sophisticated features (international roaming) and added value services at a very low cost. Infrastructure equipment interoperability was an official, though not necessarily a perceived, goal through GSM's standardization. Many proprietary system architectures and implementations are found in the MSC, the BSC, and the Abis interface between BSC and BTS. This quickly led to incompatible equipment, with the exception of the A interface (between MSC and BSC), which was stable at an early stage. Still, a network operator had one more reason to buy better infrastructure components—in particular, BSC and BTS—from a single manufacturer or a consortium. Interoperability, or at least the possibility of interoperability, is guaranteed with this sole source scenario. The situation is completely different with GSM mobile terminals, which depend on smooth interoperability with any and all infrastructure possibilities for their success. A very tight approval regime was set in place to ensure compatibility of terminals with any GSM network that conforms to the standards. We cannot tolerate any kind of misinterpretation of GSM compliance between a mobile manufacturer and an infrastructure supplier. The relevant goals for GSM type approval are as follows:

▶ To allow compatibility and interoperability of terminals with any GSM network;

▶ To guarantee a minimum quality and safety standard of the products;

▶ To make sure that electromagnetic interference/compatibility and radio emission criteria are met.

The specification that dictates GSM type approval for terminals is GSM 11.10 (ETS 300 607). This document fills more than 1,000 pages and includes several hundred test scenarios, which are referred to as *test cases*. They are split up into signaling tests, radio tests (sensitivity under various conditions, spurious emissions, transmitter performance, etc.), acoustic tests (audio quality), SIM interface tests, and other tests. Carrying out individual tests—successfully—in the proper order takes dozens of hours in the test laboratory. A full type approval testing session can last many weeks or even months. As problems are detected and tests are failed, bugs are discovered and fixed, and tests are started all over again.

2.5.2 The authorities

According to a number of *European Economic Community* (EEC) terminal directives (electromagnetic compatibility or EMC directive, low-voltage directive), which form the legal basis for European acceptance, compliance must be achieved in certified test houses. The European Type Approval Regime is managed by the *Advisory Committee for Terminal Equipment* (ACTE). ACTE accredits the test houses that carry out the tedious task of *type testing* a phone. Test houses are both laboratories and profit centers. Their investment in test equipment and expertise is recovered by collecting fees from mobile equipment manufacturers who bring in their products for approval. There are six test houses in Europe. European type approval itself is issued by one of the four national telecommunications authorities or *notified bodies* (BZT in Germany, BABT in the United Kingdom, DGPT in France, and NTA in Denmark). The certification is based on specified detailed declarations from the terminal manufacturers and on the positive test results from type testing. Applicants for type approval must have a legal entity within one of the European Economic Area's countries. Authorities in any of the MoU signatory nations will, in general, accept type approvals achieved in one of the European test houses. PCS 1900 phones from North America are certified by the *Cellular*

Telecommunications Industry Association (CTIA), which directs candidate terminals to a specific test lab in Germany.

Having explored the origins, development phases, variants, and the future of GSM, we turn to its PCS competitors in Chapter 3 before we dig into its myriad features and belowground plumbing in the remainder of this book.

References

[1]   Redl, S. M., M. K. Weber, and M. W. Oliphant, *An Introduction to GSM*, Norwood, MA: Artech House, 1995.

[2]   Mobile Data Initiative (MDI), http://www.gsmdata.com.

[3]   Gardiner, J., and S. Temple, "Personal Communications Networks: Concepts and History," Chap. 2 in *Personal Communications Systems and Technologies*, J. Gardiner, and B. West (eds.), Norwood, MA: Artech House, 1995.

[4]   GSM MoU World 1996, http://www.gsmworld.com.

[5]   ETSI, http://www.etsi.fr and http://www.etsi.org.

[6]   Arthur, W. B., "Competing Technologies: An Overview," Stanford, CA: Center for Economic Policy Research (CEPR), Stanford University, July 1987.

Contents

A look over the fence

GSM is not the only digital cellular radio standard attracting the world's PCS users. Different network architectures and radio techniques compete with and complement GSM as they strive to meet the needs of demanding PCS subscribers efficiently and reliably. As new cellular applications tried to adjust to the PCS challenges, spirited competition broke out among the supporters of the different technologies that tried to balance the wishes of the users against the reality of the networks. For instance, even before the first PCS licenses were granted, rival camps appeared throughout the North American cellular industry around nearly a dozen technologies, which quickly boiled down to three surviving standards: (1) IS-54 (TDMA) and its IS-136 PCS successor, (2) IS-95 (CDMA), and (3) GSM. The contenders argued the cost, efficiency, service quality, and reliability of each of the strikingly different technologies that were expected to respond to the PCS challenges:

1. Low-cost handsets;

2. Wireline speech quality;

3. Smooth handoffs with no roaming confusion;

4. High capacity;

5. Full data services;

6. Full and easy interconnection with the *public-switched telephone network* (PSTN) that could offer access to *intelligent network* (IN) features and *virtual private networks* (VPNs).

The advent of digital wireless systems moved cellular from a need-based market to a cost-based one with open standards, deregulation, and competition. Because cellular service costs more than wireline service, cellular subscribers purchased phones and services because they *needed* the mobility and were willing to pay for the convenience. For newer PCS networks to survive, more wireless users had to appear and costs had to come down so that people would buy handsets and service because they *wanted* them, for instance, as an alternative to the traditional wireline phone at home. The larger markets fueled even greater progress in technology that finally brought wireless communications to an international mass market.

Different technical solutions can solve the problems associated with providing wireless communications to different kinds of users with various degrees of success that depend on the specific applications and typical propagation conditions, local customs and habits, existing infrastructure, and available credit. Cordless access, for instance, can be provided through technologies such as CT2, DECT, PACS, and PHS. Wireless local loop access can be furnished by current technologies (PHS and PACS) deployed by both established and new operators who seek an efficient and quick means to increase teledensity. The next few years will see a proliferation of new technologies and techniques for *wireless in the local loop* (WLL) applications, which will transform the wireless market into one driven by culture and habits.

Cordless and WLL access can be viewed as specific and specialized extensions of PCS, which emerged from cellular technologies in response to a new generation of mobile users who continue to be frustrated by what they view as unnecessary and artificial distinctions between cordless telephones, WLL, and cellular systems. To understand GSM's place in the new wireless environment, we need to look at alternative technologies with a focus on how each of them strives to respond to the PCS challenges. Some of the systems are GSM competitors, and others enhance or complement it.

3.1 Competition or complement?

GSM technology can complement dual-mode mobile networks, and other technologies, as well as compete with them for mainstream cellular applications and PCS. The competing or complementing technologies are:

1. Old first-generation analog cellular systems like AMPS and NMT;

2. Second-generation digital systems such as the dual-mode TDMA system (IS-54 and IS-136) and the single-mode PDC system;

3. Advanced, quasi-third-generation systems, at least for some aspects such as the physical layer (e.g., CDMA [IS-95]);

4. Real third-generation systems of the FPLMTS, now referred to as the IMT-2000 within the ITU [1]. The *universal mobile telecommunication system (UMTS) term is also associated with IMT-2000 and the older FPLMTS terms as it designates the ETSI standardization effort for IMT-2000.*

To understand the relationships GSM has with other technologies, we need to examine the *applications.*

3.1.1 Cellular and personal communications

Today's market figures and the forecasts for the future clearly show that GSM will remain the prevailing worldwide standard for wireless access. Other important TDMA-based cellular technologies, such as IS-54/136 and PDC, have their own viable but smaller markets. The circumstances that favor the smaller TDMA competitors in certain applications also limit their attractiveness in others; they respond to their specific applications so well that it is not likely they will be displaced—even by GSM. Established analog NMT networks in Central and Eastern Europe do not have enough subscribers to justify the development of their own dual-mode extensions, but they can survive in the same markets with new GSM operators through their superior coverage and by adding PCS features. Although the equipment does not exist yet, SMS can be added to the analog traffic channels in the NMT specifications. Some relatively minor NMT network modifications could support new SMS-capable handsets, if they ever exist.

GSM also becomes a kind of global common denominator or international track, particularly in North America, that drives proposals to use GSM technology for PCS applications in a dual-mode terminal combined with AMPS. There are even some triple-mode proposals combining GSM/TDMA/AMPS and GSM/CDMA/AMPS. The two common technologies, GSM and AMPS, provide resources for international roaming (GSM) and national roaming (AMPS). The importance of AMPS and IS-54/136 is explained in Section 3.2.5.

The situation with CDMA is different. Because it was conceived somewhat later than GSM, and experienced some delays in its introduction to the market, CDMA technology has some catching up to do before it can meet the huge expectations set by its supporters. The current situation with IS-95-type CDMA is explained in Section 3.2.4. The CDMA air interface concept with modifications in bandwidth (wideband CDMA), modulation schemes, access techniques, system architectures, and other parameters has been proposed for some next-generation wireless personal communications systems, such as *low-earth-orbit* (LEO) and *medium-earth-orbit* (MEO) satellite systems, WLL applications, and IMT-2000 (UMTS). An interesting hybrid CDMA/TDMA system is proposed for cellular and PCS applications, and is standardized in the United States as IS-661 (see Section 3.2.6).

3.1.2   Cordless access

Cordless telephone (CT) standards exist for applications in wireless access to fixed-line networks. Such applications are found in simple home use, office use including PBX access, telepoint use (in the street or in public buildings), and for the "last mile" in the local loop (WLL). The chief standards are CT0, CT1, CT1+, CT2, DECT, *personal access communications system* (PACS), and PHS. There have not been many industry attempts to push

GSM technology into any of these applications through GSM home base stations or WLL systems. The IS-54/136 and IS-95 systems have attracted much more interest in cordless access than has GSM. There are some applications for which GSM can be a viable complement to cordless access such as the case when GSM complements DECT or PHS in a single dual-mode handset.

3.1.3 Wireless in the local loop

Even though the borders between cordless access systems, WLL systems, and PCSs will become more arbitrary and difficult to define in the future, it is convenient to treat WLL as its own application. WLL should be considered a substitute for wire and cable. Many regions of the world do not provide fixed telephone access at all, and WLL is a low-cost substitute in such places. Places in which we find adequate fixed-wire access use wireless techniques as part of the subscriber's link to the switching office. Today, the "last mile" of the local loop to a subscriber's home or office can be substituted with a wireless link that can be installed in a manner that is faster and cheaper than actually laying a copper cable in the ground or stringing it on poles. The cost of the local loop infrastructure, including installation and maintenance, accounts for up to 50% of the expenditures for a fixed-line telecommunications operator. WLL gives a subscriber a quasi-fixed-line telephone service access point, which is locally routed through a small transceiver to the backbone network. The transceiver may be attached to the outside wall of a house, or on a roof or another exposed location. The transceiver is remotely linked to radio extensions that collect the traffic from many subscribers. The equipment must meet two contradicting requirements: it has to be durable and low cost.

Cellular technology and cordless technology, both analog and digital, have been proposed and are currently used for small WLL applications. The use of cellular technologies requires some modifications in the products but, in general, we can say that WLL systems can be based on existing technology and open standards. About half of the WLL systems in the year 2000 will be based on some kind of digital cellular or digital cordless technology or, more likely, on some elaborate combination of both technologies.

Because neither of the technology categories (cordless or cellular) was explicitly designed for WLL applications, the requirements are not ideally satisfied with the best efficiency and the lowest cost. This is one reason why so much of today's WLL equipment, or that which is under development, tends to be based on proprietary technologies rather than on open standards. New WLL technologies make use of, for instance, wideband CDMA with proprietary spreading codes, or other air interfaces with unusual channel bandwidths and spacing on frequency bands not reserved for cellular service. These proprietary systems will not be widely deployed in the world, and they will not enjoy the economies of scale that would otherwise be true if the technologies were based on open standards. Still, cellular technologies such as GSM, IS-95 (CDMA), and even LEO and MEO satellites can offer the benefit of allowing WLL access to an existing infrastructure, provided that is has been deployed with the appropriate coverage and capacity.

3.2   What else is out there?

In this section we discuss some of the more important technologies that are in competition with GSM, or that can be used to complement and enhance GSM. Where there is competition, its origins are explored. Where there are enhancements, whether they are dual-mode systems and terminals or extensions to a GSM network, the interworking is explained together with the justification for the enhancements. The purpose of the following sections is to help the reader understand the broad system features and issues that relate to GSM and PCS rather than to provide a detailed treatment of the alternatives. The reader is referred to the appropriate literature and references for more details.

We start by turning our attention to today's most important digital cordless access systems: (1) Digital Enhanced Cordless Telecommunications, (2) the Personal Handy Phone System, and (3) the Personal Access Communications System. The technical features of the three systems are targeted at slow mobility, short range, high density, low cost, and high quality services. The following general parameters usually apply:

> ▶ Wideband access methods and high symbol rates can be enlisted even though they usually require expensive equalization techniques to resolve multipath components. The slow mobility and short operating range of the applications so reduce the effects of

multipath that the need for an equalizer and complex channel coding is removed.

▶ The TDD method allows the use of a single synthesizer and simplifies the radios.

▶ *Adaptive differential pulse code modulation* (ADPCM; ITU G.721 standard) speech coding provides wireline voice quality.

▶ Low transmission output power supports microcell and picocell structures with very high user traffic density.

▶ Dynamic channel allocation, in which the terminal or the base selects available channels upon call setup, makes frequency planning (reuse patterns) unnecessary. Network reconfigurations are easy and allow for an optimum use of the available spectrum.

Table 3.1 offers a technical overview and some comparisons of the technical features and parameters of the three systems. Applications include residential access (cordless telephone at home), office/campus use (PBX), WLL, WLANs (through DECT), and public wireless access (cordless mobility). Some dual-mode operation with cellular, such as GSM/DECT, PDC/PHS, and GSM/PHS are under consideration. Cordless access systems make extensive use of the existing infrastructure such as the PSTN, the *public land mobile network* (PLMN; the GSM infrastructure for DECT), or a PBX. There is no need to lay out additional backbone equipment, which is the case in other mobile radio systems.

Table 3.1
Technical Comparison of DECT, PHS, and PACS

Parameters	DECT	PHS	PACS
Origin	Europe (U.S. derivative)	Japan	North America, Asia
Original spectrum band	1880–1900 MHz (extensions expected)	1895–1918 MHz	1850–1910 MHz and 1930–1990 MHz (U.S. PCS band), unlicensed band, and others
Channel bandwidth	1728 kHz	300 kHz	300 kHz
Number of channels	10	77	200
Radio access method	TDMA/TDD	TDMA/TDD	TDMA/FDD

Table 3.1 (continued)

Parameters	DECT	PHS	PACS
Multiplex scheme	12 time slots per channel for transmitting and 12 time slots for receiving	4 time slots per channel for transmitting and 4 time slots for receiving	8 time slots per duplex channel
kHz per channel	144	75	75
Modulation method	GFSK (Gaussian-filtered frequency shift keying), FM method	$\pi/4$-DQPSK (differential quadrature phase shift keying)	$\pi/4$-DQPSK (differential quadrature phase shift keying)
Carrier bit rate	1,152 Kbps	384 Kbps	384 Kbps
Channel bit rate	48 Kbps	48 Kbps	48 Kbps
Nominal output power portable part	10 mW (max. 250 mW)	10 mW (max. 80 mW)	25 mW (max. 200 mW)
Nominal output power fixed part / cell station	10 mW	10–100 mW	up to 800 mW
Cell radius	50m–150m	100m–300m	100m–1500m
Speech coding	32-Kbps ADPCM	32-Kbps ADPCM	32-Kbps ADPCM
Designated mobility	Pedestrian	Pedestrian and low-speed vehicle	Pedestrian, low-, and moderate-speed vehicle
Channel assignments	Dynamic (available channels selected automatically)	Dynamic (available channels selected automatically)	Quasistatic automatic frequency assignment

3.2.1 Digital Enhanced Cordless Telecommunications

The DECT standard [2] was developed and described by ETSI with support from many segments of the wireless industry. As a cordless telephone concept, DECT was set to provide high-density, high-quality wireless access to a fixed telecommunications network. The chief applications were home use (*residential access*), business use (cordless *PBX access*), and outdoor use (*public access*). Additional applications that use the DECT platform have been proposed, and most have been added to the original definition.

The specifications and architecture were chosen such that (1) the system was compatible with existing analog and digital telephone systems, services, and interfaces; (2) the products could be manufactured at low cost; and (3) the system features and services were much more competitive and future proof than other wireless access technologies of the time. In contrast to the CT2 protocol that was used in sophisticated public access applications such as the failed telepoint experiment, DECT supports features such as handover, location capability (roaming), outgoing and incoming calls, and authentication and encryption [3]. The DECT standard was conceived as a universal radio access standard (mobile cordless handset to fixed radio base station) rather than a complete communications system definition like GSM. DECT, therefore, needs to be complemented with an "intelligent" network such as a PSTN/ISDN, or a PLMN (GSM). DECT supports all the features of existing analog services and the ISDN by extending them out to terminals over a wireless link; it is a wireless extension to the fixed network. DECT is an *open standard* with a number of defined interfaces for a multitude of connections to fixed networks: (1) DECT to PBX (which also has some proprietary solutions), (2) DECT to PSTN, (3) DECT to ISDN (with transparent delivery of ISDN services), and (4) DECT/data and DECT to GSM (interworking units connecting DECT fixed parts to the GSM A interface with full access to GSM network features).

More than 5 million DECT terminals were sold by 1996. Projections for the year 2000 are on the order of 30 million installed units and beyond. Simple cordless home telephone units based on DECT were very successfully sold, for instance, in Germany because they were priced only slightly above standard CT1 phones. Other countries are expected to follow the German example as soon as the prices are adjusted to appropriate levels. DECT has been formally accepted in about 40 countries, including Asia-Pacific regions (Australia, New Zealand, Hong Kong), South Africa, and some South American countries. A modified version of DECT appeared in the United States where it was proposed for PCS applications. The American DECT variation is called *private wireless telecommunications* (PWT).

For other applications such as office and public use, two major developments within the ETSI standardization process are important: the *generic access profile* (GAP) and *cordless telephone mobility* (CTM). GAP provides a basis for the interoperability of terminals and base stations that enables a multivendor environment. Mobility management features are introduced within CTM specifications that enable wireless access technologies, like DECT, to support terminal location management for complete mobility.

3.2.1.1   DECT basics and the radio interface

We will use DECT as an example of how a cordless access system is structured with particular attention to radio techniques and signaling protocols; other cordless access systems we discuss in this chapter are similar.

DECT uses the frequency band between 1.88 and 1.90 GHz, in which there are 10 frequency channels with a spacing of 1.728 Mhz (Figure 3.1). The 10 carriers (C) numbered from 0 to 9 (top to bottom), occupy the frequency band according to the following rules:

1. Center frequency: Fc = F0 − (C)1728 kHz, where F 0 = 1897.344 MHz;

2. Occupied channel: F_c − 1728 kHz/2 to F_c + 1728 kHz/2.

The access scheme is a combination of *time division duplex* (TDD) and *time division multiple access* (TDMA) with 24 time slots per frame at a repetition rate of 10 ms. The scheme implies a pulsed radio signal transmission. The TDD technique means the radio can work with a single synthesizer (local oscillator) without a duplex stage, thus simplifying designs and limiting costs. The *fixed parts* (FPs)—the base stations—make use of the first

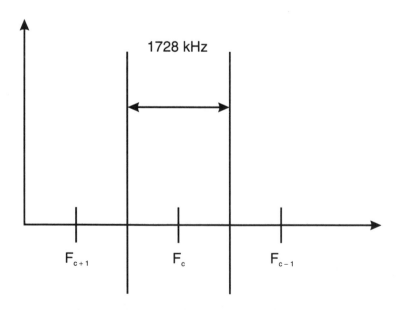

Figure 3.1   DECT channel spacing.

12 time slots, and the *portable parts* (PPs)—the handsets—use the last 12 time slots (see Figure 3.2) for transmissions on one particular frequency channel. Transmission and reception are always offset by 12 full time slots (5 ms). One full time slot has a duration of 417 μs, and 1 bit lasts 0.87 μs.

Figure 3.2 also shows the fields transmitted in a single burst for a so-called *full-slot* burst. There are also definitions for half- and double-slot transmissions. The Sync. field is used for timing and synchronization and is different for FPs and PPs. The A-field holds signaling data and the B-field carries user data that is digitized speech from a speech codec or other user data. DECT's speech codec is a 32-Kbps ADPCM (CCITT/ITU G.721) codec that yields excellent "wireline" quality. The 320 bits of capacity in the B-field, which occur every 10 ms, accommodate the 32-Kbps (320 bit / 10 × 10^{-3} sec) user data rate. Except for (parity) check bits, speech and signaling data are not error protected or channel coded. The nominal symbol rate (channel capacity) is 1,152 Kbps per radio channel and 48 Kbps per full time slot. This means that 1 bit (= one symbol) lasts 868 ns. Encryption or scrambling of payload data (the symbols) is also a DECT feature.

The modulation technique applied in DECT is a kind of *frequency modulation* (FM) called *Gaussian-filtered frequency shift keying* (GFSK). The time-bandwidth product of the Gaussian filter ($B \cdot T$), which is shaping the modulating symbols, is 0.5. Thus, the 3-dB bandwidth of this filter is 576 kHz.

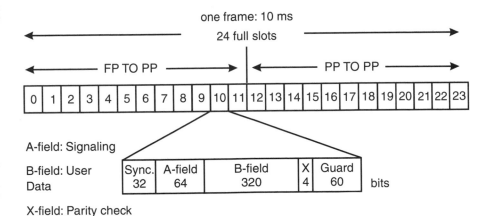

Figure 3.2 TDD/TDMA frame and slot structure for DECT.

The nominal DECT FM deviation (Δf) is ± 288 kHz. A binary symbol "0" yields a negative frequency shift, and a "1" yields a frequency deviation in the positive direction (see Figure 3.3).

The TDMA pulse shape must fit a template (Figure 3.4) that the reader may find similar to the one specified in GSM, and the pulsed RF power, also referred to as the *nominal transmitted power* (NTP), must not exceed 250 mW (24 dBm).

So, the radio part in DECT is a rather simple one; simple radio techniques are typical of cordless systems. The wideband FM (GFSK) allows some relatively wide tolerances for frequency accuracy, modulation error (FM deviation), output power, and receiver sensitivity. The building blocks in the radio are as follows:

- The *transmitter* (TX) can be a directly tuned *voltage-controlled oscillator* (VCO) without *intermediate frequency* (IF);

- Single conversion *receivers* (RX) and even some 0-Hz IF proposals and implementations can work in front of the FM discriminator;

- VCO/*phase-locked loop* (PLL);

- *Simple power amplifiers* (PA in the TX path) and *low-noise amplifiers* (LNA in the RX path) are used;

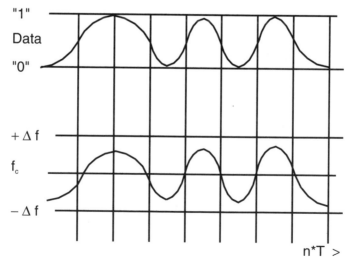

Figure 3.3 Gaussian-filtered frequency shift keying modulation.

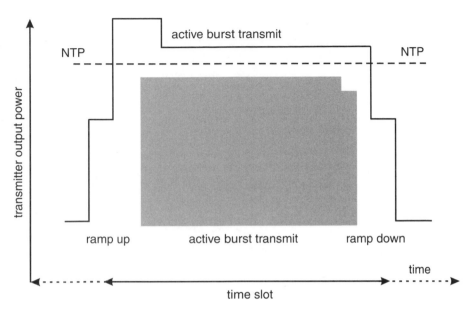

Figure 3.4 DECT RF pulse shape (transmission burst).

▶ The system needs an *antenna switch*. DECT is a TDD system; transmitting and receiving occur on the same frequency with a 10-ms time duplex offset.

DECT can be used in the North American unlicensed PCS band (see Figure 2.1), and there are proposals and standards in the United States (PWT and PWT(E) for licensed and unlicensed PCS) that use DECT baseband technology but different modulation schemes in the radio (QPSK, $\alpha = 0.5$, 1- and 1.25-MHz carrier spacing).

DECT applies a *dynamic channel allocation* (DCA) scheme, which means that the terminal scans for available frequency channels (1 out of 10) and then looks for open time slots (1 out of 24). The radio requests the optimum physical channel (frequency and time slot) when a call needs to be set up. The DCA feature is widely used in cordless systems because it has some attractive features: (1) DCA does not require frequency planning and (2) an additional *continuous dynamic channel selection* (CDCS) feature lets the portable radio scan for better speech channels after a call is set up. When it comes time for a channel change, such as would be the case when a new call is setup, the process is seamless and transparent to the user. Some users, we admit, can hear the changes. Multiple users can be served even at a very high traffic density of theoretically over 10,000 Erlangs/km^2.

An *Erlang* is a unit of *load* on the system. The load is a dimensionless number like the radian or the decibel and is the product of the number of calls initiated during a time interval (calling rate) and the average duration of a call. Let's say, for example, that 40 subscribers initiate calls during a 30-minute interval. The calling rate is 0.022 calls per second. Let the combined duration of the calls be 5,000 seconds (average call duration is 125 seconds). The load is 2.77 Erlangs (0.022 × 125). In our example, with 125 seconds being the average call duration, a 10,000-Erlangs/km^2 DECT system theoretically could handle 144,000 such calls in 30 min/km^2 (10,000: 125 = 80 calls per second, 80 × 30 × 60 =144,000).

In this perspective the potential of DECT is stunning when compared to the typical 350 to 500 Erlangs/km^2 performance in GSM 900 (up to 7,200 calls in our example) with a minimum probability of contention and blocking. High density is also ensured in DECT with the low transmit power (NTP). A mere 250 mW of power in the 1800-MHz band can extend the transmission range out to about 100m but reception conditions vary widely. In-building penetration in the 1800- to 2000-MHz frequency range is better than what is typically experienced in the cellular bands. Still, DECT, as other cordless technologies, was conceived as a low mobility system where fading and Doppler effects are not an issue. Delay spreads in most DECT applications are short (below 100 ns), which is very short indeed when compared to the 868-ns bit period. Though equalizers are not required, some DECT products employ equalizers to improve reception while extending the coverage range in a system.

3.2.1.2 DECT signaling

The signaling and call control scheme applied in the DECT standard is shown in Figure 3.5 in comparison with the *open system interconnection* (OSI) protocol layers.

The OSI structure for a layered approach to protocol stacking applies as follows:

OSI Layer 1   DECT physical layer and the physical part of the *medium access control* (MAC) layer

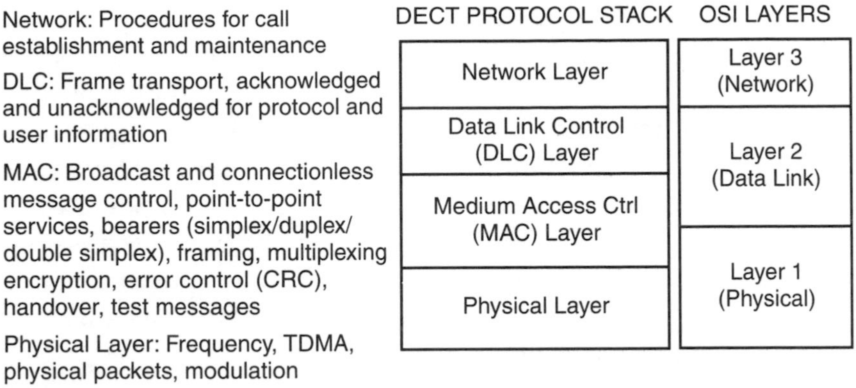

Network: Procedures for call establishment and maintenance

DLC: Frame transport, acknowledged and unacknowledged for protocol and user information

MAC: Broadcast and connectionless message control, point-to-point services, bearers (simplex/duplex/ double simplex), framing, multiplexing encryption, error control (CRC), handover, test messages

Physical Layer: Frequency, TDMA, physical packets, modulation

DECT PROTOCOL STACK	OSI LAYERS
Network Layer	Layer 3 (Network)
Data Link Control (DLC) Layer	Layer 2 (Data Link)
Medium Access Ctrl (MAC) Layer	
Physical Layer	Layer 1 (Physical)

Figure 3.5 DECT protocol stack.

OSI Layer 2   Part of the MAC layer and the data link control layer

OSI Layer 3   DECT network layer

Further aspects were considered in the DECT specifications that allow interworking with external networks: analog PSTN and PABX, digital public networks (ISDN) and PABX, GSM networks, existing and evolving telepoint networks, and *local-area networks* (LANs) and X.25 networks.

3.2.1.3 DECT dual-mode operation and interworking

As a next step in standardization, the GSM/DECT interworking profile [4]—as defined by ETSI RES 3 group (Radio Equipment and Systems) and to be adopted through SMG in Phase 2+—adds access to a GSM PLMN (cordless mobility and roaming), and the related service features, to basic quasistationary wireless access. Dual-mode terminals (GSM and DECT) can enable a wide range of new applications. All the services supplied through the backbone network (through GSM or a PSTN) are then available to the user in DECT mode, too. GSM, in particular, supplies a complete catalog of services through its own infrastructure including mobility, roaming, and billing.

One motivation for combining DECT and GSM is the additional capacity that can be achieved through DECT resources. Another one is the fact that local low mobility and short-range cordless access and PBX usage can be provided at a substantially lower cost (through fixed-line access) than high-mobility and high-range cellular services can.

3.2.1.4   Other DECT applications

Wireless access systems, like DECT, can become wireless extension for WLAN and WLL applications. The DECT channel structure and capacity is ideal for speech telephony (32-Kbps user data rate) but compromises need to be considered in ISDN data applications, which find increasing popularity in the local loops within industrialized countries; most wireless links, including cellular ones, cannot carry ISDN connections very well. The ISDN S_0 interface (2 B+D, 2×64 Kbps, and 1×16 Kbps) cannot easily be transported over present air interfaces, but DECT's data rate capacity can be enhanced by slot concatenation.

3.2.2   Personal Handy Phone System

PHS was developed in Japan and has been aggressively marketed outside Japan with considerable success [5], particularly in the Eastern Hemisphere. PHS is similar to DECT; it is a simple cordless access system for slow moving terminals (pedestrian) for both incoming and outgoing calls. Its great success in Japan (see Chapter 1) was based on high-quality service and availability, low-cost subscriptions and air-time fees, very small and convenient terminals, and superb marketing. Small terminals that weigh less than 100g and significantly lower tariffs than those typical for Japanese cellular services attracted huge masses of new customers. Delighted users purchased PHS services instead of contracting for a second (or even a first) fixed line for making and receiving voice calls. PHS is a hit in Japan where 23 MHz of spectrum is available (1895 to 1918 MHz) and is split into two parcels: 12 MHz for public use and 11 MHz for residential use (which can be shared with public operators).

3.2.3   Personal Access Communications System

PACS differs from the other digital cordless phone systems, as indicated in Table 3.1, in that it has some attributes normally seen only in cellular systems. For example, the relatively high base station power and the sensitive receivers (−101 dBm) give the handsets greater range than we would normally expect from a cordless phone system, and the mobility can extend somewhat beyond walking speeds. PACS is an adaptation of the *Wireless Access Communications System* (WACS) developed by Bellcore (Bell Communications Research) in the early 1990s. Bellcore is a commercial organization that establishes standards and tests equipment for the regional *Bell Operating Companies* (BOCs) in the United States. The system is currently championed by Hughes Network Systems under the *AIReach* name, which includes some extensions and improvements over the original proposals [6].

The system can be deployed into the North American PCS bands (see Figure 2.1) with 300-kHz FDD channel spacing aligned with 100-kHz resolution as shown in Table 3.2 [6,7]. The band edge channels are selected from the table in order to solve any guard channel problems that may arise in a particular PACS deployment. Once the edges are established only every third channel can be used, thus they are incremented every 300 kHz.

Table 3.2
North American PCS PACS Channels

Band	Channel Number	Uplink (MHz)	Downlink (MHz)
A	1	1850.1	1930.1
A	2	1850.2	1930.2
A	3	1850.3	1930.3
A	...	...	...
A	148	1864.8	1944.8
A	149	1864.9	1944.9
D	150	1865.0	1945.0
D	151	1865.1	1945.1
D	...	...	...
D	198	1869.8	1949.8
D	199	1869.9	1949.9
B	200	1870.1	1950.1
B	...	...	...
B	349	1884.9	1964.9
E	350	1885.1	1965.1
E	...	...	...
E	399	1889.9	1969.9
F	400	1890.1	1970.1
F	...	...	...
F	449	1894.9	1974.9
C	450	1895.1	1975.1
C	...	...	...
C	599	1909.9	1989.9

We will see that the physical channels of each of the disparate North American PCS schemes covered in this chapter are mapped onto the 1900-MHz PCS spectrum resource in accordance with their own rules such as the *AIReach* mapping shown in Table 3.2. Unlicensed operation is allowed in the 1910- to 1930-MHz band (see Figure 2.1). As is the case with all of these kinds of cordless systems, PACS's easy interconnection to the PSTN (switch) is evident with its P and C interfaces shown in Figure 3.6. The air interface is called the A interface. The PSTN's intelligence is used instead of building up a complete mobile telephone switching center [7,8]. A single *radio port controller unit* (RPCU) controls one or more *radio ports* (RPs, or simple base stations) through P interfaces. Costs are kept down by moving as many of the RP's baseband functions to the RPCUs as possible. In addition to handling all the 32-Kbps ADPCM voice coding and decoding, the RPCU takes care of all the call control and mobility management tasks in such a way as to hide the wireless nature of the *subscriber units* (SUs, 200-mW maximum transmitter power) from the PSTN. The P interface, which may be carried on a T1, for example, terminates in the RP base stations, which appear in several forms depending on the application. An outdoor pole-mounted RP, capable of up to 800 mW of transmitter output, can fill a WLL application with a fixed or portable SU. A smaller indoor RP, which may be screwed to a wall, can be a wireless PBX extension for SUs in an office. A common fixed subscriber instrument (telephone) can be connected to a *wireless access fixed unit* (WAFU), which operates as a stationary SU that may look like a wall-mounted RP. The WAFU has a standard connector (e.g., an RJ-11 connector) that accepts a common wired desk phone. The WAFU, and its companion RP, can find themselves in a WLL application when they replace a buried cable.

Turning to the A interface, we start with the downlink depicted in Figure 3.7. TDMA frames of only 2.5 ms are divided into eight time slots of

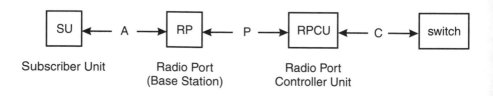

Figure 3.6 PACS network interfaces.

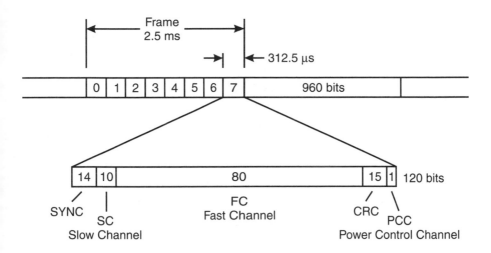

Figure 3.7 PACS downlink.

312.5-μs duration. One of the eight time slots is expanded in Figure 3.7 to show how the 120 bits contained therein are mapped into five logical channels. The bits are transmitted from the RP down to the SU in pairs that are imparted onto the carrier in symbols (2 bits per symbol) with a π/4-DQPSK modulator. This yields a frame data rate of 384 Kbps. The π/4-DQPSK modulation scheme forces a change in phase state with each symbol even if the underlying information bits do not change, thus adding a timing resource to the physical channel without the need to reserve a logical resource for the purpose. The price we pay for this benefit is that the resulting π/4-DQPSK type of modulation requires linear methods in the RF stages of the RP and SU.

The RP toggles the single bit in the downlink PCC (power control channel) to control the SU's uplink power, up or down, in 1-dB steps (+/- 0.5 dB) over a 30-dB range. If for some reason the PCC channel fails to work, the SU will transmit at full power, which is typically 200 mW. On the other end of each time slot there are 14 bits (seven symbols) reserved for the synchronization channel. These bits include a 7-bit synchronization pattern, a 3-bit frame number counter (000–111), a 3-bit time slot counter (000–111), and a reserved bit. The *slow channel* (SC) holds 10 bits of signaling information that is transported to the SU at a 4-Kbps rate (10 bits/312.5 $\times$ 10^{-6} / 8 = 4 Kbps). The *fast channel* (FC) carries the user data which can be the high-rate, high-quality 32-Kbps ADPCM speech data. Eighty FC bits per 2.5-ms frame yields 32 Kbps, which is the ADPCM rate, but PACS allows reduced user data rates of 16, 8, and 4 Kbps, which is accommodated by not using each and every frame. Consecutive frames are numbered cyclically 0 through 7, and user data are assigned a time slot in one out of eight frames (4 Kbps), two out of eight frames (8 Kbps), four out of eight frames (16 Kbps), or all eight frames (32 Kbps—speech). The downlink synchronization channel, as we saw earlier, keeps track of the current time slot and frame. The content of both the SC and the FC are protected with a 15-bit *cyclic redundancy code* (CRC) in its own logical channel.

The uplink frame structure from the SU back to the RP is depicted in Figure 3.8. The uplink structure differs from the downlink structure in only two logical channels. First, the downlink's synchronization channel is replaced with 12 bits of guard time and a 2-bit *differential decoder* (DD) channel. No signal is sent from the SU during the guard time, thus avoiding overlapping and destroying transmissions from other SUs on the same physical channel. The DD channel holds 2 bits that become a differential phase reference for the RP's receiver, hence the DD designation. Second, since the SU is the RP's slave in uplink power control, the downlink's single bit PCC is replaced by a constant "0."

Figure 3.9 shows the timing relationship between the downlink frames from the RP and the uplink bursts from the SU. The SU delays its bursts by about 375 μs with respect to the received downlink time slot.

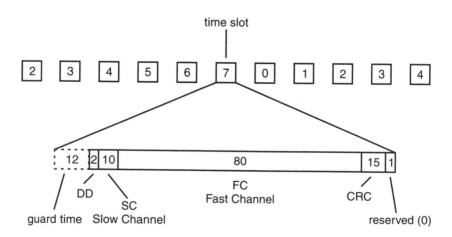

Figure 3.8 PACS uplink.

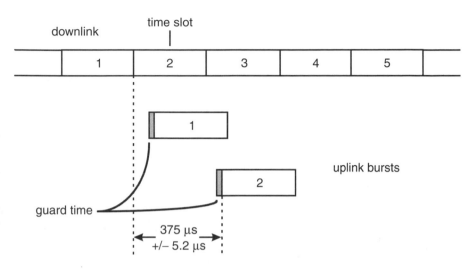

Figure 3.9 PACS TDMA offset.

This considerable delay accounts for propagation delay and allows the SP to work well without a duplexer.

PACS is a PCS insofar as it transports all the PSTN's teleservices and bearer services out to low-cost and easy-to-use handsets (SUs) with the transport techniques we explained in this section.

3.2.4 CDMA (IS-95)

The IS-95 CDMA system was originally designed as a second-generation dual-mode (with AMPS) 800-MHz band digital cellular alternative for an earlier dual-mode TDMA proposal in North America called IS-54. The system is briefly but thoroughly described in [9] and, except for the introduction of a greatly enhanced variable rate voice codec (13.3-Kbps maximum rate QCELP) in early 1997, has changed little from the description in the reference. The late industry support of CDMA must not be neglected because many PCS operators, particularly in North America, have made IS-95-based CDMA their technology of choice. Some U.S. cellular operators, as well as a few in some other countries (notably South Korea, Hong Kong, and Singapore), have opted for CDMA systems in its dual-mode configuration with AMPS. IS-95 CDMA technology will also be deployed by some operators in Japan in order to ease some regional capacity constraints in the cellular systems.

The PCS version of IS-95 (SP-3384) is almost identical to its 800-MHz cousin except for a different channel numbering scheme in the 1.9-GHz bands. CDMA channels, which are 1.23 MHz wide, are centered on relatively few 50-kHz preferred assignments (N), as listed in Table 3.3. Figure 2.1 shows the frequency ranges of the six PCS blocks referred to in the table.

The mobile station's transmitter center frequency, F_m (in MHz), is calculated from the channel number, N; thus: $F_m = 1850.000 + 0.050 \times N$, where $0 \leq N \leq 1199$. The base station's transmitter center frequency, F_b, is figured in a similar way as the mobile station case: $F_b = 1903.000 + 0.050 \times N$, where $0 \leq N \leq 1199$.

When CDMA finally matures, and when its service features, perceived voice quality, and the selection of handsets is equivalent to existing systems, then the current CDMA technology will find and maintain a strong position in the industry. Indeed, the North American PCS operators who selected CDMA and most of the Korean CDMA operators have spent astronomical sums of money deploying large systems that have done very well. The American PCS operators who adopted CDMA for their systems secured enough licenses in the PCS auctions to ensure them of virtual nationwide coverage, thus removing the urgency to install some kind of dual-mode/dual-band roaming basis in their systems and saving the time and expense that perfecting the appropriate handset would absorb. The *CDMA Development Group* (CDG) is a rather open organization of CDMA supporters that supports efforts that will surely

Table 3.3
Preferred CDMA Frequency Assignments for North American PCS

Block	Preferred Channel Number (on 50-kHz centers)
A	25, 50, 75, 100, 125, 150, 175, 200, 225, 250, 275
D	325 350, 375
B	425, 450, 475, 500, 525, 550, 575, 600, 625, 650, 675
E	725, 750, 775
F	825, 850, 875
C	925, 950, 975, 1000, 1025, 1050, 1075, 1100, 1125, 1150, 1175

hasten CDMA's maturity and importance in the future of wireless public services. The CDG's activities can be monitored on the Internet through [10]. At this writing (mid-1997), CDMA was primarily a cellular system featuring high-quality voice service with only a few PCS features (e.g., voice mail). Since installing a full catalog of network features into CDMA is just a matter of time, the IS-95 system will probably carry slightly more than half of North America's PCS traffic by the year 2000, leaving the IS-136 and GSM competitors with equal shares of the remainder. AMPS and its dual-mode IS-136 extension will continue to dominate the 800-MHz band cellular services.

Why has CDMA, in contrast to the situation in the rest of the world, captured so much of North America's PCS traffic relative to GSM? There are several reasons. First, GSM was a European invention during the late 1980s that was designed to bring a single high-quality, high-capacity cellular radio standard to a continent in dire need of one that would, among other things, facilitate international roaming. As late as 1992, when GSM was still going through its first deployment pains in Europe, North America already had a single cellular standard (AMPS) that, since it had allowed roaming all over the vast North American continent for a decade, quelled some of the urgency a search for a single high-quality cellular standard might have otherwise brought to North America. In the early 1990s there were only a few North American markets plagued with capacity problems (e.g., Los Angeles), and the industry saw a simple digital extension to AMPS as the logical and ideal entry into second-generation cellular service. The IS-54 precursor to the current IS-136 system (see Section 3.2.5) was devised for exactly this purpose. Second, GSM was rushed into European deployment by decree as the Americans selectively upgraded certain AMPS markets in relative leisure. Third, by the time the IS-95 standard was taking its present form in the early 1990s, GSM was already deployed widely in Europe and was being considered longingly by dozens of countries outside the European community. CDMA was thus as new a technology for the late 1990s as GSM was for the early 1990s around half a decade earlier. When the American and Canadian PCS licenses were granted in 1993 and 1994, the successful operators had three choices of system technologies on which to build their networks: IS-136, GSM, and CDMA. More than half the North American PCS coverage was given over to CDMA because this relatively new technology was viewed by a majority of the North American PCS markets as the technology with the least risk for the future. Indeed, when the decisions were made CDMA was not ready for full deployment, and some IS-136 and GSM supporters saw the CDMA supporters as, variously, brave pioneers or reckless fools. All PCS operators, regardless of which of the three technologies they selected, saw a need to distinguish themselves clearly from their 800-MHz cellular cousins. They understood that they had to provide truly personal services: to communicate with people regardless of where they happened to be, and to offer their subscribers a growing catalog of powerful communication management tools, many of which are explored in this book. As the newest technology available, CDMA was seen by its supporters as the most robust platform on which to hang a growing number of features, and anyone deploying it would be protected for a relatively long time from having to replace the system with something newer; that the IS-136 and GSM systems could not survive the demands the future would place on them. Fourth, the broad international appeal of GSM was met with great waves of indifference in most of the North American PCS markets. With the few important exceptions noted in Section 2.1.5, the North American PCS operators were content to cater to the typical American user; someone for whom regular trips outside the continent or country were not likely to happen. So, CDMA arrived just in time for PCS but too late to be the digital adjunct to the huge and wildly successful 800-MHz cellular system. Since the AMPS roaming mode remains intact throughout North America there is little incentive to abandon it for anything beyond a simple digital enhancement. But the AMPS operators feared the PCS operators were bent on stealing away their customers with better voice quality and advanced features; something had to be done. Thus to the IS-136 system we turn.

3.2.5 TDMA (IS-136)

The IS-136 system is an extension of the IS-54 standard (DAMPS), the derivation and technical details of which are covered in [11]. IS-136 is the first of a series of standards (IS-136A, IS-136B, IS-136C, etc.) that, after defining a new digital control resource, add functions that support PCS features. Figure 3.10 shows the overall plan. None of the IS-136 standards abandons the original AMPS control and traffic resources shown as the two upper pairs of forward (BS to MS) and reverse (MS to BS) channels in

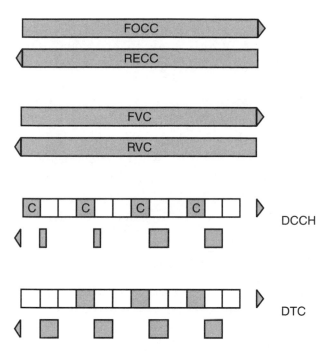

Figure 3.10 Types of IS-136 channels.

the figure. The original AMPS control channel is the *forward control channel/reverse control channel* (FOCC/RECC) pair, and the analog traffic channel is the *forward voice channel/reverse voice channel* (FVC/RVC) pair. IS-54 added the *digital traffic channel* (DTC) to AMPS, and the content of the control channel (FOCC/RECC) was modified slightly to accommodate dual-mode phones that could carry voice in digital form on the DTC. The *digital control channel* (DCCH) was added later with IS-136. The key aspect of IS-54 was its dual-mode capability; dual-mode, moreover, that was plug-and-play compatible with AMPS.

Since the DTC occupies the same narrow 30-kHz channel as the FVC/RVC, it is a short walk to simply (1) swap as many (or as few) base station AMPS channel modules with DTC-capable ones as a particular market demands, (2) install linear RF amplifiers where they are needed (the DTC's π/4-DQPSK modulation requires linear processes for it amplification), (3) load the appropriate voice codec and supporting IS-54 software in the MTSO, (4) and sell some dual-mode phones to your best customers. This is a far simpler and cheaper way for an AMPS operator burdened with capacity constraints to increase capacity than the CDMA alternative offers. Most AMPS base site equipment could easily accept digital channels when IS-54 was introduced; in some sites the conversion was as simple as unplugging an AMPS channel unit and replacing it with a DAMPS one. The usual startup problems associated with any new technology conspired with less than spectacular voice quality from IS-54's *vector sum excited linear predictive* (VSELP) codec to throw some cold water on the early market for dual-mode cellular. Nevertheless, high-quality, reliable dual-mode IS-54 phones were widely available by 1994, and the dual-mode 800-MHz cellular operators could freely turn their attention to the looming PCS threat.

The logical response to PCS was to enhance the IS-54 standard with precisely what it lacked relative to PCS: (1) specify a better voice coder, and (2) specify a DCCH with enough capacity to support a full load of PCS features.

The EFR voice coder finally appeared in early 1997 as an *algebraic code (-book) excited linear predictive* (ACELP; see Chapter 11); and the forward and reverse DCCH specifications, which are the third pair of channels depicted in Figure 3.10, borrow lots of logical channel schemes and access techniques from GSM. Figure 3.11 expands the IS-136 forward DCCH frame structure to show where all the PCS enhancements can be placed.

The forward DCCH is mapped onto the same 40-ms, six-timeslot frame structure, with the same modulation and symbol rate the DTC uses. IS-136-capable phones are obliged to seek out the DCCH from among DTCs in accordance with a search algorithm that is presently left to the various IS-136 phone manufacturers to optimize. There is help on the air interface in the phone's search for a DCCH. First, the analog control channel, which will remain in all 800-MHz dual-mode systems to accommodate older AMPS-only phones, has a pointer in its *overhead message* (OVHM) stream informing IS-136 phones of the channel number of the DCCH. The relevant OVHM symbols are not defined in the AMPS specification and are ignored by properly designed AMPS phones. Second, 11 of the 12 bits that were always set to "0" in each forward IS-54 DTC time slot are assigned the duty of a *coded DCCH locator* (CDL) channel in IS-136 [12]; see Figure 3.12.

Once the IS-136 phone acquires and then camps on a DCCH, a superframe and hyperframe structure becomes evident. Sixteen TDMA frames of 40 ms each make a superframe (640 ms), and two superframes make a hyperframe (1.28 sec). Hyperframes are numbered on the DCCH so that a

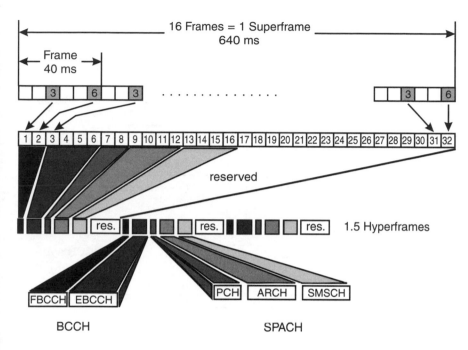

Figure 3.11   IS-136 superframe channels.

paging frame class (PFC) can be assigned to a phone corresponding to a sleep time from PFC = 1 (1.28 sec of sleep time) to PFC = 8 (122.88 sec of sleep time). Higher PFC designations with long sleep times are not intended for phones but for vending machines and other remote control applications that do not need immediate access to the radio system. Each superframe on the forward DCCH can contain up to five types of messages mapped onto the data channel shown at bottom of Figure 3.12, the DCCH downlink. The *shared channel feedback* (SCF) replaces the *slow associated control channel* (SACCH) found in the forward DTC; it controls and acknowledges shortened *random access channel* (RACH) uplink messages (not shown) from different mobiles trying to gain access to the cellular system; it is a replacement for the busy/idle bits on the AMPS FOCC. The *coded superframe phase* (CSFP) replaces the *control digital verification color code* (CDVCC) channel in the forward DTC. The CSFP field appears in the middle of each downlink DCCH and helps the phone find its place in the superframe/hyperframe structure so that it can, for example, figure out when its sleep schedule can start. The normal uplink DCCH bursts are almost identical to the corresponding uplink DTCs except for the

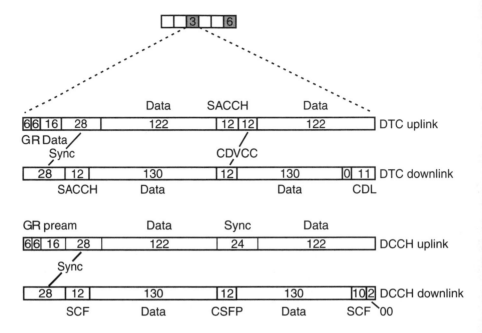

Figure 3.12   IS-136 time slots.

substitution of a synchronization pattern in place of the DTC's SACCH and CDVCC fields.

Returning to the data bits in the downlink DCCH time slots, we refer to Figure 3.11 to discover that the five possible messages are grouped into two types: *broadcast control channel* (BCCH) and SPACH. The SPACH designation refers to the three message types: (1) *SMS message channel* (SMSCH), which can send short text messages to a phone without setting up a dedicated traffic resource; (2) *paging channel* (PCH), which are the kinds of messages we normally associate with paging chores; and (3) *access response channel* (ARCH), which is the downlink's complement to the MS's RACH bursts. The SPACH messages are those directed to specific mobiles, and the BCCH messages are directed to all mobiles in the system. There are two types of BCCH messages: (1) *fast broadcast control messages* (FBCCH), which are general system messages of interest to all mobiles, and (2) *extended broadcast control messages* (EBCCH), which are those messages that are not particularly time critical such as neighbor lists. The system can adjust the number of time slots allotted to each of the message types.

With each revision of IS-136 more PCS services are added: supplementary services, short message services, data services, etc. This has the curious effect of putting PCS on the 800-MHz cellular band, thus blurring the distinction between *cellular* and *PCS* in the North American markets. The system also overlaps into simpler cordless telephone applications thanks to its support of *private system identifiers* (PSIDs) and *residential system identifiers* (RSIDs). An IS-136 mobile may register on a nonpublic system that broadcasts a PSID or RSID that matches one stored in its memory. One of the champions of IS-136 in the United States is AT&T Wireless Services, a group that, together with its affiliates, has secured enough licenses in both the 800-MHz cellular band and the new PCS bands to blanket most of the North American continent with IS-136 coverage on either the 800-MHz or 1.9-GHz bands. This organization has, moreover, succeeded in pushing key phone manufacturers into delivering dual-band/dual-mode handsets in mid-1997: IS-136/AMPS on the 800-MHz band and IS-136 only on the 1.9-GHz band, where the 800-MHz AMPS resource is the continental roaming resource. The 30-kHz-wide IS-136 channels, devoid of their AMPS possibilities, appear in the 1.9-GHz bands as either *base station* (BS) transmitter frequencies and *mobile station* (MS) transmitter frequencies.

The corresponding receiver frequencies (80.04-MHz duplex offset) and band partitions retain those shown in Figure 2.1. The mobile's transmit frequency, F_{MS}, and the base station's transmit frequency, F_{BS}, are figured from the channel number, N, as follows:

$$F_{MS} = 0.030\ 0 \times N + 1849.98 \text{ MHz}$$

and

$$F_{BS} = 0.030 \times N + 1930.02 \text{ MHz}$$

where $1 \leq N \leq 1999$. Table 3.4 shows the partitioning among the six different North American PCS bands.

The reader may see IS-136 as an elegant exercise in pragmatism, as has the likes of such giants as AT&T Wireless Services. Indeed, IS-136 is an attractive and low-cost way for 800-MHz AMPS network operators to respond effectively to new PCS services that may be licensed in their markets, particularly so that the latest version of the standard supports

Table 3.4
North American PCS IS-136 Channel Frequencies

Block	Channel Number	Frequency (F_{MS})	Frequency (F_{BS})
A	2	1850.04	1930.08
A	...	...	...
A	498	1864.92	1944.96
D	502	1865.04	1945.08
D	...	...	...
D	665	1869.93	1949.97
B	668	1870.02	1950.06
B	...	...	...
B	1165	1884.93	1964.97
E	1168	1885.02	1965.06
E	...	...	...
E	1332	1889.94	1969.98
F	1335	1890.03	1970.07
F	...	...	...
F	1498	1894.92	1974.96
C	1502	1895.04	1975.08
C	...	...	...
C	1998	1909.92	1989.96

nonpublic systems in addition to the normal public ones. There are plenty of these markets outside of North America: large portions of South America, Malaysia, New Zealand, Russia, and a growing number of fixed-terminal 400-MHz systems in, notably, Mexico and Vietnam [13].

Because most of the world's AMPS phones were designed before IS-54 and IS-136 were even dreamed of, the dual-mode operators are obliged to take special care that their systems can accommodate the older models. The operators have the unique task of assisting the industry in sorting through some rather sophisticated signaling issues that appear on the common AMPS control channel; each time a new—usually minor—signaling change is made on the FOCC, when perhaps another reserved-for-future-use bit is enlisted into dual-mode service, some older AMPS phones, which are supposed to ignore reserved bits, become confused. IS-136 also has an authentication procedure reminiscent of the GSM scheme that is complicated enough to have spawned the appearance of some dual-mode phones that could not authenticate properly in some systems. The need for the testing and verification of dual-mode phones is a special challenge in IS-136. The *Universal Wireless Communications Consortium* (UWCC) is an advocacy group for IS-136 and IS-41 (wireless intelligent network) that has, through the auspices of the CTIA, spread the different testing and verification tasks of dual-mode handsets among sundry test equipment manufacturers rather than confine the whole job to a huge system few can afford [13].

IS-136 has only one viable competitor in the 800-MHz band: CDMA. It will be interesting to watch how IS-136 fairs on the global stage with GSM, which, because it shares no compatibility with AMPS, has no place in the American 800-MHz cellular band, but gives the relatively few Americans who need it a path to global roaming with a mature complement of advanced services. Because IS-136 makes specific use of the existing AMPS roaming platform, the dual-mode appeal cannot be easily discounted even in the face of GSM. The UWCC's Internet site [13] will serve as a ringside seat to IS-136's successes, and, because of AT&T Wireless Services support, may serve as a demonstration of the importance of marketing prowess over engineering reality.

3.2.6   IS-661

Having examined some cellular and cordless telephone alternatives to providing PCS to those who want it, we take a fast trip into the future as we look at the IS-661 system developed and championed by the Omnipoint Corporation [14]. IS-661 was one of the many original PCS proposals for the North American market that submerged into anonymity when the IS-136/CDMA/GSM wars began to rage. It is not compatible with AMPS. IS-136 explicitly uses AMPS as a means to support national and international roaming at least among IS-136/AMPS systems. The IS-95 system is, as some would say, AMPS-friendly, and GSM, which is also not AMPS compatible, is deployed in too many places for anyone to ignore. IS-661 could not survive the early momentum of the three PCS survivors. It is, however, deployed in a few places (e.g., New York) alongside the AMPS and GSM systems, as a sophisticated cordless phone system and has some particularly intriguing characteristics, particularly its unusual FDMA-TDMA-TDD-CDMA air interface specification that makes it a model for future wireless access systems.

Figure 3.13 shows the IS-661 network interfaces; we are immediately drawn to the *base site controller* (BSC) box with GSM's A interface on its network side. The two systems, GSM and IS-661, indeed share the same A interface. The interface between the BSC and the BS is the *Notes* interface. The corresponding GSM interface is called the *Abis*. The air interface, which is called the U_m-*interface* in GSM, is simply called the *air* interface in IS-661. We devote most of this section to IS-661's curious air interface, which the attentive reader may find a rich playground of possibilities.

As is our custom we start with the frequency assignments in the American PCS band. Table 3.5 divides the band into its six allotments, A through F, arrayed with increasing frequency, in which we place the IS-661 channels, numbered 0 through 35. A channel is 5.0-MHz wide and is placed on 2.5-MHz centers. Because it is a TDD system, the *low* and *high* frequency assignments in the table, which are usually reserved for mobile and base station transmitter emissions, respectively, are grouped into a common spectral resource.

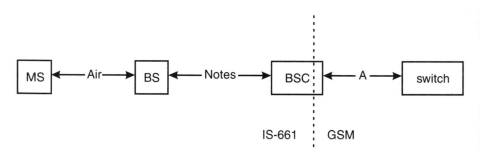

Figure 3.13   IS-661 network interfaces.

Table 3.5
IS-661 Channel Assignments

Block	Channel Number	Frequency (low)	Frequency (high)
A	0	1852.50	
A	1	1855.00	
A	2	1857.50	
A	3	1860.00	
A	4	1862.50	
A	5		1932.50
A	6		1935.00

Block	Channel Number	Frequency (low)	Frequency (high)
A	7		1937.50
A	8		1940.00
A	9		1942.50
D	30	1867.50	
D	31		1947.50
B	10	1872.50	
B	11	1875.00	
B	12	1877.50	
B	13	1880.00	
B	14	1882.50	
B	15		1952.50
B	16		1955.00
B	17		1957.50
B	18		1960.00
B	19		1962.50
E	32	1887.50	
E	33		1967.50
F	34	1892.50	
F	35		1972.50
C	20	1897.50	
C	21	1900.00	
C	22	1902.50	
C	23	1905.00	
C	24	1907.50	
C	25		1977.50
C	26		1980.00
C	27		1982.50
C	28		1985.00
C	29		1987.50

The TDMA frame structure is depicted in Figure 3.14 in which the TDD technique becomes clear. A constant sequence of 20-ms frames can appear on any of the 5-MHz-wide physical channels in Table 3.5. The frames are divided into either 25 or 32 time slots of variable duration with a 625-μs minimum in the 32 time slots per frame case. The figure arbitrarily expands time slot number 10 to show that there are guard times of variable duration between the *base station transmit time* (BS TX), the *mobile*

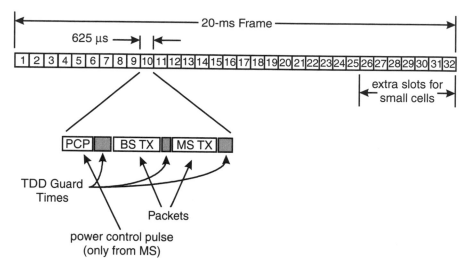

Figure 3.14 IS-661 TDMA frames.

station transmit time (MS TX), and a period reserved for a channel sounding signal (*power control pulse* [PCP]) that the mobile sends to the base station. Operation within small cells reduces the guard time requirements between the packets in each time slot, thus allowing time for 32 time slots in each frame. Proper operation in larger cells demands additional guard time, particularly between the mobile's uplink PCP packet and the base station's downlink packet, which limits the number of time slots to 25 in each frame. The PCP signal from the mobile is used by the base station to figure the propagation loss and link quality so that the mobile's power level can be set and spatial diversity control at the base site can be updated.

One time slot in each frame can support an 8-Kbps user data rate. Mobiles negotiate with the system for multiple time slots in each frame when a high data rate is desired, or they can elect to skip frames for low rate data. Because the system employs TDD, the burst data rate must be at least 512 Kbps (8 Kbps × 32 time slots per frame × 2 packets per time slot = 512 Kbps). The variable guard time requirement, and some overhead data included in each packet, forces the system to a 781.25-Kbps burst rate. The MS TX and BS TX packet possibilities are shown in Figure 3.15.

There are four types of packets: 200-bit poll packets are mapped into the uplink and downlink times during access and data rate negotiation procedures. When access is finally granted, traffic can move

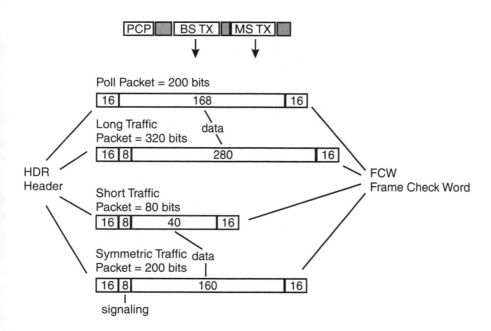

Figure 3.15 IS-661 packets.

symmetrically or asymmetrically between the MS and BS during time slot periods. The symmetric case is supported with 200-bit traffic packets in both the BS TX and MS TX times. The asymmetric case uses the 80-bit short packet and the 320-bit long packet complementary pair shown in Figure 3.15. The poll packet does not have the traffic packet's 8-bit signaling frames because the poll packet's information *is* signaling.

Figure 3.16 demonstrates the use of poll packets and traffic packets. A *general poll* message is sent from the base station in each unoccupied time slot. A mobile that desires access to the system makes its request in an uplink general poll response message packet in an idle time slot. The base station responds to the mobile's access request in the next available time slot period with a *specific poll* message, and the mobile completes its negotiations for service with its own specific poll response packet. If more than one mobile responds to the base station's general poll packet during the same time slot period, both mobiles withhold their attempts to seize a channel (a time slot in one of the 5-MHz frequency assignments). A mobile will wait a period of time determined by its identity number before sending another general poll response.

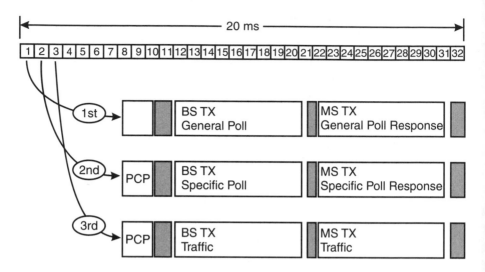

Figure 3.16 IS-661 mobile access.

The system uses FDMA and TDMA techniques to separate users, and TDD to distinguish the uplink from the downlink paths in a channel. An additional CDMA technique is applied to reduce co-channel interference and generally improve system performance and planning options. A *direct sequence spread spectrum* (DSSS) characteristic, unique to each base station, is thus applied to the transmitter emissions with a type of continuous phase modulation called *spectrally efficient quadrature amplitude modulation* (SEQAM). This additional constant envelope modulation spreads the carrier to fill a 5-MHz channel allotment without the need to employ expensive linear RF power amplification schemes. The I-Q plot for SEQAM is shown in Figure 3.17.

If we set aside the elaborate radio techniques discussed here and return to Figure 3.13, we are reminded that the purpose of the IS-661 standard is propose a means to move PCS features that reside in the intelligent network out to users who do not want to be tethered by wires. The systems compete with each other in their abilities to transport the PSTN's features out to the user on the air. There are, however, systems that are by their nature purely wireless ones; they do not have a wired operating mode. Such systems are explored in Section 3.3, and, although they satisfy applications fundamentally different from those to which cellular and PCS systems respond, there is some overlap.

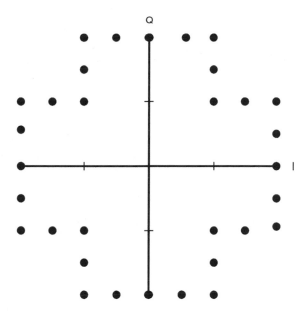

Figure 3.17   IS-661 SEQAM.

3.3   Noncellular digital trunking systems

Other wireless communication systems coexist with cellular. For instance, *private mobile radio* (PMR) and public trunking networks supply services to emergency activities (police, firefighters, ambulances) and private organizations (delivery services, taxi companies, construction companies). These systems (1) have long-range coverage, (2) tend to have a large variety of system architectures, (3) allow calls to be set up (at *push-to-talk* [PTT]) through the system very quickly (on the order of only 100 ms) as if each user always has his or her own virtual open channel, (4) allow mobile terminals the option of communicating directly with each other without using the system's infrastructure, and (5) are usually based on some proprietary or other manufacturer-specific analog technologies. Though an increasing number of these newer systems borrow cellular techniques, they differ from cellular systems in that they provide wide-area dispatch services with open channel and back-to-back working. Trunked radio systems ration out their dispatch features through a huge catalog of features not generally seen in public cellular systems such as these:

▶ User partitioning hierarchies divide the radio system users into fleets and subfleets, and sundry talk groups that facilitate a wide variety of dispatch tasks that are usually not point-to-point ones. For instance, a fire department battalion commander may want to talk privately with the fire chief in one instance and simultaneously with every member of his battalion at another time. A small town may normally partition the water and electric power services from the public safety users except during a flood or other natural disaster. The user partitioning has to be dynamically adjusted during a disaster.

▶ Data services tend to be more widely used in trunked systems than in public cellular systems. Police sometimes receive vehicle registration information, mug shots, fingerprints, and dispatch instructions as pictures or data messages.

▶ The safety of police officers can be enhanced when the dispatcher in a public safety system can gather vehicle and radio terminal location information, for instance, through an integrated GPS system in the police car, without setting up a voice call.

▶ Simple voice encryption techniques have been used in trunked radio systems in even their earliest analog forms. The need for even more sophisticated voice encryption schemes will increase in public safety trunked radio systems.

The users differ from typical cellular customers in that they access the systems very frequently (hundreds of times per day), their connect times are very short (only a few seconds), and they are tolerant of marginal voice quality. Voice traffic, particularly in public safety applications, tends to be critical information transmitted among radio terminals operated by people who receive training in the use of the radio system as part of their daily work chores. The access patterns are such that typical cellular billing is not practical.

Trunking systems vary greatly. Some of the systems are wide-area ones that sell channel time at rates much lower than cellular. The long-haul trucking industry in North America depends on these. Municipalities and electric power utilities, on the other hand, favor much smaller but very complex systems with extremely fast access times and highly specialized mobile and portable terminals. They buy these kinds of systems in one highly integrated lot: base stations, microwave links, dispatch consoles, and terminals.

There was some discussion in recent years of the eventual "death" of traditional land mobile and trunked radio because cellular was presumed to be able to assume these tasks. Though cellular systems can be designed (through packet radio techniques) to mimic many of these sophisticated dispatching systems, including the open channel feature, it is generally accepted today that large users of trunked radio will never find a place in cellular; their needs are too variable and specialized to fit into the strong point-to-point orientation of cellular radio with its emphasis on connections to the PSTN. It is more likely that the so-called software base station will create generalized RF ports or transceivers in cell sites that will carry cellular and noncellular (trunked radio and paging) traffic together. Indeed, the cellular and PCS operators may even evolve into RF utility companies who carry RF services of all kinds for a fee.

Emerging digital technologies, which were born out of the frantic spectrum-saving efforts of digital cellular through open standards, will make future trunked radio applications more economic, more compatible with each other, and much more powerful. Even more applications and services will appear in these systems. TETRA, or the trans-European trunked radio system, which is now called *terrestrial trunked radio* to remove the European bias, is already defined within ETSI [15] and features most of the logical extensions and possibilities we can predict for trunked radio. TETRA terminals have the familiar PTT switch as the embodiment of open channel working, a channel selector switch, and includes procedures that allow high-priority communications preemptive access to the radio system over more routine communications [15,16].

An American initiative called the P25 system, already deployed in a few systems in its so-called Phase 1 version, is also called the APCO 25 system, the APCO 25 project, or Project 25 [17]. The need for P25 in America comes from the same demands that are driving TETRA: a low-cost, high-quality, feature-rich, professional trunked radio system that has some kind of standard radio resource through which public safety agencies can work together in a disaster situation. Most government radio users of such systems in the United States are being forced to replace their old analog radio systems with spectrally efficient ones, thus accelerating the progress of P25 proposals through the TR 8 standards body, a part of the

TIA, which is engaged in defining a Phase 2 version of the system. Recent work in the TR 8 committees has resulted in entertaining the possibility of more than one standard that shares a common operating mode with the present proposal.

A third system is the European TETRAPOL proposal championed by Matra [18]. It is a French proposal currently used by the French National Police (ACROPOL system). Discussions are underway to examine a friendly coexistence and even a standardized system interoperation scheme (ETSI RES 6 Working Group 3) with TETRA. Table 3.6 lists a few technical features of TETRA, P25, and TETRAPOL.

The land mobile radio business will continue to grow at its own rate independently of cellular. It currently includes about 19 million units, and is growing at a 4% annual rate [19]. The fastest growing portion of the land mobile services are the trunked radio systems, which are known as *specialized mobile radio* (SMR) or professional radio systems. The typical manufacturer of SMRs for demanding public safety applications, which is the high end of the SMR market, installs between 150 and 200 base site repeaters all over the world every year. The typical SMR system has five base site repeaters, which can support about 500 terminals, and costs more than $3 million. The largest systems support more than 3,200 terminals with 20 repeaters, and cost at least $100 million.

Given these high costs, there is a market within cellular for the more casual users of trunked radio services. The likely users are found among those who would prefer to be billed for network time rather than buy a whole system, and who can accept the more generalized and much lower cost portable terminals available in cellular systems. GSM has been proposed to pick up this slack in the market through some Phase 2 and Phase 2+ additions (e.g., *closed user group* [CUG]). These new features make the equipment built around the resulting specifications fit for certain applications such as those found in some railway operations. The UIC forum, *Union Internationale de Chemin de Fer*, which is a GSM interest group of European railway operators and working groups within ETSI, is the main driver on particular GSM Phase 2+ improvements. Digital cellular systems and services can, to some degree, compete with trunking systems; see, for example, Chapter 7 on the CUG. The requirements that are compatible with modern digital cellular are listed here:

> ▶ *Coverage:* Long-range performance, achieved in traditional trunking systems by use of lower radio frequencies and higher transmission

Table 3.6
Digital Trunking Radio Systems Comparison

Feature / Parameter	TETRA	P25 (APCO 25) Phase 1	TETRAPOL
Access method	4-timeslot TDMA	FDMA/CSMA	FDMA
Channel bandwidth	25 kHz	12.5 or 6.25 kHz	12.5 or 10 kHz
Frequency band	European 380–400 MHz for emergency services; 410–430 MHz and others for civil use	U.S. public safety bands (450 and 821 MHz)	European bands (380–400 MHz) and others
Modulation scheme	π/4-QPSK	C4FM Nyquist raised cosine, cascaded with shaping, and FM modulator (6.25 kHz: CQPSK) common receiver	GMSK
Gross bit-rate	36 Kbps / 4 slots	8 Kbps	8 Kbps
Speech codec	Approx. 4.8 Kbps ETS 300 395-2 "TETRA Codec"	4.4 Kbps improved multiband excitation (IMBE) codec, 2.8 Kbps FEC	Proprietary LPC/LTP/RPCELP
Data rate	7.2 Kbps: no protection; 4.8 Kbps: low protection; 2.4 Kbps: high protection	9.6 Kbps	8 Kbps

power levels, is no longer an issue in countries where almost 100% of the area is covered by low-power cellular service access.

▶ *Services:* Plain voice telephony, of course, and simple low-rate data are already featured in all digital cellular systems. Other services (supplementary or special ones) are not always covered; group calls or direct unit calls, high data rates, preempting, and many other features are not commonly defined for all digital cellular standards.

▶ *Privacy:* Digital representation of voice and TDMA are barriers to the casual eavesdropper using FM scanners, and encryption adds yet another layer of protection against even the most tireless interception efforts.

▶ *Performance:* High voice quality, high probability of access to service, and a low number of dropped calls are not an issue with today's digital cellular technology. Some care has to be taken that all the requirements are met for particular applications all the time, for example, the highest access priority for emergency services.

▶ *Equipment cost:* The high price paid for equipment designed and manufactured for highly specialized niche markets has long been accepted and accounted for in the community of private mobile radio users. Cellular provides the alternative of a dense infrastructure and terminal costs based on mass market economies of scale.

▶ *Service cost:* The operational costs of specialized private and public networks have to be compared to the cost at which more generalized cellular service can—or better, *could*—be made available. They key here is that SMR users prefer flat monthly fees over being billed for a precise accounting of connect times.

3.4   Interference and health issues

Why do we treat the subject of health issues in a chapter entitled "A Look Over the Fence"? A discussion of the potential health risks of *electromagnetic radiation* (EMR) has been going on since the widespread use of public radio communications systems became obvious to the general population more than a decade ago. If recent court judgments in consumer complaints against all kinds of industries, particularly in the United States, are a model for the future, then the wireless industry should never return to the earlier, occasional not-in-my-yard-not-my-problem attitudes.

The discussions are often technical and disciplined; sometimes they are not. The emotive impact of perceived health risks may sometimes give way to calm investigations of *electromagnetic compatibility* (EMC) concerns, or they may, at some time in the future, boil over into jury trials during which huge awards may be made to subscribers with seemingly little connection to scientific reality. Some of the discussions are enlightening, interesting, and helpful, for instance, those concerned with automobile airbag triggering, interference with aircraft navigation systems, and interference with hearing aids and heart pacemakers; at other times, they are neither interesting nor helpful.

Recent interest in the United States concerning the possibility that cellular radio equipment may have caused brain cancer received lots of public attention and brought the matter into the offices and labs of the wireless industry without shedding much light on the subject.

Certain informal claims emerged even from within the wireless industry during the 1990s that some radio access schemes were potentially more harmful than others. Interference with devices such as hearing aids and heart pacemakers was claimed by some people sympathetic with *one* system-standard-camp to be a problem of the *other* camp. TDMA systems, for instance, have a fixed repetition rate of RF bursts that tends to lie in the audible frequency range. People who have spent lots of time around GSM phones, for example, can recognize access procedures from routine traffic handling merely from the sound the mobile station's bursts sometimes make in a public address system, a car radio speaker, or a landline telephone speaker. GSM bursts are repeated every 4.615 ms, which yields a buzzing sound in loudspeakers with a frequency of 217 Hz (1/4.615 ms). Bursts from IS-136 TDMA phones are a bit harder to discern because of their lower frequency; this rate is 20 ms or 50 Hz. The TDMA buzz can sometimes even be heard in the mobile's own speaker when shielding against such interference is inadequate. Even though CDMA uses a completely different radio access technology, it is also said to cause interference. CDMA phones, too, use bursty transmissions the rates of which are a function of the voice activity and time-varying speech coder rate. The additional amplitude modulation that is applied to the CDMA phone's transmit signal is not as regular as those from a TDMA phone; they sound more like low-frequency noise. Whatever distractions the wireless industry brought on itself with its internal finger-pointing, they helped put a handle on the more nebulous health issues.

Thus there is concern over biological effects on human bodies. There is proof, for example, that the human body absorbs *nonionizing electromagnetic* energy from an active mobile terminal's transmitter, and this energy becomes transformed into heat. Radiation can be regarded as *ionizing* or *nonionizing* depending on the wavelength. Electromagnetic radiation at a wavelength just beyond the visible spectrum is called *ionizing* because it is energetic enough to break chemical bonds within living cells. Lower frequency radiation below the visible spectrum—as in cellular radio and PCS—does not exhibit this effect but merely heats the tissues as a microwave oven or sunlight does. This heating effect is especially evident when the transmitter is held close to the body as is often the case with small handheld phones. The average temperature of the body tissue thus exposed to the transmitter's radiation will rise by a few tenths of a degree Kelvin. This means that temperature in the human head, in the ear, or in the eyes will rise. Even if the temperature change is the same in all the types of tissues and body structures, the effect of the change may not be the same. It is more difficult, for example, to cool the cornea of the eye, instead of, say, the ear, back down to a normal temperature because it does not have a cooling blood flow. Less clear are the long-term effects from prolonged exposure to low-level nonionizing radiation. What are the biological and biochemical effects of such radiation? Will there be changes in the tissues or structures in 1 year or 5 years? What about 20 years? What about damage to a fetus or to the sex cells? And then we have questions that include words with lots of emotional baggage: cancer. Is there *any* risk increase through the use of cellular phones or through exposure to this kind of radiation like from base stations? If so, under which circumstances does the risk increase? Human society has difficulty handling processes that may take years to yield a result, and society simply does not know how to deal with slow processes that, because it knows little or nothing of its workings, it feels may hurt or kill someone.

Research, to some major part funded and supported by the wireless industry, is ongoing. In the United States, for instance, an independent scientific research program is being carried out by the *Wireless Technology Research* (WTR), L.L.C. Started in 1993 and running for more than 5 years, the WTR carries out and manages research related to public health problems with the use of wireless and cellular equipment. EMC issues are addressed by the U.S. Center for the Study of Electromagnetic Compatibility at Oklahoma University [20]. Other programs are moving forward in Europe as well.

For now, as we calmly await any findings that may come from the studies, claims and counterclaims on which radio technology is safer relative to another technology, and under what circumstances phones may or may not be harmful, contribute nothing to a fruitful discussion of the subjects; they merely cloud everyone's concentration with nervous cries from confused and uneasy consumers. The unrest is fueled by the indecision of worldwide authorities to take real action, or to at least thoroughly investigate the matter and fund its research. The wireless industry should avoid emulating the habit of governments to avoid responsibility, to deny the existence of problems, and to remain silent on the issues. The persistent secretiveness of governments spawns conspiracy theories and breeds distrust.

The whole wireless industry is in a difficult situation no matter how it may respond to health concerns. Any results from the research funded by the wireless may be viewed with skepticism by those who feel the work was somehow influenced by those who pay the bills. Those who perform the research have bills to pay too, and the competition for research grants is intense enough to stir up controversy over even the most interim results in order to keep the money flow intact. Fortunately, since it has resolved to fund research rather than ignore the issues, the wireless industry has avoided the worst course of action and the appearance of avoiding social responsibility. It seems best to hold to the present course; to continue the funding while remaining open and compassionate with all who may comment on the work, and avoid being spontaneous or reckless because it strictly avoids even the vague appearance of trying to influence the outcome of the research it funds.

So, with the first three chapters of this book we have reviewed the origins and history of GSM, and defined personal communications services and the networks that furnish them. We also looked at some of GSM's responses to the PCS markets, and we saw some alternative systems and listened to some consumer questions.

In Chapter 4 and following chapters we step down into the basement of GSM; we climb around in the heating ducts and probe into the plumbing to see the mechanisms that can make it a PCN.

References

[1]   Redl, S. M., M. K. Weber, and M. W. Oliphant, *An Introduction to GSM,* Norwood, MA: Artech House, 1995, Chaps. 12 and 13.

[2]   ETSI, European Telecommunication Standard ETS 300 175, Parts 1–9, and ETS 300 176, first release in August 1992.

[3]   van der Hoek, H., "The TDMA Approach: DECT Cordless Access as a Route to PCS," Chap. 4 in *Personal Communications Systems and Technologies,* J. Gardiner, and B. West (eds.), Norwood, MA: Artech House, 1995.

[4]   ETSI, European Telecommunication Standard ETS 300 466 DECT/GSM interworking profile.

[5]   PHS International, http://www.phsi.com.

[6]   Patel, V., *AIReach*^{TM} *PACS PCS Features and Applications*, Germantown, MD: Hughes Network Systems, June 1996.

[7]   Garg, V. K., and J. E. Wilkes, *Wireless and Personal Communications Systems*, Upper Saddle River, NJ: Prentice Hall, 1996, Chap. 6.

[8]   Shih, M.-P., et al., "Development of Personal Communications Services for Taiwan Areas," *IEEE Personal Communications*, Vol. 4, No. 2, April 1997.

[9]   Redl, S. M., M. K. Weber, and M. W. Oliphant, *An Introduction to GSM*, Norwood, MA: Artech House, 1995, Chap. 13.

[10] http://www.cdg.org.

[11] Redl, S. M., M. K. Weber, and M. W. Oliphant, *An Introduction to GSM*, Norwood, MA: Artech House, 1995, Chap. 12.

[12] "Overview of the IS-136 Air Interface," Document ES-1035, revision 1.0, Kirkland, WA: AT&T Wireless Services, March 1997.

[13] http://www.uwcc.org.

[14] "Omnipoint IS-661-Based Composite CDMA/TDMA PCS System," Colorado Springs, CO: Omnipoint Corporation, 1995.

[15] "TETRA Voice + Data, Designer's Guide," version 0.0.10, draft, Radio Equipment and Systems Technical Committee of ETSI, January 1996.

[16] http://www.tetramou.com.

[17] http://members.aol.com/project25/index.html.

[18] http://www.tetrapol.com.

[19] "The State of SMR and Digital Mobile Radio: 1997," Washington, DC: The Strategies Group, 1997.

[20] Lukish, T. J., "Radiofrequency Electromagnetic Interference Risks in the Wireless Telecommunications Environment," *Proc. GSM World Congress 1996*, IBC Technical Services, 1996.

GSM services and features

The development of GSM standards and features

Throughout the remainder of this book we need to deal with phases, versions, releases, services, functions, and features; the abbreviations and fine distinctions can cloud the material. We have seen some of the terms in Chapter 2 and we will see more of them later. This chapter organizes all of GSM's services and features over time. New GSM services and features (teleservices, bearer services, supplementary services, the SIM card, and network improvements) appear in phases, versions, and releases of the standards. GSM's services and features are thoroughly explained in later chapters, but for now we can consider the following distinctions:

▶ *Teleservices* are those that people use directly; talking on the phone is an example.

‣ *Bearer services* are transport services that become useful to someone only if they attach some kind of contraption or software to his phone; Internet surfing is an example of an application that uses bearer services.

‣ *Supplementary services* (SS) are features resident in the network itself; the phone must have a way to use them. Putting a call on hold is an example of SS.

‣ The *SIM card* is the subscriber identity module; this stores subscriber-related data and identifies the subscriber in the network.

Back in 1982 when the *Groupe Spécial Mobile* was founded, the goal was to design a common Pan-European standard for mobile telecommunications to replace the chaos of sundry incompatible analog protocols that covered the continent, and which were seen as a barrier to the economic health of the European community. In the 15 years that followed, industry representatives coming from regulators, GSM operators, and manufacturers specified not only a common air interface for the future Pan-European system but a complete *public land mobile network* (PLMN) standard.

Initially, everybody wanted to introduce the best her or his own national standards could offer, be it NMT (Scandinavia), ETACS (United Kingdom), or the Net C protocol of Germany. It quickly became clear that even the most clever combination of all the early proposals could not meet the demands for the quality and features that everyone wanted, which is why the GSM standard has and still continues to evolve. The standards and features already defined for the *integrated services digital network* (ISDN) had a major impact on the proceedings and forced the participants to look to the latest technologies and radio techniques for relief. GSM eventually became a digital cellular standard based on a TDMA technique for the air interface that could offer ISDN services from a SS7 architecture [1,2].

Fueled by extraordinary efforts on specifications work, document revisions, and field tests, GSM became an elegant second-generation mobile phone system that was the topic of lectures, social conversations, and technical papers throughout the wireless industry. The sparkle was dulled as the specification process took longer than anticipated; so much longer that in countries like France, Italy, and Austria some new cellular systems based on first-generation analog standards were introduced, or adapted to exiting systems, to cope with the pent-up demand for mobile telephony. In other countries where new, competing operators had already been licensed to build and operate whatever GSM was trying to complete, the demand for operational equipment based on GSM standards became urgent. The solution was to split the standardization work into phases, which leaves us with today's legacy of three phases: Phase 1, Phase 2, and Phase 2+.

As we saw in Chapter 2, the GSM Phase 1 specifications were frozen in 1991 to allow the operators to build their first commercial GSM networks. Unfortunately, creating a stable type approval process, including equipment and test suites, for mobile phones took much longer than expected so that the actual launch of GSM was delayed further to the summer of 1992. An *interim type approval* (ITA) procedure was introduced to move *some* mobiles onto the market. All other features and changes that were to be introduced after the 1991 freeze were moved into the Phase 2 specifications. Phase 2 was frozen in October 1995, and subsequent improvements and additions will automatically make their way into Phase 2+.

The history of GSM's phases is evident in the documents that are known as either *GSM* documents or *digital cellular telecommunications system* documents. The digital cellular telecommunications system title replaces the *European* banner on Phase 2+ publications to finally acknowledge the global impact of GSM that occurred in the past 5 years. We can also refer to different versions of GSM with version numbers. The version numbers for the Phase 1 documents are 3.X.X, Phase 2 documents have version numbers 4.X.X, and Phase 2+ documents have numbers 5.X.X. ETSI could have continued to work in *phases* forever incrementing version numbers with increasing numbers of phases, but many features added these days are increasingly optional ones for the operators rather than the mandatory features of early phases, and the demand for them cannot wait for a freeze of Phase 2+ work. ETSI, therefore, now distinguishes work packages that are released on a yearly basis. Starting in 1996 the Phase 2+ documents are referred to as *Release 96*, and we just saw a *Release 97* published [3].

4.1   Phase 1

More features were specified in Phase 1 than were actually implemented or working; this situation has recently reversed itself in later releases.

4.1.1   Phase 1 teleservices

The first network services and products that *finally* made it to the market-place focused on plain voice communications, which was supported by the standard 13-Kbps full-rate speech codec. A tentative venture into auxiliary features was represented by *emergency calls,* which were accessible via the standard emergency number "112." Quick access or shortcut buttons, like "SOS," were offered too. The emergency services were available regardless of whether the subscriber was allowed access to the remainder of network services or not. Emergency calls worked, and still work today, without a SIM inserted in the handset. However, it is up to the network operator whether or not to accept emergency calls without a SIM present, because the emergency services authorities might demand an identity to call back to the subscriber. France and the United Kingdom, for example, do not allow emergency calls without a SIM present, whereas Germany does.

Even though Phase 1 specified *short message services* (SMSs) and some fax and data capabilities, these features were not available in the networks nor were they supported by the mobile phones when GSM appeared. Point-to-point mobile terminated SMS became more common only when mailbox servers started sending short notes to subscribers informing them of new voice mail messages awaiting the subscriber's attention.

4.1.2   Phase 1 bearer services

Table 4.1 lists the bearer services defined in Phase 1. We should highlight here the difference between the transparent and nontransparent services: *transparent* for GSM means that the GSM infrastructure (including mobile stations) pass on the data from data or fax terminals without any additional protocol to reduce the effect of errors introduced on the air interface. The *nontransparent* services, however, add error protection protocols that make data and fax transmissions much safer. A side effect, which is detailed further in Chapter 5, is that the effective speed of transmission is higher for the nontransparent service.

4.1.3 Phase 1 supplementary services

The first *supplementary services* (SS) were very basic in Phase 1.

Call forwarding (CF) enables subscribers to divert calls to other phone numbers. CF services distinguish unconditional forwarding of all calls from conditional call forwarding. Conditional CF applies to the subscriber who is busy, who cannot be reached, or who does not answer.

Call barring (CB) lets subscribers bar certain calls. There are five types of CB service:

1. Barring of all outgoing calls;

2. Barring of outgoing international calls;

3. Barring of outgoing international calls except those directed to the home country;

4. Barring of all incoming calls;

5. Barring of incoming international calls when roaming (to avoid hefty surcharges).

Table 4.1 lists the main features and services specified, but not necessarily implemented and working, for GSM Phase 1.

Table 4.1
Features Covered by GSM Phase 1

Service Category	Service	Comment
Teleservices	Telephony	Full-rate speech at 13 Kbps
	Emergency calls	"112" defined as a GSM-wide number to a local emergency service
	Short message service (SMS) point-to-point	Alphanumeric message exchange between two individual users via a dedicated service center
	Short message service (SMS) cell broadcast	Alphanumeric information to all mobile stations within one cell or area but not during a speech or a data connection
	Videotext access	

Table 4.1 (continued)

Service Category	Service	Comment
	Teletex transmission	
	Facsimile alternate speech and facsimile Group 3	
	Automatic facsimile Group 3	
Bearer Services	Asynchronous data	300, 1200, 1200/75, 2400, 4800, 9600 bps, transparent / nontransparent
	Synchronous data	2400, 4800, 9600 bps, transparent
	Asynchronous PAD access	300, 1200, 1200/75, 2400, 4800, 9600 bps, transparent / nontransparent
	Synchronous packet data duplex 2400 bps T/NT	2400, 4800, 9600 bps, transparent / nontransparent
	Alternate speech / data	
	Speech followed by data	
Supplementary Services	Call forwarding (CF)	All calls, calls when a subscriber is busy, not reachable, or not available
	Call barring (CB)	All outgoing calls, all outgoing international calls, all outgoing international calls except those to the home country, all incoming calls, all incoming calls when roaming

4.2   Phase 2

While Phase 1 networks were starting up and winning subscribers and admirers in Europe, curious attention came from Asia (primarily from Hong Kong and Singapore), Australia, and a wide variety of industry groups in other countries as specification work continued almost seamlessly within ETSI working groups. As the industry caught up with the Phase 1 specifications, the appetite and demand for advanced features grew.

4.2.1   Phase 2 teleservices

Phase 2 added the half-rate speech codec; Phase 1 already had provisions for implementing it on the air interface, but the delay was confined to the speech coding process itself; the algorithms were not available as they were still being improved with further testing.

The *enhanced full-rate speech codec* (EFR) was a requirement from the U.S. market for its introduction of PCS 1900. The voice quality of the EFR approaches landline quality under good radio conditions and is superior to the normal full-rate codec under nonideal conditions. At the time of this writing the GSM EFR was not deployed in GSM networks (and terminals) outside North America and is only just appearing in the North American markets where most of the PCS 1900 phones have both types of full-rate coding schemes installed; the American networks tend to favor EFR channel assignments. The EFR was originally a Phase 2+ feature, but has been moved to Phase 2 by a GSM plenary meeting to allow earlier adoption and introduction into GSM networks in Europe.

4.2.2   Phase 2 supplementary services

Major additions were made to the supplementary services, which were heavily influenced by the ISDNs.

Calling line identification (CLI) allows the receiving party to get an indication of who is calling by displaying the calling party's phone number on the caller's display. Two different services are defined for CLI:

1. The *presentation* (CLIP) service is the normal option where the caller's number is shown in the phone's display.

2. The *restriction* (CLIR) service is an option that the calling party may wish to exercise that does not show the calling party's number to the called party.

Connected line identification (COL) is the opposite of CLI; it lets the calling party see the number to which he or she is connected. This is a helpful feature when a call is forwarded to a different number. Two different services are specified for COL:

1. The *presentation* (COLP) service shows the called number to the caller.

2. The *restriction* (COLR) service lets the called party block the display of the forwarded-to-number on the calling party's terminal.

Call waiting (CW) is an indication to a *busy* subscriber that another caller is trying to reach him. *Call hold* (CH) complements the CW service in that it allows the subscriber to put active calls on hold in order to answer waiting calls or set up multiparty calls.

Multiparty communication (MPTY) lets a subscriber set up calls to more than one party.

Closed user group (CUG) is a service borrowed from the trunking systems briefly described in Chapter 3, where a group of users, such as a fleet of taxi drivers, share the radio resources with short dialing numbers of only a few digits for each other.

Advice of charge (AoC) provides a subscriber with charging information. The subscriber is not only able to check on ongoing charges (in-call) but also accumulated call charges (previous calls).

Unstructured supplementary services data (USSD) provide the network operator with a tool to define proprietary supplementary services using simple control strings that only have meaning within the home network.

Operator-determined barring (ODB) is not a subscriber-accessible service. It lets the network operator control certain feature groups for individual subscribers. Think of someone who does not pay his or her phone bills.

4.2.3   Phase 2 network improvements

Phase 2 added new mechanisms to the infrastructure to handle the radio resource, location management, call control, and higher level functions throughout the network. They are completely transparent to the user and they let the network operator use his or her resources (e.g., the radio frequencies) more effectively.

Phase 2–compatible equipment, fixed and mobile, must support the Phase 2 features and its signaling. Compatibility is controlled by a type approval regime described in Chapter 2. The first real Phase 2 mobile equipment that supported both the signaling and the services appeared at the end of 1996. Before 1996, some Phase 1 mobile phones supported many of the Phase 2 supplementary services, thus giving GSM subscribers some benefits of Phase 2 before it was officially introduced. Phase 2 features require sophisticated support from the networks, but there are many differences in support details among the networks. Confusion appears in those countries that have legal restrictions against the introduction of certain features like CLI. The reason for this was that some suppliers of infrastructure had the presentation (CLIP) implemented before the restriction (CLIR), but privacy concerns dictate that CLIP is not allowed to be introduced without CLIR. Table 4.2 gives an overview of the new Phase 2 services. The Phase 1 services are still valid but have been amended to the Phase 2 specifications. Terminal compatibility is maintained—any phase of network will accept any phase of terminal.

Table 4.2
Additional Features Covered by GSM Phase 2

Service Category	Service	Comment
Teleservices	Half-rate speech codec (HR)	Optional implementation, implying the use of dual rate (half rate and full rate) in one phone
	Enhanced full-rate speech codec (EFR)	Optional implementation, implying the use of dual rate (enhanced full rate and full rate) in one phone
Supplementary Services	Calling line identification (CLI)	Presentation and restriction of displaying the caller's ID
	Connected line identification (COL)	Presentation and restriction of displaying the called ID
	Call waiting (CW)	During an active call, the subscriber will be informed about another incoming call (offered together with call hold)
	Call hold (CH)	Put one call on hold in order to answer/originate another call (offered together with call waiting)
	Multiparty communication (MPTY)	Up to five ongoing calls can be joined to a multiparty communication
	Closed user group (CUG)	Similar to a feature known from trunking services
	Advice of charge (AoC)	On-line charge information
	Unstructured supplementary services data (USSD)	Allows operator-defined individual services
	Operator-determined barring (ODB)	Enables the operator to restrict certain features from individual subscribers

4.3   Phase 2+

As we explained in the introduction to this chapter, Phase 2+ includes all additional features specified for GSM after October 1995 (the freezing date for Phase 2). ETSI distinguishes yearly releases, thus we discuss here Release 96 and Release 97.

4.3.1   Release 96

As new *flavors* are added to GSM, such as railway operations and communications applications, special features required by particular users make their way into the specifications.

4.3.1.1   Release 96 teleservices

In the service group of the *teleservices* we find some examples. The *voice group service* (VGS) was introduced on behalf of the European railway organization and defines two services:

1. The *voice group call service* (VGCS) provides an effective way to set up a group call within a very short time to predefined groups; this is much faster than multiparty calls because multiparty calls are not predefined.

2. The *voice broadcast service* (VBS) enables the originator to make verbal announcements in a manner similar to broadcast announcements on a radio to a predefined group.

Extensions to the SMS alphabet became necessary as more Arabian and Asian countries joined the GSM community. Because these countries are already accustomed to paging services in their own languages and characters, the Roman-character-based GSM short message service is no longer acceptable. The introduction of the *Universal Coding Scheme 2* (UCS2) helps to cover Arabian and Chinese characters.

The *cell broadcast service* (CBS) was initially used for the transmission of local-area information and for tariff notifications. Operators wishing to implement new services based on CBS realized that the information capacity on a single-cell broadcast channel was not sufficient for the additional traffic. A second SMS cell broadcast channel introduced for Phase 2+ will help overcome the capacity problems.

4.3.1.2 Release 96 bearer services

Major improvements were achieved for the bearer services in Release 96. With more applications being accessible through GSM (e.g., Internet browsing), the bandwidth limitation of the 9600-bps data channel became apparent. The solution was *high-speed circuit-switched data* (HSCSD), which allows a combination of up to eight time slots for a single communications link. Combining eight time slots allows anyone to use eight times the capacity of a single slot, which could be up to 8 × 9.6 Kbps = 76.8 Kbps when using 9.6 Kbps for each traffic channel. In practice, the data rate will be limited to 64 Kbps, which is the maximum data rate on an ISDN channel.

Packet data on signaling channels (PDS) is an intermediate answer to the demand for data services. Because packet data services borrow signaling resources, the implementation effort in the networks is reduced and will be available sooner. The data rates for PDS range from 600 to 9600 bps. Modest data applications (downloading e-mail and other information packages) might be done more economically by using packet data services over the air rather than with normal Phase 1 data services and certainly more efficiently than with HSCSD.

The highest possible data rate on a single GSM data channel, defined as a time slot on an assigned frequency, is 9.6 Kbps. Implementations in the fixed network today can achieve much higher standard user data rates of up to 33.6 Kbps (PSTN) or 64 Kbps (ISDN). Work has been finalized on a new service offering 14.4 Kbps on a single time slot. The higher 14.4-Kbps rate data channel combined with HSCSD will improve the data rates when combining multiple time slots.

4.3.1.3 Release 96 supplementary services

Release 96 brought modest improvements to the supplementary services.

Explicit call transfer (ECT) allows a subscriber to pass a call on to another subscriber, a feature borrowed from the office PBX. This will be a very useful feature especially for business users, but its early introduction may be delayed until security provisions to avoid misuse are in place.

Enhanced multilevel precedence and preemption service (eMLPP) is another service introduced to meet the needs of the European railway organizations. The eMLPP service allows someone to assign priority to certain calls upon subscription. When high-priority calls meet limited resources the network will preempt other calls to free up the resources. This service is dedicated to certain user groups that need to have this priority, for example, train personnel or security services. Therefore, the "standard" users of GSM will very likely never be able to subscribe to, or make use of, the eMLPP service.

4.3.1.4   Release 96 SIM

Though we did not mention it in the Phase 1 and Phase 2 reviews, the SIM has always followed the growth of GSM services. Phase 2+ swings a bright light onto the SIM's ability to bring unique—and portable—features to GSM. We call on four examples to illustrate how the SIM can enhance the portability of features, setups, and configurations from one mobile phone to another.

1. *Barred dialing numbers* (BDN) allow a subscriber who gives away his phone—or, more precisely, gives away his SIM card—to bar certain numbers from being called. This is a complementary feature to outgoing call barring or the barring of international calls.

2. There has been a tendency in Europe to subsidize mobile equipment in order to increase the number of subscribers; the cost of the phone is seen as an artificial barrier to the regular collections of subscriber fees from people who, save for the cost of the phone, are willing to pay for them. Because the subsidy is not the same in all countries and markets, operators wanted a way to protect their investment in their subscriber base by personalizing the mobile equipment. The *ME personalization* feature not only specifies an operator lock, but also the personalization of an ME to an individual SIM as a protection against theft. There is also a corporate lock and a service provider lock, which is used depending on who subsidizes the mobile equipment. This ME personalization is often referred to as *SIM lock*, which is a misnomer since it is the phone that is locked to a SIM and not the other way around.

3. The *service dialing numbers* (SDN) feature allows the operator to program dedicated service numbers into a write-protected area of the SIM card.

4. One major improvement is the *SIM application toolkit*, which provides the operator with a variety of possibilities to create new

features. These are completely network dependent and range from basic activation of the SIM over the air to changing parameters on the SIM to the introduction of elaborate packages of value-added services such as hotel or flight bookings.

4.3.1.5 Release 96 network improvements

Phase 2+ also introduced new features to the network itself. *Network identity and time zone* (NITZ) provides a new means to transfer the network identity, the network's name, to the mobile equipment. Before Phase 2+ the identities of known networks were programmed into the mobile equipment in the form of codes or numbers representing the network. As new networks opened up it took time for new mobiles to support the new names; older phones would only display the digits of the code identifying a new network; not its name. Consider the example of when a network changes its name for commercial, legal, or marketing reasons. NITZ not only provides the means to inform the mobile equipment about new network identities in plain characters but also about the local time and the time zone in which a mobile station is roaming [4].

Customized applications for mobile network enhanced logic (CAMEL) brings IN features to GSM. By leaving the implementation of new Phase 2 and Phase 2+ features exclusively to the operators, roaming customers accustomed to a certain service feature in their home country and network could not enjoy the same feature in another country. CAMEL provides the facilities that allow roaming subscribers the same features they enjoy in the home network. CAMEL will be introduced in two phases: Phase 1 of CAMEL is part of Release 96 and Phase 2 of CAMEL is part of Release 97.

A widely discussed opportunity for the deployment of GSM access and services is *radio local loop* (RLL). Particularly in remote areas with little or no fixed line coverage at all, RLL will provide efficient and economic solutions for providing wireless coverage as a substitute to traditional fixed lines [5]. RLL provides wireline service by using the GSM infrastructure, which could include all of its entities (MSC, BSC, BTS, etc.), or only the radio parts (BSC and BTS). It is much cheaper and faster to roll out telephone services by using radio access than wiring the whole countryside. An RLL may use GSM technology but would not necessarily be operated by GSM operators. There is a goal within the GSM community not to alter the GSM implementation, particularly in the area of mobility management, just for RLL. Allowing changes would result in different dialects of GSM, which may raise the prices of equipment because of incompatibilities.

Call charges for roaming subscribers can become very expensive because calls are not necessarily routed in ideal ways. The *support of optimal routing* (SOR) feature enables GSM operators to clear most common routing problems. This will have a major impact on roaming call charges. Table 4.3 lists the additional GSM features introduced in Release 96 of Phase 2+.

Table 4.3
Features Covered by Release 96 of GSM Phase 2+

Service Category	Service	Comment
Teleservices	Voice group service (VGS)	Voice group call service (VGCS) and voice broadcast service (VBS), both defined for UIC applications
	Extension to the SMS alphabet	Allowing UCS2 with 16-bit character representation
	Second SMS cell broadcast channel	Increasing capacity for cell broadcast services (CBS)
Bearer Services	High-speed circuit-switched data (HSCSD)	Combination of up to eight time slots for a single link
	Packet data on signaling channels (PDS)	Reusing signaling resources for packet data traffic
	14.4-Kbps user data rate	
Supplementary Services	Explicit call transfer (ECT)	
	Enhanced multilevel precedence and preemption service (eMLPP)—Phase 1	Priority for certain calls defined for UIC applications
SIM Features	Barred dialing numbers (BDN)	Bars certain numbers from being called
	ME personalization	Allows restriction of phone usage to particular SIM cards (subsidy protection)
	Service dialing numbers (SDN)	Preprogrammed service numbers on the SIM, protected from erasure by subscriber
	SIM application toolkit	Vehicle for operators to define their own value added services

Service Category	Service	Comment
Network Features	Network identity and time zone (NITZ)	Simple "download" of network-related information and data to the MS
	CAMEL—Phase 1	IN features for roaming GSM subscribers
	Radio local loop (RLL)	GSM extension for fixed networks
	Support of optimal routing (SOR)	Reduced charging due to optimized routing

4.3.2   Release 97

At the time of this writing the following items were being considered for Release 97 of Phase 2+.

4.3.2.1   Release 97 bearer services

The hottest item in the group of bearer services is the *general packet radio service* (GPRS), which is a general packet service with a dynamic allocation of resources. Different channel coding (data rates of up to 21.4 Kbps per single time slot) and the option to use up to eight time slots will accommodate a wide variety of data rates up to 100 Kbps. GPRS affects the air interface and also specifies new network entities and functions; its deployment will be slow.

4.3.2.2   Release 97 supplementary services

The group of supplementary services gets more attention in Release 97.

One focus is on call completion services, *completion of call to busy subscriber* (CCBS), which enable the network monitor the called party for the caller rather than letting the caller tie up network resources trying to get through to a busy phone. When the network detects that the called party finally completed his phone call, it will first set up a call to the calling party and then try to connect the called party.

Support of private numbering plan (SPNP) will be interesting to large companies that tend to have their own internal numbering plans for outside numbers. This means that a number used internally will be mapped onto a PSTN or PLMN number inside the private switch. The advantage is that the internal number of a called party remains the same even as the *outside version* in the PSTN or PLMN changes. An example illustrates the private numbering scheme. The sales force of the X Company, a grocery supply firm, uses dedicated mobile phone numbers that are grouped according to their product ranges. This helps the staff at the home office remember the numbers: people who sell fruits and vegetables always have numbers that start with a "1," and people who sell paper products always have phone numbers that start with a "2." This handy scheme would be very difficult to implement in a real network. It would be nice if each outgoing call, identified with a private numbering plan, could eventually find its way to a matching phone number through a database stored in the network. In our example the private numbering plan only works for staff from the X Company because they subscribed to the service. X Company's customers would still have to call the individual sales managers through their normal, outside phone numbers. A subscriber may be part of up to nine different private numbering plans in a SPNP scheme.

Multiple subscriber profile (MSP) gives a subscriber the ability to have multiple profiles, including multiple subscriber numbers: one for business and one for private use with different subscription options. This is the same as having two subscriptions with one SIM. This also implies that this specific user has the choice to forward incoming calls to different numbers depending on whether it is a private or business call, or it allows the subscriber to let private calls use a different ringing tone than business calls.

4.3.2.3   Release 97 network improvements

Another improvement intended for the network in Release 97 of Phase 2+ is provision for *hot billing*. This enables network entities to service ongoing charge records much faster, thus helping fight fraud. It also allows for additional services such as prepaid SIM cards, which we already have in Phase 1 networks, but with proprietary protocols in the network equipment. The proprietary systems are replaced with a standard procedure. Table 4.4 offers an overview of additional features and services likely to be introduced in Release 97.

4.4   Conclusion

This short overview of features shows that GSM has come a long way from a simple second-generation mobile phone system—the advanced

Table 4.4
Work Items in Release 97 of GSM Phase 2+

Service Category	Service	Comment
Teleservices	Emergency call with additional data transfer	Possible combination of GPS and GSM
	Voice group service	Voice group call service—Phase 2, defined for UIC applications
		Voice broadcast service—Phase 2, defined for UIC applications
Bearer Services	General packet radio service (GPRS)	Data rates of up to 21.4 Kbps per channel/time slot, up to eight time slots used per communication
Supplementary Services	Call deflection	Divert call to any not preprogrammed number without accepting the phone call
	Completion of call to busy subscriber (CCBS)	Network monitors the "busy" called party
	Support of private numbering plan (SPNP)	Individually assigned numbers (e.g., for large companies)
	Multiple subscriber profile (MSP)	Different profiles on one subscription
	User to user signaling	Direct exchange of smaller data packages
	Enhanced multilevel precedence and preemption service—Phase 2	
Network Features	CAMEL—Phase 2	Expansion of CAMEL
	Provision for hot billing	Immediate processing of call records

features will be accepted as standard telecommunications fare in third-generation systems. One big disadvantage of GSM remains: its lack of bandwidth. A third-generation system (even when it evolves from GSM) needs to have substantially greater bandwidth that can handle up to 2-Mbps data rates. These kinds of rates will be needed to keep up with the demand for new communications services. We will note a comment by

Friedhelm Hillebrand, chairman of ETSI SMG, that GSM Phase 2+ takes GSM to a *generation 2.5 system* [6].

References

[1]   Redl, S. M., M. K. Weber, and M. W. Oliphant, *An Introduction to GSM*, Norwood, MA: Artech House, 1995.

[2]   Mouly, M., and B. Pautet, *The GSM System for Mobile Communications*, Palaiseau, 1992.

[3]   Hillebrand, F., and A. Bergmann, "The Development of the GSM Standard in 1997," *Proc. GSM World Focus 1997*, Mobile Communications International, pp. 54–56.

[4]   GSM 01.42, "Digital Cellular Telecommunications System (Phase 2+); Network Identity and Time Zone (NITZ) Service Description, Stage 1," ETSI, Sophia Antipolis.

[5]   GSM 01.50, "European Digital Cellular Telecommunications System; Radio Local Loop (RLL) Using GSM," ETSI, Sophia Antipolis.

[6]   Hillebrand, F., "The Evolution of GSM to the Third Generation," *Proc. GSM World Focus 1997*, Mobile Communications International, pp. 59–64.

GSM telecommunication services

GSM telecommunication services include all the services offered to GSM users: from basic telephony, the chief reason a GSM mobile phone might be bought in the first place, all the way to fax, data, or other more specialized services. There are two categories of services: *teleservices* and *bearer services*. We concentrate on bearer services in this chapter because teleservices and related functions, for example, speech and SMS, are explained in other chapters. We start by making a general distinction between teleservices and bearer services.

Teleservices provide a complete service to the user within the GSM network. No additional external equipment is required to complete the service. Its description includes the capabilities of any terminal equipment that provides those capabilities. Standard voice communication is an example of this where the service extends from one GSM terminal to another GSM terminal or to a fixed-line telephone. Teleservices use the higher layers in the GSM system where we find the equipment and software to handle the services. Fax service—the transport of graphic information—is a teleservice even though fax service calls on data channels (bearers) for information transport between fax machines.

Bearer services provide only the lower layers or transport mechanism between access points; they require or support some kind of *terminal equipment* (TE) at each end, for which they function as a data "pipe." The TE might be a laptop computer, a palmtop, a *personal digital assistant* (PDA), or another kind of data terminal device. Bearer services provide their own higher level communication protocols and GSM only provides the transport vehicle, which is identified by the data rates and protocols used rather than the actual service.

Figure 5.1 shows the difference between the two services. It illustrates the end-to-end teleservice connection by means of a normal phone connection, and the end-to-end bearer service connection through data connections [1]. The teleservices allow an end-to-end connection within the GSM service area, which supplies all the necessary higher layers to support the services. In contrast to the teleservice, any bearer service requires additional external equipment, which runs the necessary higher layers of the applications. These higher layers are not part of the GSM service specifications.

In Figure 5.1 we see that there are different access points for various services. An access point is a well-defined interface that provides connectivity between two entities. The standard access point for a GSM network is the U_m-interface, which is the wireless connection between the mobile station (on the left side of the figure) and the BSS. We can also see some

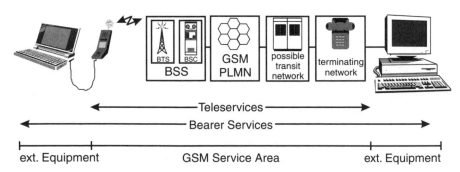

Figure 5.1   Bearer services and teleservices.

additional access points defined for user access. The first access point is the actual user interface of any mobile telephone (key pad, display, microphone, and loudspeaker) with nothing connected to it. This familiar access point is regarded as the general access point for teleservices. The second access point is the interface between the terminal equipment (the laptop at the far left of the figure) and the mobile phone. This is usually handled with some kind of specific interface for a laptop computer and is treated in Section 5.4.2. Different kinds of implementations are possible for this type of access point; each depends on its own interface.

5.1 Bearer services in GSM

Bearer services only provide *low-layer access;* lower layers are confined to GSM layers 1 through 3. Higher layer functions have to be provided by additional hardware or software. If a subscriber wants to use a bearer service along with a higher layer access, then the GSM infrastructure must support this kind of service. Bearer services are organized and described by the demands they make on the network. Bearer services (1) have different access structures (synchronous or asynchronous), (2) can use data rates up to 9.6 Kbps, (3) can have different information transfer attributes such as *unrestricted digital information* (UDI) or the standard 3.1-kHz tonal information exchange used for PSTN or ISDN, and (4) can have different access attributes describing *transparent* (T) or *nontransparent* (NT) services. All these parameters are listed in Table 5.1 [2]. Normally, someone who wants to use bearer services connects additional equipment (a portable computer is an example) to his phone. Both entities together are referred to as data terminal equipment (DTE) where the indication for the bearer services reaches out all the way to the computers in Figure 5.1.

The GSM bearer services are always circuit-switched services even when they access a packet-switched data network and are assigned numbers as shown in Table 5.1. The communication configuration is always *point-to-point.* As we take a close look at Table 5.1, we need to point out some details:

Table 5.1
List of Bearer Services

| Bearer Service | | Access | | Information | |
No.	Name	Struct.	Rate	Transfer	Attribute
21	Asynch 300 bps	Asynch	300 bps	UDI or 3.1-kHz	T or NT
22	Asynch 1.2 Kbps	Asynch	1.2 Kbps	UDI or 3.1-kHz	T or NT
23	Asynch 1200/75 bps	Asynch	1200/75 bps	UDI or 3.1-kHz	T or NT
24	Asynch 2.4 Kbps	Asynch	2.4 Kbps	UDI or 3.1-kHz	T or NT
25	Asynch 4.8 Kbps	Asynch	4.8 Kbps	UDI or 3.1-kHz	T or NT
26	Asynch 9.6 Kbps	asynch	9.6 Kbps	UDI or 3.1-kHz	T or NT
31	Synch 1.2 Kbps	Synch	1.2 Kbps	UDI or 3.1-kHz	T
32	Synch 2.4 Kbps	Synch	2.4 Kbps	UDI or 3.1-kHz	T or NT
33	Synch 4.8 Kbps	Synch	4.8 Kbps	UDI or 3.1-kHz	T or NT
34	Synch 9.6 Kbps	Synch	9.6 Kbps	UDI or 3.1-kHz	T or NT
41	PAD access 300 bps	Asynch	300 bps	UDI	T or NT
42	PAD access 1.2 Kbps	Asynch	1.2 Kbps	UDI	T or NT
43	PAD access 1200/75 bps	Asynch	1200/75 bps	UDI	T or NT
44	PAD access 2.4 Kbps	Asynch	2.4 Kbps	UDI	T or NT
45	PAD access 4.8 Kbps	Asynch	4.8 Kbps	UDI	T or NT
46	PAD access 9.6 Kbps	Asynch	9.6 Kbps	UDI	T or NT
51	Packet access 2.4 Kbps	Synch	2.4 Kbps	UDI	NT
52	Packet access 4.8 Kbps	Synch	4.8 Kbps	UDI	NT
53	Packet access 9.6 Kbps	Synch	9.6 Kbps	UDI	NT
61	Alternate speech / data				
81	Speech followed by data				

▶ Bearer services 23 and 43 actually specify two data rates (75 and 1200 bps). These services are intended for videotext such as the

Minitel service in France. In the uplink direction, 75 bps is used for the exchange of control parameters and 1200 bps is used in the downlink direction for the actual data transfer, which is specified in [3].

▶ The *packet assembler/disassembler* (PAD) access bearer services are defined for the mobile originated services only.

▶ Neither the access structure and rate, nor the information transfer method and attribute capability, have been specified for bearer services 61 and 81 because these parameters depend on the kind of service that is offered together with the speech *teleservice*. This means that the parameters may be chosen from bearer services 21 through 34 in Table 5.1.

All the offered bearer services have to be supported inside the GSM PLMN. This not only means that GSM is able to transport the information provided by the bearer service, but that it must also have the capabilities to interconnect to another network, for example, into an ISDN, a PSTN, a *packet-switched public data network* (PSPDN), or a *circuit-switched public data network* (CSPDN). These interconnections are performed by dedicated entities at the interfaces between the GSM PLMN and the other networks. Such an interfacing entity is called an *interworking function* (IWF). An interworking function is a physical interface to the MSC. Different inter-working functions are used for different purposes such as packet data or fax traffic. Upon call setup the mobile station must describe the kind of bearer service it requires from the network. Assuming the network supports the requested service, it routes the call to the appropriate interworking function, which completes the proper connections to the target network. This explains why the *multinumbering* scheme applies in most GSM networks, in which different phone numbers to an individual subscriber are allocated for speech, fax, and data services. Depending on the number, the access point for the MSC is different; each number enables an immediate connection to the proper IWF. The network operator, however, has the option to supply a *single-numbering* scheme, which leaves the subscriber who receives incoming calls with the responsibility of connecting the proper terminals and equipment to the mobile station for the different bearer services and teleservices.

5.2   Teleservices in GSM

In contrast to bearer services, teleservices require higher layer support from the GSM network. This is particularly true for SMS and facsimile Group 3. The facsimile services use the higher layers of their respective ITU-T specifications. Facsimile service are explained in more detail in Section 5.7, and SMS is covered extensively in Chapter 6. As is the case with bearer services, an interworking function is required when interconnecting with other network teleservices. Table 5.2 lists all teleservices by number [4].

5.3   Connection types in a GSM PLMN

The two previous sections of this chapter partitioned the services offered by GSM into two different categories: bearer services and teleservices. The sorting of services into categories implies that different types of connections are required for the different services. The user specifies only the service he or she wants to use, and the network supplies the resources for the requested service [5].

Table 5.2
List of Teleservices

Category No.	Name	Individual No.	Teleservice Name	Attribute
1	Speech transmission	11 12	Telephony Emergency call	
2	Short message service	21 22 23	SMS—MT SMS—MO SMS—CB	
6	Facsimile transmission	61	Alternate speech and facsimile Group 3	T or NT
		62	Automatic facsimile Group 3	T or NT

5.3.1 Lower layer capabilities

Two different lower layer capabilities are distinguished. For connections with the *transparent* attribute the portion of the connection between the terminal (TE and MS) and the IWF is used transparently without any kind of automatic procedure to repeat lost frames when errors occur on the path between the MS and the IWF. This type of connection is characterized by constant throughput, constant transit delay, and variable bit error rates. During handover and channel changes, in which the data channel is briefly replaced by signaling information, the error rate increases dramatically. The bit error rate incurred on the air interface usually requires additional higher layer protection against errors, which have to be provided by the end-to-end terminals.

For connections with the *nontransparent* attribute, error protection is applied between the terminal and the IWF. This error protection is achieved through the *radio link protocol* (RLP), which can be implemented in the terminal equipment, the mobile phone, or the terminal adaptation function. The manufacturer chooses where to provide the RLP function. The nontransparent mode is characterized by improved, stable, and constant bit error rates with a variable transit delay and throughput. During handover the excessive data loss due to signaling and channel changes is overcome by slowing down the data rate and repeating lost information. The flow-control mechanism is based on feeding back *automatic repeat requests* (ARQs) for any blocks containing errors to the data source. This mechanism asks for such a feedback channel and suitable buffering.

In addition, the *physical layer's* protection mechanisms still apply for user data. For transmission over the GSM U_m-interface *forward error correction* (FEC) through convolutional coding and interleaving provides protection for user data on the lowest level, both for transparent or nontransparent connections.

5.3.2 Connections

A connection can be established between different access points as illustrated in Figure 5.2. In part (a), the connection is between two GSM PLMN access points, which would be the case when two subscribers within the same GSM PLMN call each other. In part (b), it is between a GSM PLMN access point and a network-to-network interface, which will always be the case when the GSM PLMN subscriber calls someone outside the GSM PLMN, perhaps through a PSTN or an ISDN. In part (c), the connection is between a GSM PLMN access point and an interface to a specialized resource within the GSM PLMN. When sending a short message the interface to the service center would be such a specialized resource. Part (d) shows the connection between a GSM PLMN access point and an interface to a specialized resource outside the GSM PLMN. Access to any kind of data network (Internet or CSPDN) is an example of providing access to a specialized resource.

An end-to-end connection can reach over several network connections using transit networks for the transport of the information. The resources provided by the transit network have to be sufficient for the service offered from the GSM PLMN. It would be sufficient for a voice call from one GSM PLMN to another GSM PLMN to use an analog line, but a packet data network would be more appropriate for the transmission of a short message.

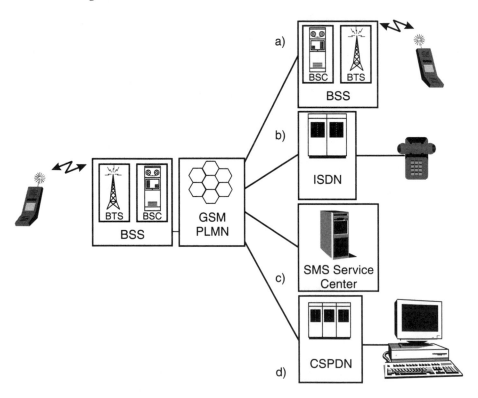

Figure 5.2   Connections in a GSM PLMN.

5.3.3 Attributes between two networks

A connection type is identified by a set of attributes, which allows communications between two networks supporting different transport mechanisms. We will discuss a few of the parameters in the following paragraphs.

The *information transfer mode* describes the means of transportation (circuit switched or packet switched) through a GSM PLMN. The value (designation) for a Phase 1 or Phase 2 GSM PLMN would always be *circuit,* since a two-way connection with a confirmation of transfer is always established. The packet mode connection is reserved for further study and might be used for dedicated *GSM packet services* (e.g., GPRS).

The *information transfer capability* describes how different types of information are transported through the GSM PLMN. There are four different categories:

1. *Unrestricted digital information* (UDI) allows digital data transmission directly into a digital network (ISDN), without converting it to an analog signal and then back to digital again, at higher rates than into normal analog PSTNs.

2. Speech information, which goes through a speech codec.

3. Group 3 facsimile follows the specifications for this service.

4. The 3.1-kHz audio (in a PLMN) allows data transmission over an analog line.

The *connection configuration* describes the spatial type of connection (point-to-point or point-to-multipoint). GSM bearer services are always point-to-point. The channel rate describes the data rate used on different channels through the network between different entities. The connection protocol indicates the appropriate protocol for each layer, for example, layer 3 protocol according to GSM TS 04.08.

Synchronous/asynchronous describes the type of transmission between two access points, which could be between the connector on the mobile phone and the interworking function. It defines how the two modems at both ends of the link communicate with each other. Modems are used in telecommunications to convert digital signals to appropriate analog

(3.1-kHz audio) or digital (64-Kbps ISDN frames in an ISDN modem) formats for transmission over the PSTN or PLMN. The modem on the receiving side does the opposite of what was done on the transmission side as it converts the signals received from the line connection back to the original data stream. Synchronous transmission means that the data are transferred continuously in step with clock signals. The additional lines required for receiving and transmitting clocks are used by the modems to identify when the individual data bits and words are received. The bits are transmitted on a fixed schedule for the synchronous format, which is 1 bit every 833 μs at a 1200-bps rate. It is very important for synchronous transmission that both ends work with the same time base, which is why clocking information is transmitted along with the actual information bits. Asynchronous transmission means data are transferred without clocking information. The data bits are transmitted as characters of 7 to 9 bits each. Each character is initiated and terminated (delimited) by start and stop bits. The nominal data rate is specified by the rate used while transmitting a single character; there can be a pause of any length between two characters. It may seem odd to consider asynchronous transmission at all when we regard the synchronous character of GSM. The reality is that there are still many interfaces based on asynchronous transmission; connecting a printer to a computer or connecting a computer to a local-area network are two examples.

The *user rate defines* the value of the user information rate at the terminal access point, which is defined by the interface between the TE and the GSM infrastructure (air interface). The user rate in GSM (Phases 1 or 2) can be 1200/75 bps or 0.3, 1.2, 2.4, 4.8, or 9.6 Kbps. The intermediate rate specifies the intermediate data rate at the A interface, which supports the two values of 8 or 16 Kbps.

Network independent clocking on Tx/Rx verifies whether such clocking is required, or not, in the transmit or receive direction, which also relates to the type of transmission. For asynchronous information transfer it is acceptable that the network uses an independent clock, whereas for synchronous information transfer it is necessary that both ends synchronize their time bases. This is done by using a control mechanism that allows the modems to delay or advance the clock or even skip or add bits.

The *number of stop bits* describes the number of stop bits included in characters used for the asynchronous type of transmission between two reference points. The *number of data bits excluding parity if present* indicates the number of data bits for the character-oriented (asynchronous) mode of transmission (7/8 bits).

The *parity information* describes the type of parity information (odd, even, none, forced to 0, or forced to 1). A parity bit can be used for error control with asynchronous characters by counting the number of 1's and 0's within the character.

The *duplex mode* includes full-duplex or semiduplex transmission. Full-duplex mode always applies to GSM.

The *modem type* describes the modem allocated by the interworking function for a 3.1-kHz audio link outside the GSM PLMN. Possible selections are V.21, V.22, V.22bis, V.23, V.26ter, V.32, autobauding type 1, or none. The specifications for these modems can be found in [3,6–10].

The *radio channel requirement* specifies what kind of radio channel is required: full-rate (bearer mobile, B_m), half-rate (low mobile, L_m), dual-rate/full-rate preferred, or dual-rate/half-rate preferred.

The *connection element* specifies the layer 2 protocol to be used: transparent, nontransparent (RLP), transparent or nontransparent but transparent preferred, or transparent or nontransparent but nontransparent preferred.

User information layer 2 protocol can be ISO 6429, code set 0; X.25; or character-oriented protocol with no flow control [11]. *Signaling access protocol* can be I.440/450, X.21, X.28 dedicated PAD, X.28 nondedicated PAD, or X.32 [12–14].

Rate adaptation is specified at the fixed reference point, which is located in the interworking function. Possible values are V.110/X.30, X.31 flagstuffing, or no rate adaptation. These are explained in the relevant specifications [15–17].

The *coding standard* is always *GSM*.

User information layer 1 protocol characterizes the layer 1 protocol to be used at the U_m-interface.

5.4   Rate adaptation

Having selected a particular service from either the teleservice or bearer service catalogs, and allowing the network to make the connections to the proper interfaces with the appropriate attributes, we now turn our attention to the communications tasks themselves.

The different interfaces in the GSM PLMN support their own particular data rates. The most common data rates we refer to are the *user* data rates, which range from 2400 to 9600 bps. These data rates are net rates as they are transmitted over the air interface (U_m); the gross data rate—after convolutional coding for FEC—always remains at 22.8 Kbps. The transmissions are furthermore classified on the air interface by different channel types: *full-rate* and *half-rate* traffic channels. This allows five different transmission rates, as indicated in Table 5.3, which shows the highest data rates that the half-rate and full-rate traffic channels can carry with either transparent (T) or nontransparent (NT) information.

The lower data rates only allow transparent transmission. A general rule applies for the transmission quality on the air interface: the lower the transmission rate, the higher the number of error detection and correction bits, and the better the data protection and transmission quality. Table 5.3 also shows the residual error rates, which are achieved on the radio interface on a channel specified for typical urban environment at a vehicle speed of 50 km/hr and frequency hopping (TU50, ideal FH). It is evident that higher rates are more susceptible to errors because there is less room for protection with channel coding. The user data rates on the radio interface do not indicate the corresponding gross data rates at the terminal equipment for data transmission or the gross data rates used in the network itself. This is an issue of capacity and economy. It is more efficient for an operator to multiplex several data links onto one physical link using a higher data rate of 16 or even 64 Kbps on the (synchronous) interface between BSS and MSC. To achieve these fixed and higher rates, special *rate adaptation* (RA) functions are specified. The RA function specified in ITU-T V.110 has been adopted for GSM. These functions are similar to

Table 5.3
Transmission Rates on Full-Rate and Half-Rate Traffic Channels

Transmission Speed (bps)	Channel Type	Lower Layer Capability	Residual Error (%)
2400	Full-rate	T	0.001
2400	Half-rate	T	0.01
4800	Full-rate	T	0.01
4800	Half-rate	T / NT	0.3
9600	Full-rate	T / NT	0.3

those used in ISDN where RA0, RA1, and RA2 are standard functions; RA1 has been adopted to GSM's needs on the air interface. The RA functions work in both directions, which means that an input data rate can also be an output data rate if the RA function is used in the opposite direction.

RA0 is used for asynchronous interfaces. Because the GSM network is a synchronous network, all asynchronous data has to be transformed into a synchronous data flow. RA0 transforms incoming asynchronous data (from the DTE) to the next (supported) higher data rate by introducing stop bits; rates of 75 and 300 bps can be adapted to a synchronous stream of 600 bps. When the incoming gross data rate is higher (up to 1% or 2.5% for nominal speeds below 600 bps) than the resulting synchronous rate, then the receiving RA0 function may delete stop bits as often as needed, up to a maximum of 1 for every 8 bits. When the incoming gross data rate is lower than the synchronous rate, then the RA0 function may insert additional stop bits. Table 5.4 shows the different possible conversions of asynchronous data into a synchronous data.

RA1 refers to rate adaptation to intermediate data rates. The input of an RA1 function is either the output of RA0 or synchronous data. These data are transformed into a stream of 8 or 16 Kbps by means of repetition. Table 5.5 shows the resulting data rates for different input synchronous rates. The table also shows how often a data block is repeated in order to achieve the intermediate data rate. The resulting data blocks (frames) not only consist of the incoming data blocks but also include some control and frame synchronization bits, which is why, for example, 4.8 Kbps is mapped onto an 8-Kbps data stream.

Table 5.4
Resulting Synchronous Data Rates From Asynchronous Data Rates

Asynchronous Data Rate (Kbps)	Resulting Synchronous Data Rate (Kbps)	Upper Limit for Asynchronous Data Rate (Kbps)
< 0.6	0.6	0.615
0.6	0.6	0.606
1.2	1.2	1.212
2.4	2.4	2.424
4.8	4.8	4.848
9.6	9.6	9.696

Table 5.5
Synchronous and Intermediate Data Rates of the RA1 Function

Synchronous Data Rate (bps)	Intermediate Data Rate (Kbps)	Duplication (times)
600	8	8
1200	8	4
2400	8	2
4800	8	1
9600	16	1

RA2 converts the output of the RA1 function to an A interface data stream of 64 Kbps, which is the basic bearer data rate for one user of one duplex ISDN channel. A basic rate ISDN channel (2B+D) consists of two 64-Kbps bearer channels and one 16-Kbps signaling channel. A 64-Kbps stream based on ITU-T V.110 is transmitted in octets: bit positions 1 through 8. RA2 places the 8-Kbps stream onto bit position 1 (of each octet) and the 16-Kbps stream into positions 1 and 2 of each octet. The remaining positions are all set to "1." The actual net data rate is not changed. It is easy to multiplex several 8- or 16-Kbps streams by positioning other streams into the remaining positions of the octets. Figure 5.3 shows the principle of RA2 with a subsequent multiplexing stage.

The *RA1'* function is dedicated to transforming data rates from the DTE to rates supported on the air interface while adding some control bits.

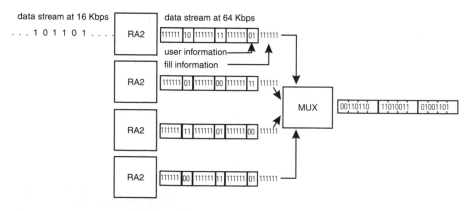

Figure 5.3 Principle of RA2 and multiplexing.

The input for RA1' is either the output of RA0 or a synchronous data stream. Table 5.6 shows the adapted rates.

To further illustrate the use of rate adaptation, we take a brief look at the overall connection from a DTE toward the MSC/IWF, as indicated in Figure 5.4. A further distinction is made for different terminal equipment such as synchronous and asynchronous, and for transparent and non-transparent services. Section 5.7 gives a similar overview for fax services.

1. DTE supporting the S interface at 64 Kbps. The mobile terminal has to reduce (RA2) the gross data rate from 64 Kbps to either 8- or 16-Kbps gross user data rate depending on the channel rate that is to be used (full-rate or half-rate). The data reduction is not achieved by coding or actual user data reduction, but simply by reducing the fill information; the actual net user data rate is still below 8 or 16 Kbps (this is the reverse process to RA2 indicated below). The combined rate adaptation RA1/RA1' transforms the 8- or 16-Kbps rate to 6 or 12 Kbps, which will be used on the air interface (full-rate or half-rate). The gross rate on the air interface is higher due to the channel coding and FEC [18]. The rate is converted again up to 64 Kbps with the help of the combined RA1'/RA1 and RA2 functions in the base station system.

2. DTE with a *synchronous* interface at various rates (e.g., 4.8 Kbps). This only requires RA1' to convert the data rate to one that can be fed into the FEC function. The rate adaptation is the same in the BSS as indicated before.

3. DTE with an *asynchronous* interface at various data rates. The main difference between this case and the synchronous case indicated

Table 5.6
Synchronous Data Rates and the Data Rate Before FEC Coding on the Air Interface

Synchronous Data Rate (Kbps)	Data Rate Before Coding on the Air Interface (Kbps)
≤ 2.4	3.6
4.8	6
9.6	12

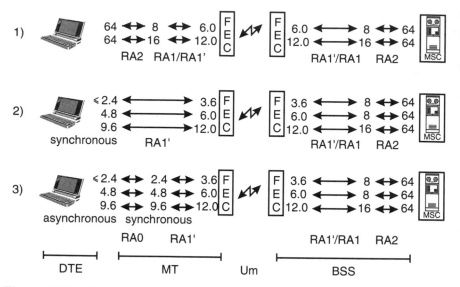

Figure 5.4 Some rate adaptation functions for transparent data transmission.

earlier is the use of RA0, which converts the asynchronous data to synchronous data. The remaining structure is the same.

The rate adaptation for *nontransparent* transmission differs from the transparent transmission for these reasons:

- Synchronous transmission is considered a special case of asynchronous transmission; it runs quite accurately at the highest allowed data rate without the need to insert start or stop bits. Both transmission types are treated equally.

- Nontransparent transmission always uses the highest rates on the air interface, which is 9.6 Kbps for full-rate and 4.8 Kbps for half-rate traffic channels.

With these two differences in mind we can describe the rate adaptation for nontransparent transmissions in the three examples of Figure 5.5, which shows three different cases for nontransparent data transmission.

1. DTE supporting the S interface at 64 Kbps. The mobile terminal will reduce (RA2) the data rate from 64 Kbps to either 8 or

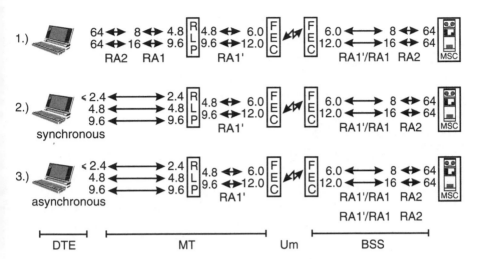

Figure 5.5 Some rate adaptation functions for nontransparent data transmission.

16 Kbps and then further reduces the rate to 9.6 or 4.8 Kbps, depending on the channel rate that is to be used (full-rate or half-rate, respectively). The RLP will add transmission security via the air interface (Section 5.5). The output of the RLP has to be adapted to the data rates available at the air interface, which is done by the RA1′ function. FEC and the rate adaptation in the BSS is the same as for transparent data transmission.

2. DTE supporting either synchronous or asynchronous interfaces. Because the RLP is used, there is no distinction between synchronous and asynchronous transmission because the RLP itself has polling capabilities that apply flow control for the user data depending on whatever errors occurred and data rate applied for asynchronous transmission. The polling capabilities are treated in Section 5.5. The RLP always uses the highest available rate, which is why the output of the RLP is either 4.8 Kbps for a half-rate channel or 9.6 Kbps for a full-rate channel.

5.4.1 Error protection

Data channels are the vehicle for the transportation of data. Forward error correction copes with the adversities on the radio channel such that channel problems do not interfere with the data communications between the two terminals. FEC can be regarded as an envelope in which user data are carefully placed for protection during transport.

Modems used for fixed-line data transmissions usually support layer 2 data protocols, which can also be used for GSM data transmission. Taking ITU-T V.42 as an example, two negotiation mechanisms for error correction are evident: (1) the *link access procedure for modems* (LAPM) allows error protection and correction (this is similar to the layer 2 LAPD$_m$ protocol); and (2) the *Microcom networking protocol* (MNP) supports different levels of implementation (MNP1 through MNP5), which specify error detection, error correction, and data compression.

For these end-to-end protocols to work, it is necessary that both ends *talk the same language*—that both ends use the same protocol. Both modems start a negotiation process over which protocol shall be used before starting actual data transfer. If one end does not support the relevant protocols, then the advantage of using these will not be available; the transmission will not run at higher gross rates and will not be able to further correct errors.

5.4.2 Terminal equipment and mobile termination

We have exposed the need for various rate adaptations along the data transmission path. On the *mobile equipment* (ME) side, the various possibilities provide room for manufacturer-dependent implementations into hardware and software, and different functions in the mobile termination. Three different types of *mobile terminations* (MT) are distinguished in GSM, and depending on the mobile termination, different terminal equipment is available.

1. *Mobile termination type 0 (MT0).* This is the simplest case; everything is incorporated into mobile equipment. For mobile speech telephony, all available mobile equipment (everything that we would regard as a mobile phone) would be a mobile termination type 0. For data applications, a mobile termination type 0 could be envisaged as a portable computer with a built-in GSM card for radio communications.

2. *Mobile termination type 1 (MT1).* The MT1 supports the ISDN S interface at 64 Kbps, which is the standard ISDN channel. Off-

the-shelf ISDN equipment (modem or fax machine) can be connected to this interface. If terminal equipment other than ISDN equipment is used in connection with a MT1, then a dedicated ISDN terminal adapter is needed. This terminal adapter would incorporate RA0 (in case of asynchronous equipment), RA1, or RA2.

3. *Mobile termination type 2 (MT2).* The MT2 supports a standard asynchronous interface with a rate of up to 9.6 Kbps. This interface is also referred to as the R interface. The MT2 incorporates all necessary data functions. It allows us to connect DTE, which follows the V-series or X-series ITU-T specifications. V.24, also referred to as RS-232, is a common example. A MT1 together with an ISDN terminal adapter will have the same physical behavior as a MT2.

The view of the mobile termination is the ideal view the fathers of GSM had in mind. The industry, however, implemented these interfaces in a more pragmatic way. Referring to a real GSM telephone, no phone supports MT1 or MT2; they all have proprietary interfaces. When we connect a PCMCIA card to a phone, what will that card be? Certainly not a MT1 or a MT2. There will be an R interface inside the PCMCIA card but it will not be accessible. The good news is that at least the connection to the PCMCIA card is somewhat standard even though it is not an S interface. So if we see a phone connected to a PCMCIA card, then we have a MT2 plus a bit more. When we take a look at the PCMCIA card market we quickly realize that many card manufacturers offer the very same card for a variety of phones stipulating that they use some sort of standard mechanism to assume control of the phone; the only difference is the cable that connects the phone to the PCMCIA card. The different cables adapt a wide variety of phones to a standard PCMCIA connection. The card manufacturers load different software into the PCMCIA cards that adapt the card and the sundry phone-specific cable to each of the different phones.

5.5   Radio link protocol

The success of the radio interface's efforts at transporting 9.6 Kbps on full-rate channels and 4.8 Kbps on half-rate channels is not very impressive. It is not a particularly efficient use of the spectrum to leave it to the two terminals to cope with the error correction tasks. First of all, the terminals have to repeat large blocks of information even if only one important bit is exchanged incorrectly. Secondly, the inherent signaling needs to span the connections (TE-MS-BSS-MSC-IWF-...-TE) rather than merely from the mobile phone to the mobile services switching center (MS-MSC). This signaling burden consumes a lot of time and system capacity. A new protocol was introduced to counter these inefficient efforts. It overlays the lower level functionality and handles the traffic between the switching center and the mobile station. This protocol is called the *radio link protocol* (RLP). The communication for RLP actually takes place between the *terminal adaptation function* (TAF) on one side and the *interworking function* (IWF) on the other side.

The purpose of the RLP for data services is similar to that of layer 2 for signaling messages: it makes sure that data will eventually arrive at the other end of the air interface without being altered or lost. When data are lost, the RLP makes sure that the data are repeated. This ensures that the repetition rate is minimized to recover the actual lost data, and that the quality is increased. Information transmitted with RLP is not transparent to the GSM network; data transmission using RLP is regarded as nontransparent data.

5.5.1 Frame structure

In a manner similar to routine signaling information, the RLP makes use of a frame structure using blocks of 240 bits, which are encoded before they are transmitted over the radio channel. The channel coding and the interleaving structure for RLP follows exactly the same structure as for the 9.6-Kbps data channel, which is described in [19]. The 240 bits of a RLP frame are subdivided into the following information, as shown in Figure 5.6.

The *header* fulfills the same purpose as the L2 control bits for the signaling frames. The control bits are used to establish or release a RLP link, to indicate that no link is established, and to count the frames. The window size, which indicates the maximum offset between sending a frame and receiving a confirmation of its delivery, is, in the RLP case, up to 62 (0–61), compared to 1 for layer 2. In theory, the window size for layer 2 is 8, but it is defined as 1 for the *service access point identifiers* (SAPI) 0 and 3 on the air interface. The use of larger windows for other SAPI values is

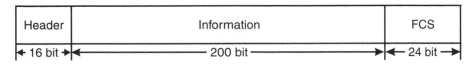

Figure 5.6 Frame structure of a RLP frame.

reserved for further study. Finally, the control bits in the header are used to signal the presence of erroneous frames that need retransmission.

The *information* field contains the actual user data that is to be transmitted.

The *frame check sequence* (FCS) is a sequence used to detect transmission errors in the header or the information field. A detected error will cause the protocol to request a repetition of the affected frame. The bits of the FCS are derived from a cyclic code, which combines the 200 information bits using a specified polynomial that detects errors. The result of the FCS will have a fixed value in the absence of errors. The FCS has the same significance as the check sequence for the class Ia bits used for speech transmission, the only difference being that for data traffic the frame will have to be repeated rather than discarded. In addition, there are the normal error detection and error correction resources through the GSM layer 1 channel encoding (convolutional coding and interleaving), that take place immediately before the RLP frame is sent over the air. Figure 5.6 shows the structure of a RLP frame.

5.5.2 Control of RLP

The control information of RLP, located in the header, is limited to the establishment and release of a RLP link, simple error protection, and setting up some link parameters [20]. The combination of the control information and user data in one frame makes more efficient use of the single frame and reduces the need to send pure supervisory frames during the actual data transmissions.

The RLP has two modes. One is the *asynchronous disconnected mode* (ADM), which indicates that no RLP link is established. A prerequisite for any RLP action is an established data channel on an established radio link such as a 9.6-Kbps channel on a full-rate TCH. The other mode is the *asynchronous balanced mode* (ABM), which initiates the exchange of numbered frames thus acknowledging the transmission of information.

The control information is identical to normal layer 2 control information and will, therefore, only be described briefly. Further information can be found in [19]. Figure 5.7 shows the structure of the 16-bit header for different frames (U, S, S+I).

There are two control parameters: (1) the *command/response bit* indicates whether a frame is a command or a response; and (2) the *poll/final bit* indicates whether the polling of frames is taking place or if the sent frame is the final frame.

The following unnumbered frames (U) are indicated:

▶ The *set asynchronous balanced mode* (SABM) command initiates a RLP link for numbered information exchange or resets a link, which also resets all internal variables. Depending on the data flow, the command may come from the mobile station side (the originator depends on where the RLP is implemented; it may be the ME, the TE, or a TAF) or it may come from the network side (IWF).

▶ The *unnumbered acknowledge* (UA) response is used to acknowledge a SABM or DISC command.

▶ The *disconnect* (DISC) command is used to release a RLP link.

▶ The *disconnect mode* (DM) response indicates that the RLP link is in disconnect mode. If DM is used as an answer to a SABM command then this indicates that the receiving entity is currently unable to establish a link.

	C/R	X	X	1	1	1	1	1	1	P/F	M1	M2	M3	M4	M5	X
U	C/R	X	X	1	1	1	1	1	1	P/F	M1	M2	M3	M4	M5	X
S	C/R	S1	S2	0	1	1	1	1	1	P/F	— N (R) —					
S + I	C/R	S1	S2	— N (S) —						P/F	— N (R) —					

S1	S2	
0	0	RR
0	1	REJ
1	0	RNR
1	1	SREJ

	M1	M2	M3	M4	M5
SABM	1	1	1	0	0
UA	0	0	1	1	0
DISC	0	0	0	1	0
DM	1	1	0	0	0
NULL	1	1	1	1	0
UI	0	0	0	0	0
XID	1	1	1	0	1
TEST	0	0	1	1	1

Figure 5.7   Structure of the 16-bit header.

▶ The *null information* (NULL) frames are sent when no information is being sent. This is similar to the fill frames for LAPDm. This is the case in the disconnected mode when no UI, TEST, or XID frames have to be sent; in the balanced mode in the reset state; and when no unnumbered frames have to be sent.

▶ The *unnumbered information* (UI) carries information that will not be acknowledged.

A major difference with the $LAPD_m$ implementation, which is used for signaling on the air interface, is the *exchange identification* (XID), which is used to negotiate (and renegotiate) RLP parameters via unnumbered information. This procedure may be initiated either by the mobile station side or by the IWF. It is done by simply issuing the exchange information command including a proposal for a new parameter set. If the other entity accepts these new parameters, then it will issue a XID response confirming these values. Otherwise it will respond with its own proposal for new parameters, which starts a negotiation process in only one direction —either increasing or decreasing—depending on the parameter. Default values apply when a RLP link is set up.

The following parameters are contained in a XID command:

▶ The *RLP version number* indicates the current version number. At the time of this writing the current version was "0." It is mandatory that future versions are downward compatible.

▶ The *window size* is defined differently for the MS-IWF and IWF-MS directions. This allows the accommodation of different processing speeds in either direction. The purpose of this is to limit outstanding frames and avoid a bottleneck effect.

▶ The *acknowledgment timer* specifies within which time frame each entity has to respond to a previously sent frame.

▶ The *retransmission attempts* specify how often a frame that did not properly arrive at its destination has to be repeated.

▶ The *reply delay* defines how long (in steps of 10 ms) each entity may wait until it sends a reply to a command.

The parameters and the default values are shown in Table 5.7.

Table 5.7
Parameters of XID Frames and Their Ranges and Default Values

Parameter Name	Range	Units	Sense of Negotiation	Default Value
RLP version number	0..255	—	↓	
IWF to MS window size	0..61	—	↓	61
MS to IWF window size	0..61	—	↓	61
Acknowledgment timer	0..255	10 ms	↑	480-ms full-rate 780-ms half-rate
Retransmission attempts	0..255	—	↑	6
Reply delay	0..255	10 ms	↑	< 80-ms full-rate < 80-ms half-rate

The *test pattern* (TEST), which is used as test information, can be used in balanced mode or in disconnected mode.

The supervisory frames (S) are used alone with a receive sequence number, or combined with information (S+I), along with the send and receive sequence number. These will only be transmitted in the asynchronous balanced mode. The frame counting allows a window size of up to 62 (0..61). It is, however, possible to diminish this figure at the network operator's option. The supervisory commands are:

- *Receive ready* (RR) indicates that frames up to the minimum window size have been received and that the transmitting entity can start transmitting frames again.

- *Reject* (REJ) indicates that frame counting is out of sequence, which means that one or more frames are missing. This condition is cleared upon receipt of the correct frame numbered with the last properly received frame, on time-out, or on resetting (SABM).

- *Receive not ready* (RNR) indicates that the receiving entity is not able to receive any further frames at the moment.

- *Selective reject* (SREJ) is used to indicate the loss of a specific frame. This is far more efficient than a single reject command, which would repeat all frames from the indicated one all the way through the ones that have already been sent. Selective reject requires only one single frame and accepts the ones sent afterward as being correct. As in the REJ case, this condition is cleared upon receipt of the correct frame numbered with the indicated last properly received frame, on time-out, or on resetting (SABM).

RLP also supports *discontinuous transmission* (DTX) on the air interface, which occurs when nothing has to be sent, and therefore provides a considerable savings in battery capacity and a reduction in uplink interference from all the mobiles in a sector or a cell. RLP may indicate to the lower layers (layers 2 and 1) that DTX may be invoked, in which case the RLP will send null frames for ADM or S frames for ABM.

5.5.3 Error recovery

The RLP needs to cope with varying propagation conditions on the air interface and clear the errors caused by the loss of portions of frames or whole frames. If only parts of frames are destroyed, the frame itself will probably still arrive at the receiving end but the information within this frame will be corrupted. This error can be detected by checking the FCS field. It is up to the receiving entity to deal with these problems. If whole frames are destroyed, then there will be a numbered sequence error. This will also be the case when the partial destruction of a frame "hits" the sequence counter. The RLP will be able to recover from this situation by issuing REJ or SREJ recovery commands requesting that the transmitting entity repeats the missing frames. There is a third mechanism to cope with loss of frames, which works from the transmitting station and is called *P/F bit recovery* or *checkpointing*. Checkpointing requests the receiving station to transmit its latest status. The transmitting station will not respond to any REJ or SREJ command during this procedure.

5.5.4 RLP summary

This section on RLP discussed the mechanisms and parameters employed to transport data over the air interface. It should be clear that data transmission using the RLP is rather fast because the amount of data to be retransmitted in response to errors is less than in more general procedures used by the terminal equipment in the case of transparent transmission.

Quality is further enhanced because the amount of repeated data are diminished, thus reducing the opportunities for additional errors. The use of RLP depends on whether it is supported by both the TE and the network.

The most important characteristic of the nontransparent feature is that it incorporates an ARQ technique for error protection. The consequence of this is the variable speed on the path between the MS and the IWF that indeed demands flow control, delays information transfer once errors occur, and, in the case of good transmission quality, speeds up the transfer of information.

5.6   Access to different networks

The previous five sections of this chapter confined themselves to the different tools and resources provided by GSM to support data transmission. GSM data traffic is certainly not limited to the GSM networks and infrastructure. It is time to consider how different types of other networks can be accessed through a GSM PLMN [21,22].

5.6.1   Transmission into the PSTN

The most common network interconnection is to the analog PSTN, which was originally designed for voice traffic with a bandwidth of 3.1 kHz from 300 to 3400 Hz. Today, the PSTNs not only carry most of the world's voice telephony but also data, fax, and other services. Data and fax services impart digital modulation onto the analog telephone lines with appropriate modems [23].

The GSM PLMN requires an interface in order to interconnect properly with the PSTN. This interface is incorporated into an IWF, which, in this case, is a relatively simple audio modem that converts the PLMN's digital information (data) into analog modulated signals compatible with the PSTN. Figure 5.8 shows a connection into a PSTN. The audio modem is part of the GSM PLMN and is attached to the MSC. The end-to-end transmission is a virtual digital transmission in which the transmission via the analog PSTN is the major part of most circuits even as GSM network

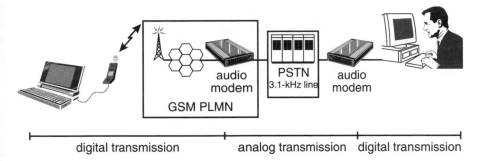

Figure 5.8 Transmission from a GSM PLMN into PSTN.

operators try to optimize the routing into the PSTN in order to limit the length of the leased lines from the fixed network operators.

Because the modem on the GSM side is installed in the IWF at the MSC, thus making it part of the network, the GSM network operator can decide which data services he wants to offer to his subscribers. The number of options is limited to only a handful. Table 5.8 lists the different data rates, the modes of transmission, and the respective modem types to be used in the IWF. The modem type corresponds to the ITU-T V.xx specifications. The ITU-T specifies the interface between the analog modem and the DTE, which in our case (see Figure 5.8) is a desktop computer, and for which the relevant connection specification is V.24 [24].

Table 5.8
Analog Modems Supported by GSM

Data Rate (bps)	Transmission Mode	Modem Type
300	Asynchronous	V.21
1200	Asynchronous, synchronous	V.22
2400	Synchronous	V.22bis
1200/75	Asynchronous	V.23
2400	Synchronous	V.26ter
4800	Synchronous	V.32
9600	Synchronous	V.32

5.6.2   Facsimile transmission

One of the key applications for GSM, born of the *mobile office* concept, is facsimile transmission. Even though facsimile transmission belongs to the *teleservices*, it still uses data channels and is, therefore, covered in this section. GSM supports fax services according to the analog Group 3 specification. The generic GSM fax adapter supports the connection of an off-the-shelf fax machine, which may require an audio modem as shown in Figure 5.9(a). This kind of implementation is unusual because it is an expensive exercise to install two analog modems at the user side just to convert digital data to analog and then back to digital again. Dedicated solutions are available today that convert the digital data (from a fax application in a PC) into a digital stream, which fits the requirements of the GSM terminal as shown in Figure 5.9(b). Dedicated solutions in the earlier days of GSM were fulfilled with PCMCIA cards that contained all the circuits and software for the data and fax protocols. During 1997 more and more companies offered solutions that were connected directly to a PC's serial interface; the data and fax protocols were implemented in dedicated SW packages running in the computer itself.

Only the analog PSTN is considered in the configurations of Figure 5.9, but there is no difference when dialing into the digital ISDN

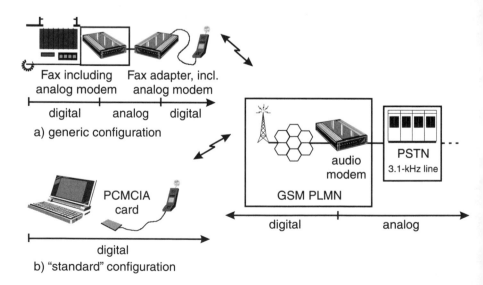

Figure 5.9   Implementation of fax services.

since the audio modem, which is located in the IWF in the PLMN, is simply shifted toward the fax machine on the ISDN side.

A more elaborate view of fax transmission is offered in Section 5.7, where we take a closer look at the implementation in GSM, and where the difference between bearer services for data and the teleservice for facsimile support becomes obvious. GSM provides the framework for supporting the transmission of data services, and it also provides higher layer support for facsimile services.

5.6.3 Transmission into the ISDN

Even though GSM was specified around ISDN services, protocols, and architectures, there is still a need for interworking between the two systems. This need arises because some of the resources need to be adapted to the needs of a mobile network. The main difference lies in the data rate, which is 64 Kbps for ISDN, and a maximum of only 9.6 Kbps for GSM. Because both networks are digital, there is no need for an analog modem. However, due to the different data rates, there is a need for rate adaptation between the two networks. Note that for voice signals the rate adaptation is simply done by unpacking the 13-Kbps speech data to 104 Kbps and then packing them again to 64 Kbps according to the A-law method, which gives lower audio levels higher significance [23]. Figure 5.10 shows the interconnection between a GSM and an ISDN network. ITU-T V.110 specifies the protocol for the data exchange.

5.6.4 Transmission into the PSPDN

Even though GSM is a circuit-switched network it provides access to a *packet-switched public data network* (PSPDN) such as the Internet. The

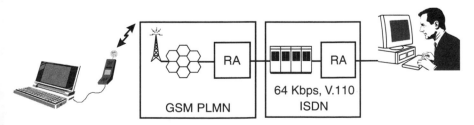

Figure 5.10 Interconnection with an ISDN.

PSPDN can be accessed in various ways [25]. Fixed users of a PSPDN might have a direct connection into the PSPDN using a dedicated access protocol (e.g., X.25). This is not appropriate for the GSM user since the subscriber is always connected to a mobile network, which acts as a transfer network. This leaves the mobile GSM user with the following options.

Access to the PSPDN using the PSTN is the most obvious option. A disadvantage is that the user has to identify him or herself with a separate number for verification. In standard packet data networks the user is automatically identified by the terminal number/identity, which cannot be transmitted through the PSTN. The GSM user must also subscribe separately to the PSPDN. The PSPDN may be accessed either by using a PAD or through a *packet handler* (PH) inside the PSPDN. The access through the PAD is intended for terminals that do not support a packet protocol and is only supported in the mobile originated direction. An example of a PAD protocol is X.28, which is an asynchronous protocol. Access to a PH is accomplished through X.32, which is a packet protocol. Figure 5.11 shows the principle of acquiring access through the PSTN.

The more elegant access to the PSPDN is the dedicated direct access from the GSM network. There are two possibilities: (1) access through a PAD or (2) direct access through the X.32 protocol into the PH of the PSPDN. The advantage of these configurations is that the GSM network (or better, the IWF) knows that the terminating network is a PSPDN and is, therefore, also able to pass the user identity onto the packet network, which makes the assignment of a second "subscriber" number redundant.

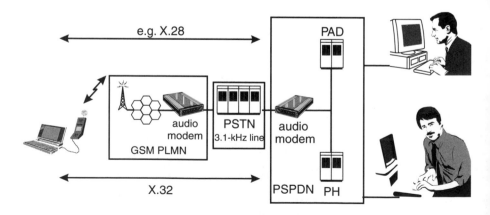

Figure 5.11   Access of GSM subscriber to a PSPDN through the PSTN.

In this case the GSM operator is the PSPDN's "customer," and the GSM operator can pass the connection charges onto the mobile customers. Figure 5.12 shows the principle of dedicated access into PSPDN. In the figure we assume that there is no dedicated modem used for interconnection and that access is subject to agreements between the GSM and the PSPDN providers.

Another possibility would be access through the ISDN network, which although defined for GSM Phase 1, has never been put into service. It is now suppressed for Phase 2 and will probably never be used.

5.6.5   Transmission into the CSPDN

CSPDNs are dedicated data networks using circuit-switched transmission, which is the same transmission mode used in GSM and ISDN. CSPDN access is always synchronous with the standard GSM data rates of 2400, 4800, and 9600 bps. Because ISDN already supports access to CSPDNs, GSM inherited the specifications (X.71) for interconnecting to a CSPDN [26]. Again, there are two ways for GSM to interconnect with a CSPDN:

1. The first possibility is access via the ISDN, which transports the data and interfaces to the CSPDN. If the destination network is a CSPDN then the IWF for interconnecting to ISDN is rather simple since it only has to perform rate adaptation according to ITU-T X.30. The structure is shown in Figure 5.13(a).

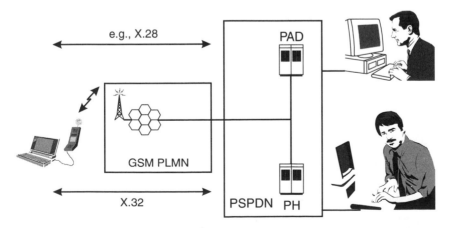

Figure 5.12   Dedicated access of a GSM subscriber to a PSPDN.

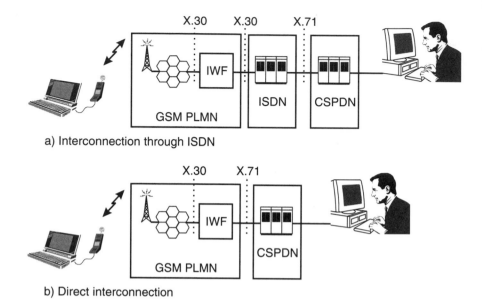

Figure 5.13   Interconnection of GSM and a CSPDN.

2. The second possibility is the direct approach since GSM provides
    direct access to the CSPDN using the protocols specified for ISDN.
    The IWF simply has to perform rate adaptation to the ISDN for-
    mat (X.30) and then convert it to the CSPDN format (X.71). The
    structure is shown in Figure 5.13(b).

5.7   Fax services

Fax over cellular has been possible on analog systems for as long as mobile
phones have supported an interface for a "second" analog phone connection, which could be an answering machine, an actual second phone, or a
fax machine. The fax machine may be a portable PC with an analog fax
modem. The fax machine transmits signals over the air exactly as it would
transmit them over a normal PSTN line using analog tones. There are
no particular, dedicated error protection and error correction schemes
applied to these fax signals for transmission over the air interface, which
means that the connection between the mobile and the base station needs
to be a good one in order to ensure the quality and speed the user is accustomed to in the fixed network. Fixed-line fax machines cope with possible

errors by means of repetition, which reduces transmission speed. Since the radio path is much more susceptible to errors than the fixed wire network, this dependence on the repetition method significantly reduces transmission speed in cellular networks.

Digital cellular has the advantage that the data coming from the fax machine need not be converted into a tone sequence representing the black dots that appear on a white piece of paper; the information remains digital. Because digital data *can* be error protected, keeping the entire process in the digital domain enhances the transmission quality even when faxes go over the air. GSM goes beyond this simple principle by specifying two different modes of fax transmission.

1. *Transparent fax* makes use of the standard Group 3 fax protocol and protects the data over the air interface with normal forward error correction techniques. See [19] for more details.

2. *Nontransparent fax* not only uses the standard Group 3 fax protocol but adds another "layer 2" (the radio link protocol) on the air interface to ensure delivery at the other side of the mobile connection.

As is the case for normal fax machines, GSM allows for transmissions in both directions: (1) mobile-originated fax traffic where a fax machine, or a computer configured as a fax machine, connected to the mobile station is the originator, and (2) mobile-terminated fax traffic, where a fax machine (or a PC) connected to the mobile phone is the receiver. The mobile-originated case does not usually require specific preparations beyond the network's ability to support fax service. The mobile-terminated case, however, often requires the subscriber to get a second phone number dedicated to fax service so that the GMSC can be addressed directly with the applicable fax IWF. The second phone number will generally entail an additional fax subscription charge to support the extra infrastructure.

5.7.1 End-to-end view via the GSM infrastructure

This section covers fax transmission with special regard to those aspects that relate to GSM. More detailed information on fax transmissions can be found in [27]. Before going into further detail let's take a brief look at how a fax connection is made via the GSM and other networks. Figure 5.14 shows an end-to-end connection. It does not matter in which direction the fax is sent, and we will assume that the fax is originated in the fax machine (laptop computer) connected to the mobile station shown on the left side of the figure.

The fax is originated in the computer, which sends its digital data to the mobile phone via a modem or other hardware, which is often a PCMCIA device (see Section 5.7.2) simulating modem functionality. The data are sent over the air interface in either transparent or nontransparent mode to the BSS and to the MSC. The MSC will transfer the data link to the proper IWF. The different IWF possibilities account for the different needs of the various networks such as pure digital transmission for the ISDN network, or tone generation for the standard PSTN. The fixed network, the ISDN, or another PLMN subsequently transports the fax information to the fax terminal on the right side of the figure, which in our case is a PSTN fax machine. Both fax terminals communicate with each other using the standard fax protocol defined in ITU-T T.30 throughout the whole fax connection [28]. The encoding and decoding of the pages is specified in ITU-T T.4 [29].

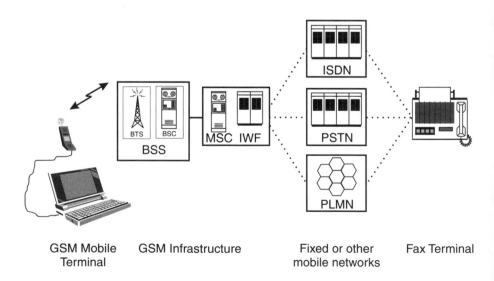

Figure 5.14   Network infrastructure supporting GSM fax service

5.7.2 Configuration at the mobile station

In the previous section we split the fax functions between the GSM mobile station and the computer. The implementation of fax into GSM requires additional functionality in order to connect a basic PSTN two-wire fax machine to a GSM PLMN.

The *fax adapter* converts the analog signals coming from the two-wire fax machine into a serial digital stream having the ISDN-specified R interface as the output. This is the standard *modem* functionality that includes some basic understanding of the T.30 fax protocol. A fax adapter is always required. It may, however, be possible to use a GSM-type fax where the fax functionality and the fax adapter are combined, which is the case for most PCMCIA fax cards supporting GSM phones. This also makes the most sense because the separate use of the fax adapter would mean that two real modems—the one from the fax machine and the one from the adapter—communicate with each other merely to convert the analog data back into the digital format used on the GSM side. Thus we have today's reality in applications that make full use of the PC; no real modems are used.

Depending on the type of mobile station, an additional *terminal adaptation* might be necessary in order to connect the fax adapter to the mobile terminal. The input to the terminal adaptation is the R interface and the ISDN-specified S interface is the output.

This leaves us with the question of how the three different kinds of mobile stations described in Section 5.4.2 are connected to the fax adapter.

1. *Mobile termination type 2 (MT2).* The R interface of the MT2 is connected directly to the fax adapter. Depending on if the fax adapter is integrated into the fax machine or not, there will be two boxes (devices) or only one. Figure 5.15 shows the different configurations for MT2 types of connections.

2. *Mobile termination type 1 (MT1).* Because the MT1 supports the S interface it requires an additional terminal adapter to convert the asynchronous serial data stream into a synchronous serial data stream at the phone side. The fax adapter and the terminal

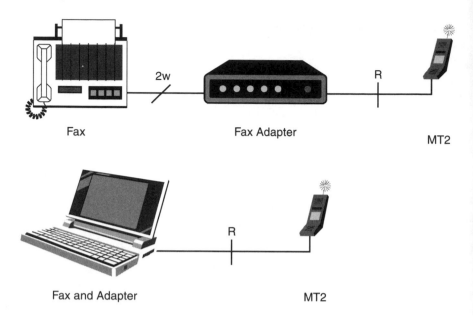

Figure 5.15   Configuration for MT2.

adapter may be incorporated into one unit providing several choices, which are indicated in Figure 5.16.

3. *Mobile termination type 0 (MT0)*. MT0 does not support a terminal interface at all, thus leaving two options. Either the MT0 is a voice-only GSM mobile phone, or it possesses all the data and fax capabilities indicated in Figure 5.17. The Nokia 9000 Communicator is an example of this.

We should note here that these three types of mobile terminations are rather more academic than real (with the exception of the Nokia 9000), since the available implementations construct the interfaces within the products (e.g., within the PCMCIA card) leaving the accessible interfaces as proprietary ones. The same has already been stated in Section 5.4.2.

5.7.3   Transparent fax service

The main differences between fax transmissions via a fixed network and the mobile GSM network can be traced to the air interface where the influence on transparent fax service is not particularly significant [30]. To

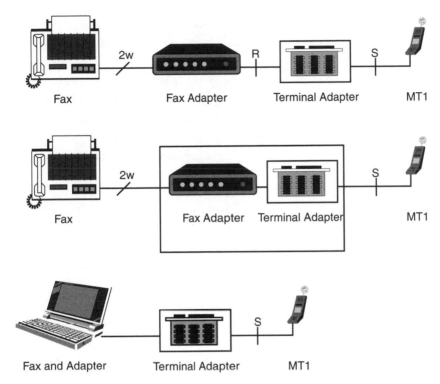

Figure 5.16 Configurations for MT1.

MT0

Figure 5.17 Configuration for MT0.

explain this, we need to take a more detailed look at the path through the GSM network, which includes rate adaptations and their consequences.

The end-to-end signal is transferred with the T.30 protocol, which specifies the ITU-T fax protocol and is not further detailed here.

Figure 5.18 shows the information transfer flow through the GSM network as it would apply to an MT2. The fax data and its signaling arrive at the fax adapter via the T.30 protocol. The fax adapter, together with the attached interface circuit, transforms this signal into one compatible with the R interface, which is a synchronous signal. RA1 adapts the speed to one that fits the data rate on the U_m-interface after encoding (e.g., 3.6, 6.0, or 12.0 Kbps). FEC is the protective envelope for the encoded data on the air interface. The data are unpacked again (decoded) in the *base station subsystem* (BSS) and go through further rate adaptation until they reach a data rate of 64 Kbps on the A interface on the MSC side of the BSS. This allows easier multiplexing of several (either 4 × 16-Kbps channels or 8 × 8-Kbps channels) data or fax transmissions onto one channel on the A interface. In the *mobile switching center/interworking function* (MSC/IWF) the rate is reduced again in order to match the input rate of the fax adapter, the interface toward the fax machine, which may appear in the fixed network as if it were attached to a 3.1-kHz audio line.

Fax communication is controlled between the two fax adapters in Figure 5.18 in accordance with the T.30 protocol. The other layers below the T.30 layer are simply used to ensure good transmission quality and the best transport mechanism through the GSM infrastructure. This mechanism is similar to the layered structure on the U_m-interface. At the beginning of each fax setup both fax machines negotiate the transmission speed over the whole transmission path. This is reflected in the speed over

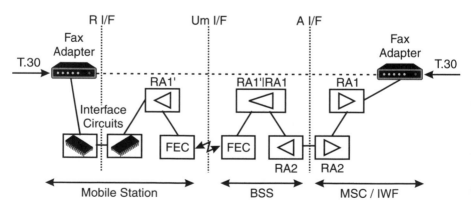

Figure 5.18   Path through the GSM network.

the air interface, which will be 2400, 4800, or 9600 bps for a full-rate channel. If the quality drops drastically during the transmission, then it is up to the fax machines to renegotiate the transmission speed. Transparent fax service is characterized by constant speed, constant transit delay, and variable error rates.

5.7.4 Nontransparent fax service

The nontransparent fax service is characterized by variable speed, variable delay, and a low error rate. As was the case with the transparent mechanism, setup and error control are the responsibility of the two fax machines. The error correction mechanism is defined in the ITU-T specifications. The chief constraint for data and fax transmission over the GSM infrastructure, as we should be well aware by now, is the air interface. All other components in the communications link have the same constant transmission quality that is found in the fixed network. Fixed-line transmissions, too, are often routed through radio links (satellite and directional microwave transceivers) and fiber optic cables, but these are not narrowband mobile links, which are more hostile to orderly data flow than fixed radio links and fiber optic cable. As long as the transmission quality is good enough, the difference between transparent and nontransparent fax service will be negligible. However, if the line quality of any segment in the link degrades, then the nontransparent fax service has the advantage of the RLP in which only corrupted data frames are repeated rather than a whole fax page. This means that the RLP has to correct errors even though the overall error control remains with the two fax machines communicating via the T.30 protocol. This repeating mechanism may cause problems when the repetition of frames on the air interface leads to a kind of "traffic jam" in the fax adapter, which will call on a buffering function to overcome the problem. It also asks for communications between the RLP and the fax adapter as shown in Figure 5.19. This communications link copes with the influence of the RLP and the air interface on the T.30 protocol, which is not designed to handle the delay introduced by these two influences [31].

The use of various data rates for nontransparent service is restricted. It is highly recommended that only full-rate channels at the highest possible speed—9600 bps—be used. The RLP will maintain the quality on the

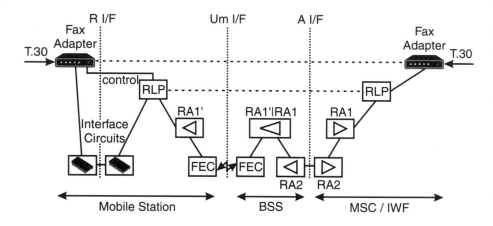

Figure 5.19   Path through the GSM network.

air interface and the fax machines will negotiate the overall transmission speed. The nontransparent service is characterized by a reduced error rate due to the radio link protocol, but variable overall transmission rates and transit delays. Even though there is additional protection through the RLP, the actual control is left with the two fax adapters and their connected fax machines via the T.30 protocol.

5.7.5   In-call modification

The *in-call modification* (ICM) feature applies only to teleservice 61 (alternate speech/facsimile Group 3). This service is used when the receiving fax machine is operated in manual mode, which is when the user has to answer the call and then switch the fax machine into its receiving mode. The ICM feature has to be handled via the user interface of the mobile station where the user switches from a voice call to a fax transfer. This is indicated to the network via a layer 3 channel modify message indicating that the channel type will be switched to a data (or fax) channel. The user can start sending the fax message after the modification is completed.

We did not mention this in the sections on data transmission, but we should point out here that the same mechanism is also provided for data transmission. The two processes are (1) speech followed by data and (2) alternate speech/data (bearer services 61 and 81).

5.8 Connecting a mobile station to external devices

We have seen many applications that connect a laptop computer, a fax machine, or other *data terminal equipment* (DTE) to a GSM mobile station for the exchange of data with dedicated data networks or to send faxes anywhere in the world. The only problem related to these applications is that today's mobile user no longer accepts the need to carry around bulky portable computers simply to accomplish routine nonvoice communications. The smaller palmtop computers or pocket organizers are preferred to help the mobile manager get organized, stay in touch, get informed, and send messages and documents anywhere. GSM supports these mobile office applications in two ways:

1. Retrieving and sending short messages through a mobile phone with a portable device. This makes the SMS more acceptable and easy to use [32].

2. Remote control of a mobile phone through an external device, which allows the user to organize his work and control the mobile phone as a single unit [33].

The advantage of the connection specifications is that they make special hybrid solutions (e.g., connecting the Nokia 2110 mobile phone to the HP 700LX organizer) obsolete. Standard connection specifications allow the user to mix and match many palmtops with a whole catalog of phones. It is still up to the manufacturers to support these interfaces with their products and to define the extent to which specific support is offered. A manufacturer may decide, for example, to support only a subset of features from one of the specifications.

Some of the very small computers support a PCMCIA interface, which can enable a connection to a GSM mobile phone. But PCMCIA cards tend to be expensive and they consume a considerable amount of power from the palmtop computer, thus limiting their operating time. Figure 5.20 shows a mobile phone connected to a notebook computer, using standard GSM commands via the serial interface.

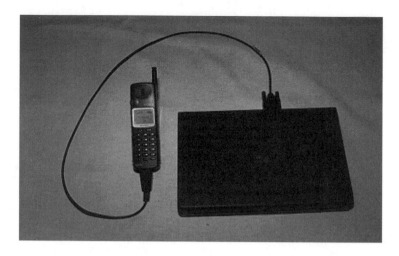

Figure 5.20   A mobile phone connected to a notebook computer.

Sadly, the specifications do not standardize the physical connector between the mobile phone and the palmtop, which means that the user is still burdened with conducting cautious experiments in search of the correct cable to be fitted between her mobile phone and her data terminal—a tedious exercise vigorously pursued by those who find value in the connections.

5.8.1 Application for short message services

As new mobile phones become smaller and lighter, and as SMS becomes more popular, maintaining a measure of convenience and utility with today's tiny keypads and displays becomes a challenge. The little displays are difficult to read, and typing a short message is even harder since there are a minimum of three letters allocated to each key. The tedium becomes intolerable for many users who need to distinguish capital and lowercase letters and add some special characters.

Relief comes by exporting the display and the keypad to a palmtop or an organizer. The telephone and the organizer are connected together with a cable or, soon, with a wireless, infrared link, which is already included in most palmtop computers. Figure 5.21 shows the user interface of a dedicated SMS software displaying how a message can be

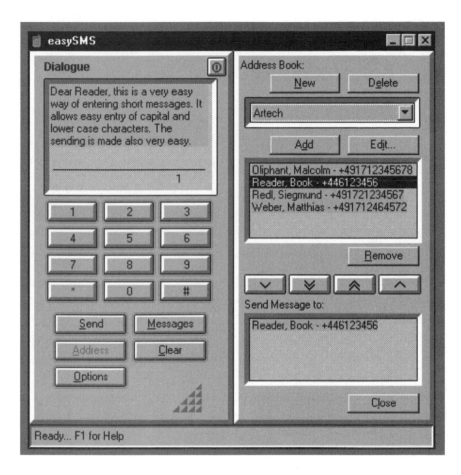

Figure 5.21 Display of short message software on a computer.

composed for sending by a mobile phone. The advantage of this solution
is that the user can use the normal keypad with lowercase and capital
character. Good software also allows the user to have his own address
book, and maintain a list of different GSM networks and the access numbers to them.

The *terminal equipment* (TE) supports the following features, which
are available for short messages and cell broadcast messages handled by a
mobile terminal (MT).

▶ *Initialization of short message services and cell broadcast message sessions.*
The MT indicates when at least one new message has arrived.

▶ *Request for a list of messages.* The MT divides the messages into blocks of five messages and transfers the header information for each of the five messages to the TE.

▶ *Transfer of a specific message.* The MT sends the full message, including the header, to the TE.

▶ *Transfer of all messages.* The MT sends the first message, and all subsequent messages, until the last message is transferred.

▶ *Diversion of incoming messages.* The MT sends all incoming messages directly to the TE.

▶ *Request of indication of incoming message.* The MT indicates each newly arrived message to the TE. A message is identified by its header.

▶ *Request of transfer into MT.* This command allows the TE to submit a short message to the MT. The TE specifies whether the message should be stored in the MT, sent over the air interface, or both. The MT returns an indication of the success or failure of the operation. This is only possible for point-to-point short messages since the cell broadcast is only defined for the downlink direction.

▶ *Deletion of messages.* The TE can specify that messages should be deleted from the memory in the MT (either in the ME or SIM). An individual message is identified by the message header.

▶ *Termination of short message services and cell broadcast message sessions.* A session can be terminated when the MT has been switched off.

The user's acceptance of this application depends on good user interfaces offered by the terminal equipment makers. Special care should be taken to provide easy access to the functions such as special icons for list or message retrieval. Once messages have been loaded into the TE, the message header should be displayed while the message itself is made available upon selection of its respective header. A possible user interface could be one similar to the user interfaces designed for e-mail tools.

5.8.2 Remote control of mobile equipment

Where the short message access application allows only access to short message service and cell broadcast service, the remote control application allows complete control of a mobile phone. To support this feature both the mobile terminal and the terminal equipment must support the *AT command set* specified for GSM. This is not necessarily limited to GSM but can also be used for other cellular systems. The use is similar to the AT commands known from normal terminal equipment connected to data networks. The need for this remote control arises for mobile terminals, without keypads and displays, dedicated to data services, or for the activation of the GSM terminal by external events such as a theft alert in a car, or when an accident is detected by sensors in the car in combination with the airbag sensors.

Why did GSM implement the AT command set rather than the relevant ITU-T specification V.25bis? The ITU-T specification was the standard for PSTN modems for many years, but its details were ignored. Hayes came up with a much more versatile command set for their successful modem products, and the AT (for "attention") command set was born. Everyone in the data communications business copied the Hayes protocol without bothering with the V.25bis standard. If a MT2 was built strictly to the GSM Phase 1 specifications, which defined the use of V.25bis, there would be no way to select options for the data service (data rate, modem type in the IWF, transparent or nontransparent, etc.) via the R interface; it would be necessary to select these from the MMI of the phone. This would be a highly user-hostile situation, because most users would not have a clue as to what selection to make. Hence, manufacturers chose to implement the familiar AT command set instead, which allows software in PCs to set up all aspects of the data capabilities of the phone. This is a much more versatile option that works with most standard communications software, which supports the AT command set.

The interface for the AT command set operates over an ITU-T V.24 specification. The standard Windows terminal program allows anyone to control a V.24 interface. To support V.24 the mobile equipment must either connect to, or include, a *terminal adapter* (TA). The TA converts the AT *commands* and AT *responses* into mobile equipment control and response strings. The principal structure is displayed in Figure 5.22. It is up to the mobile equipment manufacturer to decide if the TA is implemented inside or outside the phone. The manufacturer may also decide if the TA and the TE are connected with a cable or with an infrared link.

The functions of the AT command set are split into five functional blocks:

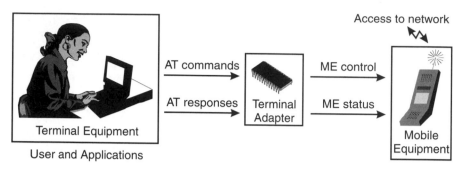

Figure 5.22 Setup of mobile equipment (ME), terminal adapter (TA), and terminal equipment (TE).

1. *General commands:* The general commands allow access to the ME identification including such information as the manufacturer (full name plus optional additional information provided by the manufacturer), the model name, revision identification, and the *international mobile equipment identity* (IMEI). They also allow the user to select a character set such as the code page from the PC character set. The TA will convert the character strings between the TE and the ME.

2. *Call control commands:* The commands in this functional block are related to call control, which are (1) dialing from the phone book, (2) hanging up a call (hook on), (3) the call mode that allows voice/fax or voice/data calls, (4) selecting the bearer service, (5) setting the RLP parameters, and (6) setting call control parameters such as how often an incoming call should ring the phone before the call is automatically answered, how long the phone should wait for call completion before giving up and disconnecting, and how long it should wait when the pause identifier is set.

3. *Network service related commands:* The first type of these commands displays the subscriber number in the PLMN and the services related to this number (e.g., voice or fax). Another type of commands are those for network registration status (e.g., registered, not registered, searching for networks, and whether operator selection should be done automatically or manually). Still other types of service related commands are facility lock (barring incoming and/or outgoing calls) and phone locking (to the SIM card—theft prevention). We also include commands that change

the password and support CLIP, COLP, CUG, call forwarding, call waiting, call hold, and call transfer. The following supplementary services are *not* supported: multiparty, advice of charge, and call completion to a busy subscriber.

4. *Mobile equipment control and status commands:* These commands include (1) phone activity status (ready, unavailable, ringing, call in progress, and asleep); (2) PIN entry (detection of whether the PIN or PUK is requested); (3) battery charge condition; (4) signal quality (in terms of level measured in dBm and quality measured in seven levels); (5) control mode of the mobile (only through the keypad, only from the TE, or from both); (6) emulation of the ME's keypad; (7) display control, which allows the TE to write into the ME's display; (8) indicator control (battery level 0–5, signal level 0–5, service, ringer, message received, call in progress, transmit activated by voice activity, and roaming indicator); (9) phone book control: read, write, find (SIM fix dialing/last dialing phone book, ME/SIM phone book, and TA phone book); and (10) clock reading or setting (if available), and setting the alarm.

5. *Mobile equipment errors:* These are errors that are returned when a function is not supported, when problems occur in the phone (SIM not inserted), or upon invalid data entry.

The functions included with the remote control of mobile phones are very universal. Unfortunately, universality also means increased complexity and user unfriendliness, which places a challenge on the engineers developing palmtop software to create a user interface that is attractive and easy to use.

5.9 Future developments

This concluding section examines some future developments and enhancements for data and fax services. Even though their standardization is discussed within the industry, some or all of these features and functions may not necessarily survive to be final standards or products. Some of the ideas such as *high-speed circuit-switched data* (HSCSD), *general packet radio service* (GPRS), and *packet data on signaling channels* (PDS) will certainly appear as formal standards and will eventually appear in future products.

5.9.1   High-speed circuit-switched data

Several of today's circuit-switched teleservices, some new teleservices, and new bearer services and related applications (video) that are already standardized for fixed networks require higher data rates than GSM can offer today. Latest coding techniques (see Section 5.9.4), which are used in the end-to-end path between both terminals, will not provide the necessary enhancements. The only feasible solution that is consistent with the current air interface specification is to increase the data bandwidth by spreading the data transmission over *several* time slots up to the maximum of eight on a channel. The services will be defined for both transparent and nontransparent data transport modes.

Some companies have already successfully demonstrated multiple time slot transmissions. For instance, Nokia, during the Telecom exhibition in Geneva in 1995, used two time slots for HSCSD data. In this case a "standard" mobile station was used with modified software in both the mobile equipment and in the infrastructure. Because this service has great potential to become very important in the near future, we will take a closer look at its implementation.

In principle, HSCSD is a vehicle for higher speed data than a normal bearer service. It not only influences the end-to-end communication and the signaling but also the physical part, which we examine in this section [34,35].

5.9.1.1   Implementation on the air interface

HSCSD combines between two and eight time slots of one channel on the air interface (U_m) for each direction. Assuming that the highest rate of 9.6 Kbps is used, this allows data rates of 19.2 (two time slots) up to 76.8 Kbps (eight time slots) for user data rates. The user data rate, however, is limited to 64 Kbps, which is exactly the amount of data for an ISDN channel, and what can be transported on one channel on the interface between the BSS and MSC. If data rates higher than 9.6 Kbps on one time slot are achieved in the future, then the number of allocated time slots will decrease. Figure 5.23 shows how two and three time slots are combined. For a single time slot application the time duplex offset

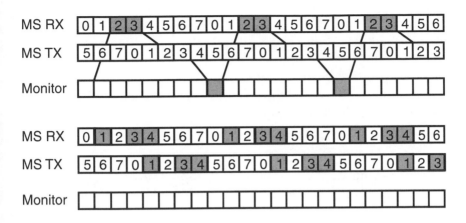

Figure 5.23 A combination of two and three time slots.

between a mobile station's reception of a burst (time slot) from the base station, and the MS's next transmission (a burst), is separated by three time slots [19]. In Figure 5.23 we realize that a combination of two channels still allows the mobile station to monitor neighboring cells in the idle times during a frame. If we increase the application's demand to three channels (time slots), then the implementation becomes more difficult, especially if the channels are not linked, but are instead spread over the whole TDMA frame of eight time slots. The design of a mobile station will be significantly affected by these multiple slot applications.

When frequency hopping is used in a network supporting HSCSD, the same hopping sequence will apply to all channels used by one mobile station. The same applies for the training sequence, which is located in the middle of the transmitted information bursts, and enables the receiver's equalizer to cope with the different propagation conditions on the channel. The ciphering keys used for each individual traffic channel forming one HSCSD channel are different, but they are derived from the same K_c stored on the SIM.

In many bit-hungry applications the user requests big chunks of information from a server. This means that the uplink traffic (commands/requests/selections) from the mobile station requires relatively little bandwidth compared with the high transmission bandwidth required in the downlink toward the MS. GSM also covers these cases, allowing symmetric and asymmetric configuration of HSCSD. *Symmetric* means that the same number of time slots (channels) are allocated for the uplink and downlink directions. *Asymmetric* means that the number of slots allocated for the uplink and the downlink is different. As a rule the allocated number of channels in the uplink direction should not be greater than in the downlink direction. An example of an asymmetric configuration might be one time slot allocated in the uplink and two or three in the downlink direction.

5.9.1.2 Classes of mobile stations

As shown in Figure 5.23, the number of channels used for an application influences the design of a mobile station. It is easy to understand that, especially when high numbers of slots are required for a channel, the mobile station may need to receive and transmit at the same time, which is not the case with current GSM phones. Today's practice takes advantage of the time gap of three time slots between receive and transmit schedules. To account for the possibility of receiving and transmitting at the same time, which is the lack of TDD, GSM distinguishes two classes of HSCSD mobile stations. Class 1 HSCSD phones will have a maximum sum of uplink and downlink allocations of five channels or slots. Class 2 phones will cover the range from 6 to a maximum of 16 channels (slots), which is eight channels in each direction. The class 2 MS will need to be able to transmit and receive at the same time.

Besides the time slot allocations there are some additional requirements for the air interface timing in a mobile station, such as how fast (measured in units of time slots) the MS must be able to transmit after the last reception (t_t), and how fast the mobile station must be able to receive after its last transmission (t_r). The different requirements are reflected in different *multislot classes*, which become part of the MS's negotiations with the network during a call setup. Table 5.9 summarizes the various parameters.

5.9.1.3 Network architecture supporting HSCSD

One bottleneck in GSM is the bandwidth on the U_m-interface, which is a 22.8-Kbps gross rate for single time slot speech channels. The communications between different parts of the network is based on higher data rates (64 Kbps is typical) and poses no limiting effect to the data rate. Indeed, several channels from the U_m-interface are mapped onto one channel on the interface between the BSC and the MSC (A interface). One exception to this rule is the interface between BS and the BSC (Abis

Table 5.9

Parameters for Multislot Transmission

MS Type (class)	Multislot Class	Maximum Number of Slots			Minimum Number of Slots	
		Rx	Tx	Sum	T_t	T_r
1	1	1	1	2	2	2
1	2	2	1	3	2	1
1	3	2	2	3	2	1
1	4	3	1	4	1	1
1	4	2	2	4	1	1
1	6	3	2	4	1	1
1	7	3	3	4	1	1
1	8	4	1	5	1	1
1	9	3	2	5	1	1
1	10	4	2	5	1	1
1	11	4	3	5	1	1
1	12	4	4	5	1	1
2	13	3	3	6	1	1
2	14	4	4	8	1	1
2	15	5	5	10	1	1
2	16	6	6	12	1	1
2	17	7	7	14	1	0
2	18	8	8	16	0	0

interface), which has been adapted to the data rate on the U_m-interface and, therefore, in most cases supports submultiplexed 16 Kbps for one U_m channel. Multiple channels on the U_m-interface are mapped onto multiple channels on the Abis interface. The data rate on the A interface is limited to one circuit, which is normally a 64-Kbps resource. Figure 5.24 shows the infrastructure required for HSCSD.

Packing data on various channels requires a *combining and splitting function* in both the mobile station and the IWF. The combining and splitting function is not done in the BSC where the multiple channels from the Abis are multiplexed onto one circuit on the A interface. The reason for keeping the combining and splitting task out of the BSC is to maintain compatibility with single-slot applications. It should not matter if there is a multislot or a single-slot application anywhere in the path from the BTS

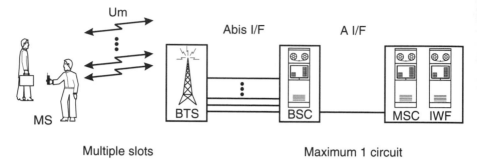

Figure 5.24   Infrastructure for HSCSD.

to the MSC. Figure 5.25 makes this situation clearer. When a high data rate stream is fed into the user's side of the MS, it is then split into an appropriate number of lower rate channels appropriate for the U_m- interface. Moving to the right side of Figure 5.25, the standard *rate adaptation* (RA), the *forward error correction* (FEC), and the *multiplexing* (MUX) are performed exactly as for normal data channels at rates of 2.4, 4.8, or 9.6 Kbps. Only in the IWF will these multiple channels be combined back into one single high data rate stream that matches the one originally fed into the MS. The experienced reader may have noticed that we have just described a transparent data service but the mechanism also applies for the nontransparent data services.

The radio link protocol has been adapted to allow frame counting for multiple time slots. An important fact to be noted here is that each individual channel is treated similarly to the single slot services. It is left to the mobile station and the IWF to handle HSCSD.

5.9.1.4   Services supported by HSCSD

HSCSD should be regarded more as a transport vehicle than as a dedicated service. For this reason HSCSD supports both transparent and nontransparent services at rates higher than 9.6 Kbps. Another prerequisite for mobile stations beyond its ability to support the modifications to the air interface required for multislot allocations is the support of a minimum data rate of at least 4.8 Kbps. It would be rather wasteful to use multiple time slots for a basic baud rate of only 2.4 Kbps.

The BSS will try to reach but not exceed the requested data rate on the air interface by selecting a specific number of channels. For example, if a

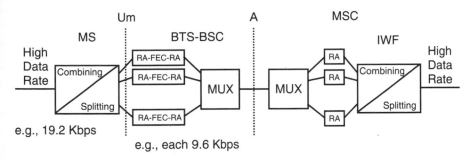

Figure 5.25   Location of the combining and splitting functions in the MS and the IWF.

data rate of 19.2 Kbps is requested, and the mobile station supports 9.6 Kbps (per time slot), then the BSS would allocate two channels. In the unlikely case where the MS only supports 4.8 Kbps (per time slot), the BSS would allocate four channels.

If the requested user rate is not a multiple of a supported rate, then the BSS has to select the next higher rate and then fill the excess bit capacity with fill bits. This is the case if the requested user rate is 14.4 Kbps and two TCH/F9.6 channels are used. Depending on the network loading conditions, the network might allocate different resources during a data call, which means the rate on the user interface might change during a call.

Charging for HSCSD is determined by the number of channels that are used for an outgoing call. When the number of channels varies during a call, the charge rate would be adjusted each time the allocated number of radio resources changes. Calling parties from the PSTN will not incur higher charges.

5.9.2   General packet radio service

All standard Phase 1 and Phase 2 GSM data services are based on circuit-switched services. Even those services that access packet data networks are circuit-switched in GSM. The SMS provides a packet transfer mode, but for evolving mobile packet systems the capacity of SMS is insufficient. GPRS provides a solution to a growing demand for packet-oriented data services. Because this service also requires higher data rates, the average size of data packages tends to grow with time and GPRS allows for data rates of up to 170 Kbps. To implement this service, a completely new network architecture is required [36,37].

5.9.2.1   Services offered by GPRS

GPRS is a complete set of new bearer services. In contrast to existing circuit-switched mode transfer, GPRS provides packet mode transfer. Two types of services are possible.

Point-to-point (PTP) service allows the exchange of information between two dedicated endpoints or users. The PTP service has two services, the *connectionless network service* (PTP-CLNS) and the *connection-oriented network service* (PTP-CONS). The connectionless service is characterized by unconfirmed, nonguaranteed delivery of data packages similar to a pager. The connection-oriented service is characterized by confirmed, guaranteed delivery.

Point-to-multipoint (PTM) allows the transmission of information from one user to many others. This is similar to electronic mail, which also allows this PTM service. The PTM service includes a total of three services: (1) the *PTM multicast* (PTM-M), (2) the *PTM group call* (PTM-G), and (3) the *IP multicast* (IP-M). PTM-M allows the user to send messages within a geographical area defined by the sender of a message. PTM-G permits the user to send messages to a predefined group within a specific area. IP-M defines a service that uses the *Internet protocol* (IP) and addresses a group in which there is no geographical limitation since the message can be spread throughout the whole Internet.

5.9.2.2   Network architecture for GPRS

As indicated before, the GPRS service requires some new network entities. Actually the GPRS structure is an overlay to the GSM infrastructure and a mobile phone supporting GPRS can be considered a dual-mode phone. Figure 5.26 shows the infrastructure required for this service.

The interface between the MSC and the BSS serves as a simple signaling interface. The main GPRS functionality lies in the two additional network entities: the *serving GPRS support node* (SGSN) and the *gateway GPRS support node* (GGSN).

The SGSN is the service node supporting the mobile station for GPRS, including functions such as:

▶ Authentication and authorization;

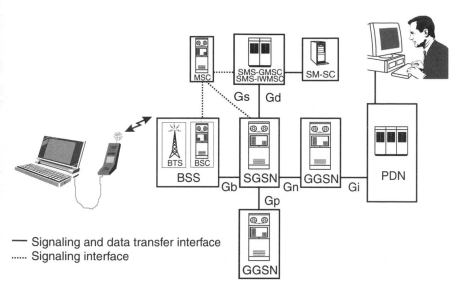

Figure 5.26 Network architecture necessary to support GPRS.

- Admission control;

- Charging data collection;

- Packet routing and transfer;

- Mobility and logical link management.

The GGSN is the interface toward the *packet data network* (PDN); it is also the interface to other PLMNs. Data are actually exchanged between the SMS entity and the GPRS part of the network as well. However, the packet-oriented SMS service should also be performed by the GPRS service once it is available. The implementation of GPRS foresees that mobile stations supporting SMS should receive short messages via GPRS.

5.9.2.3 Implementation on the radio interface

Besides the addition of new infrastructure, new access techniques on the radio interface are required. Space limits us to only scratching the surface of the subject for which detailed information can be found in [38].

A complete new set of control channels is defined for GPRS, which have functions similar to those of their cousins in the standard GSM implementation. One difference is that the capacity will be allocated dynamically; that is, if there is additional need for capacity then the network will recognize this and allocate more resources (channels or time slots).

GPRS distinguishes four *coding schemes* (CS) that have different levels of error protection ranging from high protection down to no protection at all. Table 5.10 summarizes some parameters for the different coding schemes numbered CS-1 through CS-4. The code rate indicates the redundancy added to the actual data bits where 1/2 means that for each data bit another bit is added, and 2/3 means that for two data bits one protection bit is added. The table also indicates the actual data rate. A mobile phone may use up to eight time slots simultaneously.

5.9.2.4 Introduction of GPRS

Due to the complexity of GPRS and the changes required to the network and the phones, this feature will be introduced gradually in two phases. Phase 1 of GPRS enables the use of PTP services. Phase 2 will add the PTM services to the feature list.

5.9.3 Packet data on signaling channels

GPRS is a long-term project because it requires major changes to the GSM network architecture. An intermediate relief for packet data demand can come from the implementation of PDS. This service can be used for short data packages of smaller size, as is the case for fleet management where, for instance, only the parameters of a location have to be transmitted. The advantage over the existing packet service (SMS) is that it does not have the signaling overhead. It can be considered an improvement to the already defined packet mode access. PDS is easy to implement because it

Table 5.10
Parameters for Coding Schemes

Coding Scheme	Data Rate (Kbps)	Code Rate
CS-1	9.05	1/2
CS-2	13.4	≈ 2/3
CS-3	15.6	≈ 3/4
CS-4	21.4	1

can be used on all existing dedicated channel types (SDCCH or SACCH) and it can also be used in parallel with speech. The data rates on PDS range from 600 to 9600 bps [39,40].

5.9.3.1 Services offered by PDS

PDS is a bearer service that offers two services: *PDS service 1* (PDSS1) and *PDS service 2* (PDSS2). PDSS1 needs a subscription to a GSM network, whereas PDSS2 does not require a subscription but may only be subject to the usage by an application. The difference between these two services is that the first one provides more options, such as keeping a connection while performing a handover when driving from one cell to another, and allowing mobile originated and mobile terminated traffic. Due to the lack of mobility functions, a handover for the PDSS2 will cause a connection to break; only mobile originated traffic is allowed. A new mobile identity is introduced for both services: *anonymous*. This anonymous identity allows access to certain applications where users can retrieve data from a service offered by a value-added service supplier or the operator.

5.9.3.2 Network architecture for PDS

Regarding changes on the infrastructure side for PDS, Figure 5.27 gives an overview of where and how the two services interface to the network. PDSS1 does not need any additional functionality but simply interfaces to the MSC. PDSS2 requires an additional support node called the *PDSS2 support node* (PDSS2-SN). The PDSS2-SN interfaces directly into the BSC via a dedicated interface: Ap. This explains why handover or mobile terminated calls are not possible; there is no dedicated link through the MSC that would supervise and handle the mobility management function. This implies that the call charges will not be collected by the GSM operator but by the provider for the application using the PDSS2.

PDS makes use of the standard signaling channels on the air interface, either on a SDCCH or SACCH. If a TCH is to be used then it will be used in a signaling-only mode. The only change for signaling is the introduction of new parameters identifying the service.

All of this indicates that PDS is a service that will be available within a short period of time because it does not require major implementations in the network. PDS will be an interim service for the GPRS service, which may use the same applications but at much higher data rates.

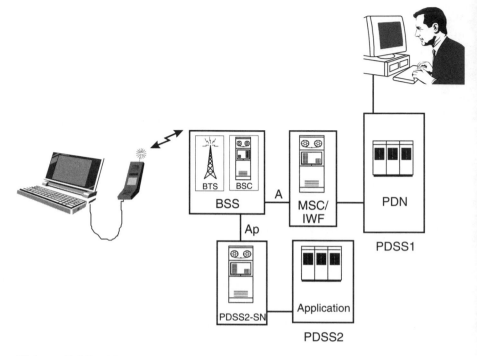

Figure 5.27   Network architecture for PDS.

5.9.4   The 14.4-Kbps user data rate

Enhancing coding of user data is one way to increase the data rate on the whole path from one terminal to the other terminal. It has been proven that the redundancy on the air interface can be reduced, thus allowing us to increase the user data rate to a level of 14.4 Kbps, which is already a widely used rate for fixed-line services. This user data rate can be achieved by modifying the coding for the traffic channel. Depending on the traffic channel, two different rates are available for full-rate TCH (14.4 Kbps) and for half-rate TCH (7.2 Kpbs). These data rates also have an influence on all data services, specially the HSCSD, which can achieve 28.8 Kbps with only two time slots.

5.9.5   Facsimile enhancements

ITU-T is discussing enhancements to facsimile Group 3 that will result in major improvements in resolution and transmission speed. People will get used to a better fax quality from the PSTN/ISDN and will expect the same quality in GSM. Such enhancements will have to find their place in GSM by integrating the relevant standards into the GSM specifications.

5.9.6 General bearer services

Phase 2+ will introduce new bearer services that allow a more flexible allocation of data rates, but also support new data transmission techniques such as HSCSD or GPRS. The general bearer services support new information transfer capabilities into the fixed networks such as V.120 rate adaptation, which is more commonly used in countries outside of Europe. Table 5.11 gives an overview of these supplemental services. The fixed network data rates of these bearer services depend only on the selected information transfer type, and the support on the air interface when rates beyond 9.6 Kbps are specified in the HSCSD specifications. The highest rate for the fixed network side is 28.8 Kbps when 3.1-kHz audio is selected. The maximum rate is 56 Kbps for V.110 UDI, X.31 flagstuffing UDI, and V.120. The highest data rate of 64 Kbps is available only for bit transparent mode as shown in Table 5.12 [41].

Table 5.11
Overview of General Bearer Services

Bearer Service		Access		Information Transfer
No.	Name	Struct.	Rate	
20	Asynch general	Asynch	Up to 64 Kbps	3.1-kHz audio, V.110, X.31 Flagstuffing, V.120, or bit transparent
30	Synch general	Synch	Up to 64 Kbps	
40	General PAD access	Asynch	Up to 64 Kbps	
50	General packet access	Synch	Up to 64 Kbps	

Table 5.12
Overview of Attributes Depending on Information Transfer Protocol

Information Transfer	Fixed Network Data Rate (Kbps)	Access Structure	Attribute
3.1-kHz audio	0.3, 1.2, 1.2/0.075, 2.4, 4.8, 9.6, 14.4, 19.2, 28.8	Asynch	NT or T
3.1-kHz audio	1.2, 2.4, 4.8, 9.6, 14.4, 19.2, 28.8	Synch	NT or T

Table 5.12 (continued)

Information Transfer	Fixed Network Data Rate (Kbps)	Access Structure	Attribute
V.110	0.3, 1.2, 2.4, 4.8, 9.6, 14.4, 19.2, 28.8, 38,4 Kbps	Asynch	NT or T
V.110	0.3, 1.2, 2.4, 4.8, 9.6, 14.4, 19.2, 28.8, 38,4, 48 (only T), 56 (only T)	Synch	NT or T
X.31 flagstuffing	2.4, 4.8, 9.6, 14.4, 19.2, 28.8, 38,4, 48, 56	Synch	NT
V.120	1.2, 2.4, 4.8, 9.6, 14.4, 19.2, 28.8, 38,4, 48, 56	Asynch	NT
V.120	2.4, 4.8, 9.6, 14.4, 19.2, 28.8, 38,4, 48, 56	Synch	NT
Bit transparent	56, 64	Synch	T

5.9.7   Emergency call with additional data transfer

The emergency call service enables a subscriber to get in touch with a local emergency service simply by dialing, for example, "112." This service is mandatory for all GSM networks. The emergency services features supported through GSM improve as GSM services develop. There would be significant time savings and less confusion if, along with the speech service, data could be transmitted indicating the location of the mobile phone. Applications for this could be special modules in cars that connect a GSM terminal with a *global positioning system* (GPS) receiver. The receiver would provide exact location data that could be included with each emergency call. In the future it will even be possible to initiate emergency calls either manually, which will be useful for taxi drivers, or automatically when a driver is injured in an accident.

References

[1]   GSM 02.01, "European Digital Cellular Telecommunications System (Phase 2); Principles of Telecommunication Services Supported by a GSM Public Land Mobile Network (PLMN)," ETSI, Sophia Antipolis.

[2]   GSM 02.02, "European Digital Cellular Telecommunications System (Phase 2); Bearer Services (BS) Supported by a GSM Public Land Mobile Network (PLMN)," ETSI, Sophia Antipolis.

[3] ITU-T Recommendation V.23, "600/1200 Band Modem Standardized for Use in the General Switched Telephone Network."

[4] GSM 02.03, "European Digital Cellular Telecommunications System (Phase 2); Teleservices Supported by a GSM Public Land Mobile Network (PLMN)," ETSI, Sophia Antipolis.

[5] GSM 03.10, "European Digital Cellular Telecommunications System (Phase 2); GSM Public Land Mobile Network (PLMN) Connection Types," ETSI, Sophia Antipolis.

[6] ITU-T Recommendation V.21, "300 Bit Per Second Duplex Modem Standardized for Use in the General Switched Telephone Network."

[7] ITU-T Recommendation V.22, "1200 Bit Per Second Duplex Modem Standardized for Use in the General Switched Telephone Network and on Point-To-Point 2-Wire Leased Telephone Type Circuits."

[8] ITU-T Recommendation V.22bis, "2400 Bit Per Second Duplex Modem Using the Frequency Division Technique Standardized for Use on the General Switched Telephone Network and on Point-To-Point 2-Wire Leased Telephone-Type Circuits."

[9] ITU-T Recommendation V.26ter, "2400 Bit Per Second Duplex Modem Using the Echo Cancellation Technique Standardized for Use on the General Switched Telephone Network and on Point-To-Point 2-Wire Leased Telephone-Type Circuits."

[10] ITU-T Recommendation V.32, "A Family of 2-Wire, Duplex Modems Operating at Data Signaling Rates of Up To 9600 Bit/S for Use on the General Switched Telephone Network and on Leased Telephone-Type Circuits."

[11] ITU-T Recommendation X.25, "Interface Between Data Terminal Equipment (DTE) and Data Circuit-Terminating Equipment (DCE) for Terminals Operating in the Packet Mode and Connected to Public Data Networks by Dedicated Circuit."

[12] ITU-T Recommendation X.21, "Interface Between Data Terminal Equipment (DTE) and Data Circuit-Terminating Equipment (DCE) for Synchronous Operation on Public Data Networks."

[13] ITU-T Recommendation X.28, "DTE/DCE Interface for a Start-Stop Mode Data Terminal Equipment Accessing the Packet Assembly/Disassembly (PAD) in a Public Data Network Situated in the Same Country."

[14] ITU-T Recommendation X.32, "Interface Between Data Terminal Equipment (DTE) and Data Circuit-Terminating Equipment (DCE) for Terminals Operating in the Packet Mode and Accessing a Packet Switched Public Data Network Through a Public Switched Telephone Network or an Integrated Services Digital Network or a Circuit-Switched Public Data Network."

[15] ITU-T Recommendation V.110, "Support of Data Terminal Equipment (DTE) With V-Series Interfaces by an Integrated Services Digital Network (ISDN)."

[16] ITU-T Recommendation X.30, "Support of X.21, X.21bis, X.20bis Based Data Terminal Equipment (DTE) by an Integrated Services Digital Network (ISDN)."

[17] ITU-T Recommendation X.31, "Support of Packet Mode Terminal Equipment by an ISDN."

[18] GSM 04.21, "European Digital Cellular Telecommunications System (Phase 2); Rate Adaptation on the Mobile Station–Base Station System (MS–BSS) Interface," ETSI, Sophia Antipolis.

[19] Redl, S. M., M. K. Weber, and M. W. Oliphant, *An Introduction to GSM*, Norwood, MA: Artech House, 1995, Chap. 5.

[20] GSM 04.22, "European Digital Cellular Telecommunications System (Phase 2); Radio Link Protocol (RLP) for Data and Telematic Services on the Mobile Station–Base Station System (MS–BSS) Interface and the Base Station System–Mobile Services Switching Center (BSS–MSC) Interface," ETSI, Sophia Antipolis.

[21] Mouly, M., and B. Pautet, *GSM System For Mobile Communication*, France: Palaiseau, 1992.

[22] GSM 09.01, "European Digital Cellular Telecommunications System (Phase 2); General Network Interworking Scenarios," ETSI, Sophia Antipolis.

[23] GSM 09.03, "European Digital Cellular Telecommunications System (Phase 2); Signaling Requirements on Interworking Between the Integrated Services Digital Network (ISDN) or Public Switched Telephone Network (PSTN) and the Public Land Mobile Network (PLMN)," ETSI, Sophia Antipolis.

[24] ITU-T Recommendation V.24, "List of Definitions for Interchange Circuits Between Data Terminal Equipment (DTE) and Data Circuit-Terminating Equipment (DCE)."

[25] GSM 09.05, "European Digital Cellular Telecommunications System (Phase 2); Interworking Between the Public Land Mobile Network (PLMN) and the Packet Switched Public Data Network (PSPDN) for Packet Assembly/ Disassembly Facility (PAD) Access," ETSI, Sophia Antipolis.

[26] GSM 09.04, "European Digital Cellular Telecommunications System (Phase 2); Interworking Between the Public Land Mobile Network (PLMN) and the Circuit-Switched Public Data Network (CSPDN)," ETSI, Sophia Antipolis.

[27] McConnel, K., D. Bodson, and R. Schaphorst, *FAX: Facsimile Technology and Applications Handbook*, 2nd ed., Norwood, MA: Artech House, 1992.

[28] ITU-T Recommendation T.30, "Procedures for Document Facsimile Transmission in the General Switched Telephone Network."

[29] ITU-T Recommendation T.4, "Standardization of Group 3 Facsimile Apperatus for Document Transmission."

[30] GSM 03.45, "European Digital Cellular Telecommunications System (Phase 2); Technical Realization of Facsimile Group 3 Transparent," ETSI, Sophia Antipolis.

[31] GSM 03.46, "European Digital Cellular Telecommunications System (Phase 2); Technical Realization of Facsimile Group 3 Nontransparent," ETSI, Sophia Antipolis.

[32] GSM 07.05, "European Digital Cellular Telecommunications System (Phase 2); Use of Data Terminal Equipment–Data Circuit Terminating Equipment (DTE–DCE) Interface for Short Message Service (SMS) and Cell Broadcast Service (CBS)," ETSI, Sophia Antipolis.

[33] GSM 07.07, "European Digital Cellular Telecommunications System (Phase 2); AT Command Set for GSM Mobile Equipment (ME)," ETSI, Sophia Antipolis.

[34] GSM 02.34 "Digital Cellular Telecommunications System (Phase 2+); High Speed Circuit-Switched Data (HSCSD)—Stage 1," ETSI, Sophia Antipolis.

[35] GSM 03.34 "Digital Cellular Telecommunications System (Phase 2+); High Speed Circuit-Switched Data (HSCSD)—Stage 2 Service Description," ETSI, Sophia Antipolis.

[36] GSM 02.60 "Digital Cellular Telecommunications System (Phase 2+); General Packet Radio Service (GPRS); Service Description Stage 1," ETSI, Sophia Antipolis.

[37] GSM 03.60 "Digital Cellular Telecommunications System (Phase 2+); General Packet Radio Service (GPRS); Service Description Stage 2," ETSI, Sophia Antipolis.

[38] GSM 03.64 "Digital Cellular Telecommunications System (Phase 2+); General Packet Radio Service (GPRS); Overall Description of the General Packet Radio Service (GPRS) Radio Interface, Stage 2," ETSI, Sophia Antipolis.

[39] GSM 02.63 "Digital Cellular Telecommunications System (Phase 2+); Packet Data on Signaling Channels Service (PDS)—Stage 1," ETSI, Sophia Antipolis.

[40] GSM 03.63 "Digital Cellular Telecommunications System (Phase 2+); Packet Data on Signaling Channels Service (PDS), Service Description, Stage 21," ETSI, Sophia Antipolis.

[41] GSM 02.02, "Digital Cellular Telecommunications System (Phase 2+); Bearer Services (BS) Supported by a GSM Public Land Mobile Network (PLMN)," ETSI, Sophia Antipolis.

Further Reading

GSM 05.02 "Digital Cellular Telecommunications System (Phase 2+); Multiplexing and Multiple Access on the Radio Path," ETSI, Sophia Antipolis.

GSM 07.01, "European Digital Cellular Telecommunications System (Phase 2); General on Terminal Adaptation Functions (TAF) for Mobile Stations (MS)," ETSI, Sophia Antipolis.

GSM 07.02, "European Digital Cellular Telecommunications System (Phase 2); Terminal Adaptation Functions (TAF) for Services Using Asynchronous Bearer Capabilities," ETSI, Sophia Antipolis.

GSM 07.03, "European Digital Cellular Telecommunications System (Phase 2); Terminal Adaptation Functions (TAF) for Services Using Asynchronous Bearer Capabilities," ETSI, Sophia Antipolis.

GSM 09.02, "Digital Cellular Telecommunications System (Phase 2); Mobile Application Part (MAP) Specification," ETSI, Sophia Antipolis.

GSM 09.06, "European Digital Cellular Telecommunications System (Phase 2); Interworking Between a Public Land Mobile Network (PLMN) and the Packet Switched Public Data Network/Integrated Services Digital Network (PSPDN/ISDN) for the Support of Packet Switched Data Transmission Services," ETSI, Sophia Antipolis.

GSM 09.07, "Digital Cellular Telecommunications System (Phase 2); General Requirements on Interworking Between the Public Land Mobile Network (PLMN) and the Integrated Services Digital Network (ISDN) or Public Switched Telephone Network (PSTN)," ETSI, Sophia Antipolis.

ITU-T Recommendation V.26bis, "2400/1200 Bit Per Second Modem Standardized for Use in the General Switched Telephone Network."

ITU-T Recommendation V.27ter, "4800/2400 Bit Per Second Modem Standardized for Use in the General Switched Telephone Network."

ITU-T Recommendation V.29, "9600 Bit Per Second Modem Standardized for Use on Point-to-Point 4-Wire Leased Telephone-Type Circuits."

ITU-T Recommendation V.32bis, "A Duplex Modem Operating at Data Signaling Rates of Up To 14400 Bit/S for Use on the General Switched Telephone Network and on Leased Point-To-Point 2-Wire Telephone-Type Circuits."

ITU-T Recommendation X.21bis, "Use on Public Data Networks of Data Terminal Equipment (DTE) Which Is Designed for Interfacing to Synchronous V-Series Modems."

ITU-T Recommendation X.29, "Procedures for the Exchange of Control Information and User Data Between a Packet Assembly/Disassembly (PAD) Facility and a Packet Mode DTE or Another PAD."

ITU-T Recommendation X.75: "Packet-Switched Signaling System Between Public Networks Providing Data Transmission Services."

Short message service

The GSM *short message service* (SMS) provides a unique means of bringing short alphanumeric (letters and numbers) messages or other kinds of information to the user of a mobile telephone. The messages appear on the phone's display, which is usually an LCD. GSM distinguishes two different types of short messaging:

1. A dedicated service, between two parties, which requires the establishment of a (dedicated) point-to-point bearer link, thus the name *point-to-point SMS*;

2. Broadcast service between the network (through one or more base stations) and all users within a cell or service area. This is called *point-to-omnipoint* or *cell broadcast SMS*.

6.1   Short message
service: point-to-point

The point-to-point service may be compared to the familiar paging service, but it is much more comprehensive. Traditional paging services deliver messages through an operator or computer, whereas GSM additionally allows direct (bidirectional) messaging without operator interaction. Sophisticated two-way paging (Motorola's FLEX family of protocols is an example) has recently appeared in the paging and messaging industry that closely mimics GSM's SMS, thus assuring us the tiny pagers we know so well are not likely to disappear, even though they do not have GSM's voice services. Nevertheless, GSM's SMS allows anyone to send alphanumeric messages to a mobile user at any time. Short messages might be sent because the recipient did not respond to a call, or has not switched on his mobile phone, or is busy on another phone, or cannot be reached for some other private reason. The "store-and-forward" mechanism behind SMS makes sure that when the recipient's phone cannot be immediately reached, the message is eventually delivered when the receiving subscriber's terminal is back in service or otherwise ready to receive messages.

Another difference between SMS and the common one-way pager is that with traditional paging services one never knows exactly if or when messages really arrive at the addressee, whereas with GSM, one can receive confirmation of the arrival of each message. In the unfortunate case where a message does not arrive within a specified time, the sender will be notified. There is an optional service for notification of when a message actually does arrive at the mobile phone.

Point-to-point short messages use a dedicated link, as in a voice call, on a duplex radio channel. This has the advantage that the network and the mobile station actually talk to each other, thus error handling mechanisms make sure that a message arrives properly at the recipient's mobile station and is stored properly therein. If this is not the case or if the mechanism fails for any reason, an error message will be created by the mobile station notifying the network of the problem together with the reason why a message was not received correctly. Since this message link uses a different *service access point identifier* (SAPI) than speech, it is even possible to send a message to a phone that is already carrying normal voice traffic for the recipient.

Another advantage over, for example, leaving voice mail messages is cost: when compared to a voice mail service, for which the message is stored during a charged voice call and has to be retrieved by yet another charged voice call, direct short messaging is by far the most economical way of getting quick and short information to the right destination.

The term *short message* comes from the relatively short length of each individual message, which is limited to 160 characters (mapped onto 140 bytes) for each single message. The technical specification for point-to-point SMS can be found in [1].

6.1.1 Implementation of point-to-point SMS in the network

To clearly describe the message flow, the relevant network structure is depicted in Figure 6.1 and is described as follows. The *short message entity* (SME) is, in general, the device that originates a short message. It might be:

- *A mobile station.* In this case we observe the transmission of a message between two mobile stations. By the way, the two mobile stations do not necessarily have to be roaming in the same network; SMS is transparent to user locations.

- *A computer directly connected to the service center (SC).* This is a service provided by the operator, which can be used simply by dialing in and requesting a message to be sent to a specific subscriber.

- *A device or computer located in the fixed network.* The special computer allows someone to dial directly into the SC in order to dispatch a message.

- *A computer that delivers updated news services.* This is an example of a service the user subscribes to and to which she makes some kind of request. This should not be confused with cell broadcast, which is always without request.

The SC handles all functions related to point-to-point SMS. The SC receives a message from a SME and forwards it to the addressee. If it cannot reach the addressee, it stores the message up to some maximum time specified by the user or the operator. If necessary, or if requested, the SC

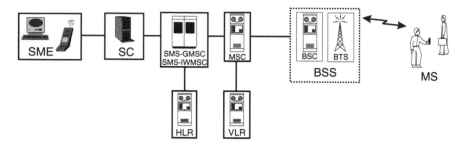

Figure 6.1 Network entities involved in the provision of short messages.

notifies the sender about the successful or unsuccessful delivery of a message. The SC, by definition, is not part of the GSM PLMN. In actual practice it is integrated into or connected to the MSC.

When SMS services first started it was common for there to be only one service center per operator. However, with the increasing number of applications for short messages, an operator would need several service centers to handle all of the messages. This is, for example, the case if one center were to handle the SMS for standard text messages between two subscribers, and another center handles a kind of service where the user sends a message to the SC requesting the latest result of a Formula 1 car race or the latest stock exchange information. This second set of examples is similar to faxback services.

It is also possible, on the other hand, for one SC to serve several PLMNs. This would be the case, for example, when two operators in one country share an SC, or two operators in different countries share an SC. One service center might be connected to several MSCs.

The *short message service–gateway mobile services switching center* (SMS-GMSC) is a dedicated function within a MSC that allows it to receive a short message from the SC, investigate the attached *home location register* (HLR) in order to find out where the recipient mobile station is currently located, and then deliver the short message to the visiting MSC and, eventually, to the mobile station. The SMS-GMSC function is used when a mobile station is the recipient of the message. This is represented in Figure 6.1 by a mobile-terminated short message, which would flow from left to right in the figure.

The *short message service–interworking mobile services switching center* (SMS-IWMSC) is another dedicated function within a MSC that allows the SC to receive short messages from the MSC. This would be the case if the mobile station is the originator of a short message. It then submits the message to the recipient's service center. The SMS-IWMSC is used for mobile-originated SMS.

SMS-GMSC and SMS-IWMSC are not only responsible for the forwarding of short messages, but in general for the complete exchange of information between the network and the SMSC, which includes, for example, the exchange of status messages, as described later.

The *home location register* (HLR) stores the identity and user data of all the subscribers belonging to the area of the related MSC. For SMS, the relevant data in the HLR is (1) the current location (MSC) of the recipient mobile station, and (2) whether it is switched on and available to receive messages.

The *mobile services switching center* (MSC) is a complete exchange system. With all of its registers, it is able to route calls from the fixed network, via the mobile network, to an individual mobile station. The MSC has connections to other entities of the PLMN, which allow it to gather information about the current whereabouts of a mobile station; the location is needed in order to forward messages.

The *visitor location register* (VLR) contains all relevant data on visiting mobile stations of the serving MSC. In the case of SMS, the relevant datum is the more precise paging area where the recipient mobile station is located.

The *base station subsystem* (BSS) is used as a transport vehicle for SMS over the radio (U_m) interface.

The *mobile station* (MS), in the particular case of SMS, is either a sender or a recipient of a short message. The mobile station would also inform the network about abnormal (erroneous) situations, such as memory overflow.

6.1.1.1 Functions and parameters of point-to-point SMS

Short message service is based on functionality that is distributed over the service center, the HLR, the VLR, and the MSC with a number of parameters. Depending on the direction of transmission, the data frame containing the message accommodates different parameters. There are, of course, other message frames supporting different functions of the SMS, such as a status report and a command. Here we discuss only the frames carrying the actual user data, which will appear on the display of a mobile station when a mobile terminated SMS transaction is completed success-fully. In the other direction we have mobile originated message services, which are initiated by messages generated in the mobile phone.

The messages described here are higher layer messages. Additional information must be added to the messages before they can be trans-ported on the air interface. Further details are found in [2].

SMS-MT message frame   Figure 6.2 shows the structure and format of a *short message–mobile terminated* (SM-MT) message frame with the different fields and parameters. The *reply path* (RP) flag indicates that an answer for this specific message is paid by the originator of the message, and the recipient can answer free of charge. Sometimes it is sufficient to get a simple *yes* or *no* reply to a short message, such as, "Should we meet today at 12:00?" The reply path option has been defined for this purpose. This gives the recipient the ability to send a short answer without having to pay anything for this service. In GSM SMS, this option is invoked by the sender of the short message.

The *status report indication* (SRI) shows that a status report will be sent back to the originator (SME) of this short message. This means that the report of proper delivery from the mobile station will be passed on to the originator as confirmation of delivery of the short message.

The *more-messages-to-send* (MMS) flag indicates that, for this particular recipient, several messages are waiting to be delivered in the SMSC. This

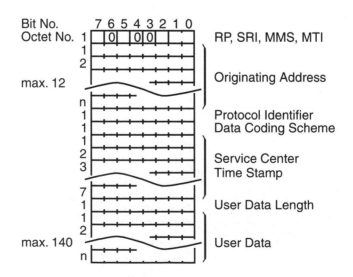

Figure 6.2   Format of an SM-MT.

can be the case when a mobile station was switched off for a long time. The basic principle behind this is that all messages awaiting delivery should be delivered in one go once the signaling channel link has been established. This not only reduces signaling effort and traffic (air time) on the radio interface, but also in the network itself since all the authentication and ciphering procedures are only performed once.

The *message type indicator* (MTI) is a 2-bit value describing the kind of message contained in the frame. The message type is related to:

▶ Delivering a short message from the network to the mobile station, and the corresponding confirmation from the mobile station;

▶ Submitting a short message from the mobile station to the network, and confirmation of the network;

▶ Status report from the network to the mobile station about the whereabouts of a previously sent message;

▶ Issuing a command from the mobile station to the network.

This last command issued by the "subscriber" enables the MS to invoke an operation or inquiry at the SC, which could be the deletion of a previously sent short message or the cancellation or initialization of a status report request.

Table 6.1 lists the different values for the message type, the direction, and their meanings.

Table 6.1
List of Values for the Message Type

Value	Direction	Message Type
00	Network–MS	Delivery of a short message from the network
00	MS–network	Report on delivery of short message from the mobile station
01	Network–MS	Sending of a status report from the network
01	MS–network	Sending of a command from the mobile station
10	Network–MS	Submission of a short message from the mobile station
10	MS–network	Report on delivery of short message from the network
11	Any	Reserved value

The *originating address* (OA) contains the address of the sender of a specific message. If the sender was another mobile station, then the originating address is the actual phone number of a specific mobile station. Many mobile phones nowadays realize that this address is a phone number, and allow the user to select this number in order to call the number directly, or respond with another short message. The first option is especially helpful if the user wants to talk to the person who sent the message rather than respond with another short message.

The *protocol identifier* (PID) instructs the service center in which format or protocol the short message should be delivered. There are two major distinctions. One is the transfer of a short message between two mobile phones within the GSM infrastructure, which requires no special treatment because it uses only the higher layers. The other is the use of telematic interworking when a short message has to be "translated" into another format. The user of a mobile telephone is entitled to set up the protocol identifier. Note, however, that not all networks support the different protocols since these always require additional hardware in the network. See Table 6.2 for the different values for the protocol identifiers in telematic interworking. As should be obvious from the list, it is, in principle, possible to transfer a short message into a message for any kind of device.

Table 6.2
Values for the Protocol Identifiers Used for SMS

Value	Telematic Device
0	Implicit, the device type is specific to this service center or can be deduced from the basis of the address
1	Telex (or teletex reduced to telex format)
2	Group 3 telefax
3	Group 4 telefax
4	Voice telephone (i.e., conversion to speech)
5	ERMES (European Radio Messaging System)
6	National Paging System (as it is known to the service center)
7	Videotext (T.100/T.101)
8	Teletex, carrier unspecified
9	Teletex in PSPDN (public-switched packet data network)
10	Teletex in CSPDN (cellular-switched packet data network)
11	Teletex in analog PSTN (public-switched telephone network)

Value	Telematic Device
12	Teletex in digital ISDN (integrated services digital network)
13	Universal computer interface (UCI) as specified in ETSI DE/PS 3 01-3
14–15	Reserved (two combinations)
16	Message handling facility that is known to the service center
17	Any public X.400-based message system
18–23	Reserved (six combinations)
24–30	Values specific to each service center; the usage is based on mutual agreement between the short message entity and the service center (there are seven combinations available for each service center)
31	GSM mobile station. The service center converts the short message from the received alphabet coding to any alphabet decoding supported by that mobile station (e.g., the default one)

The *data coding scheme* (DCS) informs the mobile station about which decoding is used for the alphabet used in the actual message. Currently, two alphabets are available: (1) the default alphabet or (2) 8-bit data. These are specified in Section 6.1.2. In addition to the alphabet, the DCS also specifies the class of the short message.

The *service center time stamp* (SCTP) indicates when a specific message arrived at the service center. This time is also delivered to the recipient. Each mobile phone user can see, by this time stamp, when the received message was actually sent, or, to be more precise, when the service center received it. From this time stamp, one could actually evaluate how long it took the network to deliver the message. The format of the time stamp is in the Year:Month:Day:Hour:Minute:Second:Time zone order. Each value is represented by two digits.

The *user data length* (UDL) indicates the length of the following data in characters. The length is, therefore, dependent on the coding scheme. If the 8-bit coding scheme is used, then the UDL is identical to the number of octets. If the default (7-bit coded) GSM alphabet is used, then the number of octets is less than the number of characters.

The *user data* (UD) contains the actual encoded message with a maximum length of 140 octets. Section 6.1.2 describes the GSM alphabet and the coding of the individual characters. This reveals the mystery of why a short message contains up to 160 characters, but can be carried by a maximum of only 140 octets. In Section 6.1.3 a complete message will be shown along with the coding of the user data.

SMS-MO message frame   Figure 6.3 shows the structure and format of a *short message–mobile originated* (SM-MO) message with the different fields and parameters. The RP indicates that the sender of the message is willing to pay for the response.

The *status report request* (SRR) indicates that the sender of this short message wants to have a confirmation that the message was delivered. If the message was delivered within the validity period, then the sender will get a positive confirmation along with the time when the message was delivered. All of this will appear within a status report explained in the paragraph about the message type. If the message could not be delivered, then the sender would be notified about this fact along with a cause value.

The *validity period format* (VPF) describes whether the validity period is present in this message, and if it is present, whether it is represented as a relative or an absolute value.

The *message type indicator* (MTI) is identical to the one described in the chapter on supplementary services, Chapter 7.

The *message reference* (MR) allows the mobile station to number the short messages that have been sent. This MR is used when a status report is issued to the mobile station by the service center.

The *destination address* (DA) indicates the recipient of this short message. The address is entered as a normal phone number. The PID and the DCS are, again, identical to the mobile-terminated message.

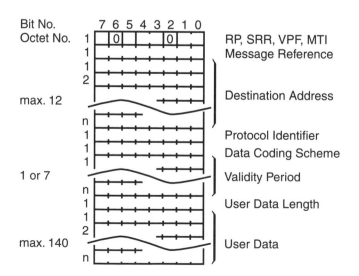

Figure 6.3   Format of an SM-MO.

The *validity period* (VP) indicates how long a specific short message is valid, and how long the SC should, therefore, try to deliver the message. Let's say we sent an urgent short message indicating an appointment for this evening. It would not make much sense to try continuously to deliver this message tomorrow morning. The validity can be represented in two different formats: the relative validity and the absolute validity.

In the case of the absolute validity, the format is identical to the SC time stamp. The absolute validity specifies the "expiration date" of a message. In the relative representation of validity, the resolution of time frames decreases with increasing time. The resolution is 5 minutes for validity periods up to 12 hours. From 12 hours to 1 day the resolution decreases to 30-minute steps. For validity periods between 1 day and 1 week, the resolution decreases even further to a whole day; and from 1 week up to 63 weeks (more than a year) it decreases to 1-week steps. In reality, however, the long validity periods with low-resolution expiration indications are mostly theoretical curiosities because network operators tend to limit validity times to a maximum of several days. Doing this limits the amount of expensive network resources required for storing messages.

The UDL and UD are similar to the mobile-terminated case.

Additional functionality in the network   So far we have only considered specific messages that were sent out to mobile stations or received from them. We have not yet explored the details of how the messages are sent and how the network ensures their proper delivery. These details become important, for example, when mobile phones are switched off or move out of reach of the network.

Consider a message that is waiting in the service center for delivery. When the *priority* indicator for such a message is set, then the service center continuously tries to deliver the message even though an error, such as "no memory available," is returned from the target mobile, the mobile station is temporarily absent, or the HLR does not confirm that an error condition is cleared. Sending the short message more often increases the chances of early delivery; there are more opportunities to detect the return of a mobile station to service, or that the user has finally decided to clear his short message memory, either in the phone or in the SIM card.

When the SMS could not be delivered, the *messages waiting* element is available in the VLR as the *mobile-station-not-reachable* flag, and in the HLR as *message-waiting-data* field, which is further subdivided into different conditions. The implementation in the VLR is mandatory, and it is optional in the HLR. If a delivery attempt of a short message was unsuccessful, then the network will return some diagnostic results. If the mobile station is not available, then the *mobile-station-not-reachable* flag is set in the VLR. If the mobile station becomes alive again, or becomes otherwise accessible by the network, the VLR will notify the SC of this happy event through the HLR. Other error cases will generate an entry into the optional *message-waiting-data* field in the HLR. If this is not present, the option to include it in the HLR has not been exercised, and then the SC would simply receive a failure indication. The SC, then, is obliged to cope with failure indications itself. Even though it is optional, the assistance of the HLR is a more effective way to deal with erroneous message deliveries. As soon as the HLR detects that a problem is resolved, it notifies the SC with the *alert-SC* command. In the *message-waiting-data* field, there is sufficient room to enter the addresses of multiple service centers that have to be notified when problems are resolved.

The *mobile-station-not-reachable* flag and the *message-waiting-data* field will only be cleared if all messages have been sent successfully to a mobile station. This data management scheme makes sure that all messages arrive at the mobile properly.

The service center can also send a status report to the originator of a message. The status can be simply that a message was successfully delivered to the SME. The request for this status message can be invoked by the originator with a dedicated request for this status report within the short message. On the other hand, a status report can also be issued when a message could not be delivered to the recipient. The implementation of this status report capability is optional, and is left entirely to the operator.

6.1.1.2 SMS MT

In the case of mobile-terminated SMS, where the message will arrive in the mobile station, the following steps will take place:

1. The short message is generated in the SME. It does not matter whether the SME is another mobile station or an entity within the PLMN.

2. The short message is sent from the SME to the service center. The service center controls further handling of the message. It checks

the validity period, and it adds the time stamp showing when the message was received. If necessary, it transforms it to another protocol, it checks whether there are more short messages to be sent to the same addressee, and it verifies the priority level of the message.

3. The SC notifies the SMS-GMSC that there is a message waiting for a mobile station. If the mobile station does not belong to this MSC, then the MSC will pass this notification on to the "home" MSC of the recipient.

4. The SMS-GMSC asks the HLR in which MSC area the mobile station is currently located (visited MSC). If the mobile station is currently not available, it is switched off, or some other problem occured (see Section 6.1.4); then the SC will be notified, and the HLR will set a flag that indicates that the attempted delivery was not successful. This situation will only be cleared when the delivery is finally successful, or the validity period expires.

5. With the MSC's knowledge of the precise whereabouts of the recipient mobile station, the SMS-GMSC passes the short message on to the visited MSC.

6. The visited MSC checks with the VLR to determine in which location area the mobile station is currently located.

7. The BTSs in the location area, in which the mobile station is registered, page the mobile. The currently visited BTS establishes a channel and delivers the message to the mobile.

8. The mobile terminal notifies the user that a short message has been received. The notification is some kind of audible alarm, or a visual signal that may be viewed on the display.

Figure 6.4 shows the display of a mobile station that has received a short message. A problem that often occurs in mobile stations is that the size of the display is not big enough for viewing complete messages. The mobile station, therefore, has to insert line feeds into the received messages and provide a means for the user to scroll through messages. Besides the message itself, there is lots of other information available. The date at

Figure 6.4 Example of a short message on the display of a mobile phone.

which the message has been sent, here 03.04.1997 at 11:29, and from which mobile phone number (+491712464572).

The most common mobile-terminated short message is a notification to the user that one or more voice messages were received, recorded, and then stored in the user's mailbox during the time the user had left the phone off, was busy, or was not answering calls.

6.1.1.3 Packing a short message on the radio channel

Short messages can be sent to, or received from, a mobile station irrespective of the state or condition of the mobile station. Assuming the mobile station is switched on and connected to a network, two states can be distinguished: (1) the *idle* state occurs when the mobile is simply monitoring the common control channels (PCH, FCCH, SCH, and BCCH), and (2) the *active* state occurs when a traffic channel (TCH/SACCH) is assigned to the mobile station for communication [3].

In the idle state, short messages are sent on a dedicated signaling channel (SDCCH), which happens to be the fastest way of transmitting short messages, since each signaling channel supports the transmission of 184 data bits (23 octets of 8 bits) within approximately 240 ms. In the active state, the radio channel is used for the *traffic channel* (TCH). In this second and less optimal case, the short message has to use the resources of the *slow associated control channel* (SACCH), which only supports 168 bits within 480 ms.

If a state change happens while a short message is being sent, this will take the signaling resource away from the short message. A state change from idle to active means that the signaling channel is needed for setting up a call, and it will eventually be replaced by a traffic channel. A state change from active to idle results in disconnecting the signaling resources. In these cases, short message transactions will fail, and the service center will be notified so that it can send the message later.

6.1.1.4 Classes of SMS

When specifying SMS the intention was to define and distinguish different classes of messages depending on their importance and the location where they should be stored. The classes of messages are restricted to the case where the recipient of a message is a mobile phone. There are four classes.

▶ *Class 0:* Short messages of this class are to be displayed on the screen of the mobile station immediately, and an acknowledgment to the service center will be sent as well. This message must not necessarily be saved in the mobile station or on the SIM card. It is, however, acceptable for these messages to be stored first in the mobile equipment or on the SIM card, if such has been selected by the user (through the user interface or menu). In this case, the message is to be treated as if there was no message class. (Note that if the user deliberately decides to store messages in a certain place, this overrides the distinction of message classes.)

▶ *Class 1:* Short messages of this class should be stored by the mobile equipment in the memory of the mobile station. If this is not possible, perhaps because there is no memory available or all available memory is already full, then the message may also be stored in a different place such as on the SIM card. Again, the user might be able to select the location where she wants to store the message.

▶ *Class 2:* This message class is reserved for Phase 2 mobile equipment and carries SIM-specific data. The mobile station must ensure that the message is transferred to the SIM data field before sending an acknowledgment to the service center. If this is not possible, then an error message will be sent to the SC. An application for this class of SMS is over-the-air activation, which actually modifies SIM fields.

▶ *Class 3:* Messages of this class are meant to be sent so that they can be forwarded to an external device such as a palmtop device or personal digital assistant. The mobile station will send an acknowledgment to the SC regardless of whether the message could be transferred to external equipment or whether the external equipment was connected at all.

Some mobile stations do not have SMS memory space. This is not as big of a problem as it may seem because most of the newer SIM cards have large memories. This underlines the reason behind putting as much information as possible on the SIM card. When the SIM card is moved from one mobile station to another all the phone's settings, phone memory, and all the messages follow the card.

6.1.1.5 SMS MO

For the SMS mobile-originated case, the message flow is in a direction opposite to that in the mobile-terminated case:

1. The short message is generated in the mobile (or transferred to it from an external device). Some details need to be attended to, as explained below.

2. The short message is sent from the mobile station to the MSC.

3. The MSC passes the short message on to the SMS-IWMSC of the addressed service center.

4. The SMS-IWMSC hands the short message over to the service center. From here onward, it is the responsibility of the service center to handle the short message.

The first-time user of SMS in GSM Phase 1 systems needed to know lots of little parameters and details. The details were neither convenient to handle nor particularly easy to comprehend and remember. The simplest of these parameters is certainly the recipient's phone number or access code. After this comes the task of entering the message into the handset, which can be quite awkward if one is confined to the standard keypad of the mobile telephone. Some better solutions are available these days that allow for easier entry of text. But, if after entering the message text, the user thinks the message is ready to be sent, he will be surprised or annoyed when some kind of error message is displayed. This is because he still, at least, has to enter the service center address, which is when the trouble really starts. What is the SC address? Where does he get the address? How does he enter it? What will the next surprise be? The service center address, which is a phone number, is unique to each operator, and has to be obtained by the usual means: examining the "welcome kit" and other subscriber instructions, or calling the operator's help desk.

The next hurdle is the protocol identifier, which is usually set to the "0" default value. This is the implicit identifier for the addressed service center (see Table 6.2). But if one wants to transform a message into a voice message because the recipient does not have a GSM mobile phone, then more difficulties arise. Some mobile phone manufacturers list the codes in the user manual, or in the user interface of the telephone itself. Compared to these hurdles, the validity period seems to be a trivial matter, except that not all phones can accept fractional entries of days.

That was GSM Phase 1, and the past is welcome to keep it. All this is made much easier with Phase 2 mobile equipment and Phase 2 SIM cards. With Phase 2 SIM cards, the operator merely has to specify default values for the service center, the protocol ID, and the validity period. Phase 2 mobile stations will read this information from the preprogrammed SIMs and use it.

Finally, the user has to select whether she wants to get the status report confirming if a message was delivered to the SME.

It is, in principle, possible to send a short messages from any PLMN to any other PLMN; the only prerequisite, of course, is that both networks support SMS. Some exceptions to the rule include competing networks in a country, which for some reason (e.g., legal restrictions) are sometimes not allowed to support transmissions from one network to the other. The way around this kind of artificial restriction is that the sender could use a service center outside his own network. A user who wishes to send a short message through his home system, say, in Germany, to a recipient who can only receive messages through the other German GSM operator, could do so through a service center in the Netherlands or even Singapore. However, if the operator in the Netherlands or in Singapore detects increasing traffic of this kind, which cannot be charged to anybody, the operator is likely to want to restrict this possibility.

6.1.2   Alphabet of SMS

Currently there is only one alphabet specified for SMS [4]. It supports a total of 28 characters, which covers all European languages, including special characters such as Å, å, Ä, ä, Ç, Æ, æ, Ø, ø, etc. It does not, sadly, fully support the Greek language. It does not even attempt to support the Cyrillic characters, Asian characters, or Arabic characters. Because the languages that make exclusive use of non-Roman characters are becoming increasingly important, there are efforts to specify the use of them in the future.

As many as 128 different characters can be coded into binary patterns of only 7 bits length ($2^7=128$). If the text of one message consists of a maximum 140 octets (1 octet is 8 bits), it is possible to pack 160 characters into the available space. The digits for the individual characters are packed together by simply completing octets from the left (see Figure 6.5). The remaining space is filled with zeros. See, also, the message format immediately following this section. The appendix shows the coding of the default GSM alphabet.

6.1.3   Example of a SMS-MT message frame

To make the SMS messages easier to understand, let's take a brief look at a very common message: a notification from the network mailbox that a new voice message has been stored, and the subscriber should call his voice mail system to retrieve the message. This example (see Figure 6.6) shows the complete message frame along with the coding for the various parameters. Once again, we explain the significance of the parameters as we go through the message.

▶ *Octet 1:* Bit 7, the 8th bit of the octet, specifies whether the reply path is set, which, for this example, is not necessary. The same applies for the status request (bit 5), and the more-messages-to-send flag

bit number	7	6	5	4	3	2	1	0
Octet 1	2g	1a	1b	1c	1d	1e	1f	1g
Octet 2	3f	3g	2a	2b	2c	2d	2e	2f
Octet 3	4e	4f	4g	3a	3b	3c	3d	3e
Octet 4	0	0	0	0	4a	4b	4c	4d

Figure 6.5   Message octets.

Bit	7	6	5	4	3	2	1	0	Comment
Oct.									
1	0	0	0	0	0	0	0	0	RP=0, SRI=0, MMS=0, MTI=0
2	0	0	0	0	0	1	1	1	originating address length indicator, 7 octets
3	1	0	0	1	0	0	0	0	ext. field, TON, numbering plan identification
4	0	1	0	0	1	1	0	0	Number: 4 +
5	0	0	0	1	1	0	0	1	1 9
6	0	0	0	1	0	1	1	1	1 7
7	0	0	1	1	0	0	1	1	3 3
8	0	0	0	1	0	0	0	1	1 1
9	0	0	0	0	0	0	0	0	Protocol Identifier
10	1	1	1	1	0	0	0	1	DCS: CG = 15, bit 3 res., def. alphabet, class1
11	0	1	1	1	1	0	0	1	Service Center Time Stamp: Year = "79"
12	0	1	1	0	0	0	0	0	Month: 60
13	0	0	0	1	0	0	0	0	Day: 10
14	0	0	0	1	0	0	0	1	Hour: 11
15	0	0	0	1	0	0	1	1	Minute: 13
16	0	1	0	1	0	0	0	1	Second: 51
17	0	0	0	0	0	0	0	0	Time Zone: 00
18	0	0	1	0	0	1	0	1	User data length in characters: 37 characters
19	0	1	0	1	0	0	0	0	P
20	0	1	1	1	0	1	1	0	l
21	0	0	1	1	1	0	0	1	e
22	0	0	1	1	1	0	0	0	a
23	0	0	1	0	1	1	1	1	s
24	1	0	0	0	0	0	1	1	e
25	1	1	0	0	0	1	1	0	_, c
26	0	1	1	0	0	0	0	1	a
27	0	0	1	1	0	1	1	0	l
28	0	0	0	1	1	0	1	1	l
29	1	0	0	1	0	1	0	0	_
30	0	1	1	1	1	1	1	1	y
31	1	1	0	1	0	1	1	1	o
32	1	1	1	0	0	1	0	1	u, r
33	0	0	1	0	0	0	0	0	_
34	1	1	1	1	1	0	1	1	v
35	0	0	1	1	1	0	1	1	o
36	0	0	1	1	1	1	0	1	i
37	0	0	1	0	1	1	1	0	c
38	1	0	0	0	0	0	1	1	e
39	1	1	0	1	1	0	1	0	_, m
40	1	1	1	0	0	0	0	1	a
41	0	0	1	1	0	1	0	0	i
42	0	0	0	1	1	0	1	1	l
43	0	0	0	0	0	1	0	0	_
44	0	0	0	0	0	0	0	0	@
45	1	0	0	0	1	0	0	1	_
46	0	1	1	0	0	1	1	0	", 3
47	1	0	1	1	0	0	1	1	3
48	0	1	0	1	1	0	0	0	1
49	0	1	0	0	1	1	0	0	1
50	1	1	1	0	0	1	0	0	"
51	0	0	0	0	0	0	1	0	.

▓ predefined fixed values

Figure 6.6 Detailed content of a SMS message.

(bit 2). The message type indicator (MTI, bits 1 and 0) describes the message as one that is delivered from the service center to the mobile station. Bit numbers 3, 4, and 6 are preset to a value of zero.

▶ *Octets 2–8:* The address field contains the length indicator for the address part of the message in octet 2. The following octet specifies the type of number, which in our example is an international number (even though it is possible for the number destination to be in the home PLMN). The numbering plan identification in our example is set to unknown (unknown = 0). Some other numbering options are the ISDN and data numbering plans. Octets 4 through 8 hold the actual number, where the single digits are represented by semi-octets, and the most significant number is held in the earliest (lower numbered) semi-octet. In our example the number is +491713311.

▶ *Octet 9:* The protocol identifier is set to the default value, which means that the message will be sent as a normal short message.

▶ *Octet 10:* The data coding scheme distinguishes the coding group (bits 4–7) and the alphabet type. The alphabet type is set to the default, which for GSM means 7-bit coding of the characters. In our example, not only is the alphabet type set, but the message class, which is one (bit 0–1), is also set. This means that the message will be stored in the mobile equipment.

▶ *Octets 11–17:* The service center entries are in semi-octets, where one octet (two semi-octets) is used for each of the year, month, day, hour, minute, second, and the time zone. The lower semi-octet is always the most significant digit, which is the reason that in our example the year "97" is entered as "79." Thus the date and time shown in the example is the "1st June 1997 11:31:15."

▶ *Octet 18:* The user data length specifies the number of characters contained in the following message. If the data coding uses 8-bit coding, then the user data length equals the number of octets. If the data coding applies 7-bit coding, then the number of octets is less than the number of characters.

▶ *Octets 19–51:* These fields are filled with the 7-bit coding of the message, which is *Please call your voice mail @ "3311"*. This is a total of 37 characters, which use 33 octets. Figure 6.7 shows how this message is displayed on the screen of a mobile station.

6.1.4 Problems that can occur while sending short messages

GSM SMS is subject to the same delivery problems found in any messaging system; there is always a chance that a message will not make it to its final destination. The cause of the problems may be either permanent, and no solution is possible; or the cause may be temporary, and a solution is possible. Once the delivery problem is solved, the short message may eventually be sent successfully by the SC.

The permanent problems with no solution are these:

▶ Unknown subscriber (no such number);

▶ No SMS subscription (you get only what you pay for);

▶ Illegal subscriber (authentication for delivery failed in call setup);

▶ Illegal equipment (IMEI was blacklisted).

Permanent problems cannot be overcome by the system. The subscriber will get a negative confirmation for the delivery of the message, along with the reason why the system was not able to deliver the message properly. In the common case of the unknown subscriber, the sender simply needs to check the address (phone number) of the person who should

Figure 6.7 Display of a mobile station with a mobile-terminated short message.

receive the message. If the network subscriber requires a dedicated subscription to SMS, then the user must register this service. For an illegal
subscriber, or the legitimate user trying to use illegal equipment, the bad
news is he will probably never be able to use SMSs unless he abandons his
errant habits.

Here is a list of temporary problems that have solutions:

- Call barred (while barring SS and during operator-determined barring, the recipient MS has no access to service);

- VPLMN does not support SMS;

- Absent subscriber;

- Error in MS (e.g., no memory available).

The temporary problems can be overcome either by the network or
by the user. If the addressee does not want to be disturbed (i.e., call barring has been selected), then the system is temporarily not able to deliver
the message. As soon as call barring is disabled, the HLR and VLR will
notice this, and the HLR will inform the service center that the user is
again available to receive messages. The same applies to an absent subscriber, who may only seem to be absent because he is driving through a
tunnel. As soon as he becomes alive (to the network) again, the HLR and
VLR will take appropriate action. If there is an error in the mobile station,
such as would be the case when memory is exhausted, the system will
have to try again and again until the user clears the memory in his mobile
station such that new messages can be delivered.

6.1.5 SMS and supplementary services

Most of the supplementary services available thus far exclude any applications to SMSs. The only exceptions are the barring services, with which
the user can specify if he wants to prevent incoming calls or if he wants to
prevent outgoing calls. If one of these services is activated for any reason,
it will also apply to SMSs. If a short message cannot be delivered because
the addressee has barred his mobile station, then the sender of the barred
message will receive an error message with an indication of the barring.

6.1.6 Use of additional devices for SMS

The access to and use of short message services is not limited to a mobile phone. Because the keypad of the telephone is very small, entering long messages might be tedious and annoying. In addition, the display may be too small. Tiny displays make it difficult to edit and read long messages. The industry is reacting to the requirement for better access to SMS. Two useful accessories are presented here that point to solutions.

6.1.6.1 External devices

Anyone who has ever tried to enter a short message of not so short a length, even a simple sentence of five or six words, on the keypad of a phone will share the view that this is not very practical. The awkward keypad is the reason why most of the SMS is of the mobile-terminated variety, which is used by an operator to notify the customer that there is a new message waiting for her in her voice mailbox.

To make the mobile-originated service more attractive, some of the mobile manufacturers have taken the approach of bypassing the keypad of the phone to a palmtop or laptop computer. The external device is connected to the telephone via a dedicated cable, which, typically and unfortunately, is different for each mobile and computer manufacturer. In most of the cases, it is possible to control remotely the mobile phone from the external device. The external device can dial a number and can gain access to the menu and the stored numbers in the handset's phone book. Many of the external devices have their own copies of the handset's phone book, which is handy when the user is using his palmtop for other uses that have nothing to do with communications. The external devices feature a complete alphanumeric keyboard, which, even if it is a tiny version of a standard desktop keyboard, is much easier to use for creating longer messages. It is also easier to read long messages on the larger displays that are a part of the devices. Figure 6.8 shows the setup of a palmtop together with a GSM phone and a longer message on the display of the palmtop computer.

Today, there is even a specification for a standard protocol between external devices and mobile stations. Not only is the sending and retrieving of short messages from the phone covered, but also the complete remote control of a GSM terminal. The specification makes the mobile

Figure 6.8 A GSM phone connected to a palmtop.

phone appear to be something like the modem one often finds attached to a home computer. Further details are given in Chapter 5.

6.1.6.2 Direct access to the SMS service center

Before the days of alphanumeric paging, SMS, and voice mailboxes, there was a long and indeterminate delay associated with getting word to someone about anything of importance. Indeed, if the message had any kind of time sensitivity at all (e.g., flight schedule changes), an array of alternative communications channels had to be tried before the message could be successfully delivered. The advantage of SMS is that someone can receive this kind of information on her mobile phone in almost any circumstances, even during a meeting without disturbing others. A message can just pop up on the mobile phone's display without so much as a beep.

There is an increasing number of dedicated software and hardware solutions on the market that allows a computer, equipped with a generic modem, to dial directly into the service center of an operator. This kind of message, the reader will recall, is an MT type of message, which should be delivered to a specific user within the network. The message is sent from the computer to the service center along with all the other necessary parameters and delivery instructions in a manner similar to mobile-originated messages: receiver address, validity period, and protocol ID. The appropriate service center address and other parameters are selected by the computer's software. Writing the message on a computer is simpler and faster than waiting for a message center operator and then dictating the message.

Figure 6.9 shows the infrastructure used for this application. The personal computer uses the normal PSTN line to access the service center. From the SC onward toward the recipient's mobile station, the delivery is performed according to the standard SMS mobile terminated case.

6.1.7 The future

In the previous paragraphs we have examined the application and implementation of the short message services, but as with all other features of GSM, these services are undergoing improvements and further development, which will appear in GSM Phase 2+.

One important improvement is *concatenation of short messages*. One of today's limitations is the restriction on the length of messages, which is limited to only 160 characters. This is the original standard length for pocket pagers. Because it is becoming much more common to use external devices attached to a mobile station (PCs, PDAs, etc.), the possibility of exchanging much longer messages is highly desirable. It is for this reason that specification work was initiated for specifying the concatenation of messages. This feature allows anyone to create, transmit, receive, and display messages with a length of multiples of 160 characters, thus giving an e-mail flavor to SMS.

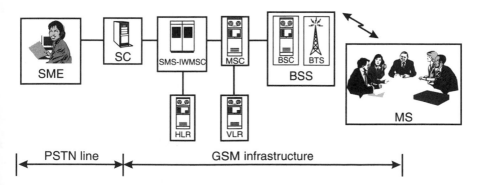

Figure 6.9 Infrastructure used for direct access to the SMS service center.

An additional possibility to increase the number of characters transmitted in one short message is the use of compression algorithms. The detailed specifications can be found in [5]. The principle behind compression is that characters used more often will be coded with smaller number of bits, whereas characters used more seldom will be coded with a larger number of bits. Note that the same principle has already been applied to the Morse alphabet. Along with the coding, puncturing of abundant characters (e.g., spaces at the beginning and end of a message and keywords from dedicated dictionaries) will enhance the throughput rate. Because the patterns of a language vary from language to language, sets of parameters are put together to match these requirements. Vodafone, a GSM operator in the United Kingdom, announced in March 1997 in a press release that this compression technique will allow the transmission of up to 240 characters [6]. However, this possibility depends heavily on support of mobile manufacturers to implement the relevant specifications.

Another improvement is the *development of additional alphabets*. Currently the standard GSM alphabet is limited to only Roman characters, including a few Greek characters. Because GSM systems are enjoying explosive growth and great popularity in Asian and Middle Eastern countries, an Arabian and Chinese character set will be added. For this purpose a dedicated 16-bit coded alphabet will be used: *Universal Coding Scheme 2* (UCS2). With 16 bits, up to 65,536 different characters can be distinguished and displayed.

Special SMS-message-indication is a third improvement. Many mobile-terminated messages are still originated by a voice mail system. The presence of a new stored voice mail message is indicated as just another new message along with all the other messages. Users prefer a specific indication that there are messages waiting in the voice mail system. The *special SMS-message-indication* feature allows the mobile station to distinguish messages from a specific resource, such as the voice mail server, the fax server, or the e-mail server. Upon receipt of a voice mail notification, the mobile station will take special action, such as displaying a dedicated icon, or storing the message in a special location (directory).

Forwarding of short messages allows a subscriber simply to transfer a received short message to another subscriber.

The networks will experience some improvements for SMS as well. *SMS mobile busy* is a condition that quite often occurs. Because a HLR notifies all connected service centers when a mobile station becomes available again, it is a simple task to make the information available anywhere. Service centers that encountered a SMS mobile busy condition will get special notification when the mobile station is again available to receive short messages. This procedure will speed up message delivery and also save network resources.

There are not only new developments for SMS itself, but also innovative applications that improve and promote the use of the SMS. Data services, such as SMS, can take easy advantage of latency in the networks. Operators will welcome anything that uses the network's latency. Anything that promotes the use of nonvoice services or anything that increases the mix of services in the networks will make them more efficient. Because the SMS-MT allows the transport of information or data directly to the MS, or the SIM card, it can be used for the SIM application toolkit (see Chapter 9). This application is so transparent and hidden that the user does not even realize that her mobile station received a short message. These kinds of SMS-MT transactions are only used by the GSM network operator to update parameters in the mobile station or the SIM.

An application of SMS-MT transaction is as a housekeeping function. As more and more countries and operators adopt the GSM system, mobile station manufacturers are experiencing difficulties keeping the network name list in the mobile terminal up to date. To overcome this new problem, mobile-terminated messages could be used to update the network list in the phone. It could be possible for a new network to update newly roaming users with the necessary parameters, which would allow the mobile station to display the network name in addition to the five-digit code.

6.2   SMS cell broadcast

The *cell broadcast* (CB) feature is very similar to the *radio data system* (RDS) in a car radio. The RDS in car radios displays certain information on the display of the car stereo such as the identification of the station. The cell broadcast feature is even more similar to Videotext service on some of our home TVs, where a user can select certain pages to be displayed.

There are also some differences. For one, the cell broadcast is not necessarily sent nationwide or even within the entire coverage area of a network operator, but can be refined and subdivided into so-called cell broadcast areas, where one area might include only one or a few cells. The data that are delivered by SMS-CB can, for instance, carry traffic information, weather forecasts, tariff or other location-sensitive information, or even commercial advertisement.

In the few cases in which it was used during the first years of GSM services, cell broadcast was used only to display zone information. The zones were those that were used to determine different tariffs for local or long-distance calls.

A single CB message contains 82 octets, which, if we are using the default character set as described for the point-to-point service, yields a maximum of 93 characters per message. The different types of messages are sent on so-called channels that the user can select for display on his mobile station. The messages may also carry a language indicator, and since the user can select his desired operating language on the phone, the mobile station will only display messages sent in the language that was selected.

We saw earlier in this chapter that the point-to-point service is a very secure service in terms of making sure that a message arrived at the intended recipient. The sender can take measures to make sure that his message arrives at the recipient, and be notified of delivery failures. There is no such assurance with the CB service, because it is a messaging service to all users within a cell, an area, or a network. The CB service is the GSM version of advertisements pasted in the subway station or notices posted in the news papers. The sender, which in the CB case is always the network operator, can never be sure if messages arrive. SMS-CB is more of an added-value type information service, than a substitute for something else [7].

6.2.1   Implementation of CB in the network

The *cell broadcast short message service* (CBS) is transmitted to any mobile station able to read the *cell broadcast channel* (CBCH) on the radio channel. The presence of the CBCH for an individual cell is indicated within the general system information on the *broadcast control channel* (BCCH). The CB messages occupy a complete signaling channel, in which individual messages are sent sequentially. The CBCH uses the structure of a *stand-*

alone dedicated control channel (SDCCH), which may be either a SDCCH (4), which is in combination with the BCCH, or a SDCCH (8). With the BCCH information the mobile station knows on which frequency and channel CB messages are sent. Before a message finally arrives at the mobile station, it must find its way through the network. The path is herewith described together with references to Figure 6.10.

The cell broadcast short message is generated in the *cell broadcast entity* (CBE). The functionality of the CBE is not described in the GSM specifications; it is entirely up to the manufacturer of this entity to supply and document the necessary functionality. This CBE includes all aspects of formatting the cell broadcast message as well as splitting a message into various pages, which will eventually be transmitted on one channel.

The *cell broadcast center* (CBC) actually handles all the GSM-related functions of CBS. The CBC may be connected to one or more CBEs, and is also connected to one or more *base station controllers* (BSCs), which will serve part or all of the operator's coverage area. The CBC coordinates the formatting and organization of the messages it receives from the CBE into GSM form:

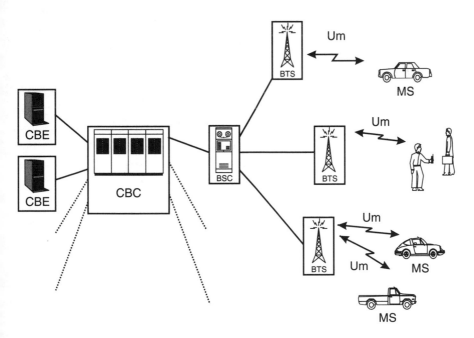

Figure 6.10 Network architecture for the CBS.

▶ It allocates serial numbers for individual messages, which allows the mobile station to distinguish whether a certain message has been updated or merely repeated.

▶ It modifies or deletes messages that are already stored in the BSC, which is the case for weather forecasts and other messages that retain relevance for short periods of time. When the weather changes or when a commercial message has expired, they should be cleaned out of the system or overwritten.

▶ It initiates messages for a BSC, sets the language, determines the area where a certain message is to be sent, and sets up the starting time for transmission.

▶ It determines at which rate certain messages have to be sent.

▶ It determines when a message should no longer be transmitted.

A BSC is connected to only one CBC, whereas it may be connected to several base stations that belong to its area. The BSC takes care of the radio part of transmitting CBS. The BSC's responsibilities, among many other functions, are as follows:

▶ Receive the messages from the CBC, and interpret these along with the commands.

▶ Store the messages as long as they are to be transmitted.

▶ Route the messages to the appropriate base stations. It is possible that different BTSs connected to one BSC belong to different areas where different CBS messages (e.g., traffic information) have to be transmitted.

▶ Schedule the messages according to the repetition rate specified by the CBC.

The base station has only one simple task concerning the CBS: it must transmit the message at the time specified by the BSC.

6.2.2 Contents of a cell broadcast message

A cell broadcast message contains 88 octets of data, 82 of which contain the coding for the actual message displayed for the user. If the default alphabet with its 7-bit code is used, it is possible to transmit 93 characters in one single message. One single message is divided into four blocks, of 22 octets each, plus one octet of control header, which are all transmitted consecutively on the air interface within the cell broadcast channel. A mobile station must discard any corrupted message, which would clearly be the case if all four blocks were not successfully decoded. The 88 octets of the cell broadcast message shown in Figure 6.11 contain the following data: serial number, message identifier, data coding scheme, page parameter, and user data.

The *serial number* occupies two octets. The two highest bits (bits 6 and 7) of octet number 1 form the *geographical scope*, which describes the area where the cell broadcast message is to be displayed, and the display mode as revealed in Table 6.3. The immediate display mode means that the message is to be displayed directly on the mobile station's display, whereas the normal display mode leaves it to the user to select the display of broadcast messages or not. The *message code* differentiates between messages from the same source and type within the same message identifier. The allocation of those codes is up to the operator. The *update number* marks whether a message has been updated since it was read from the mobile station. Normally a new message has the update number of "0000." Each update of a message will increment the number by 1.

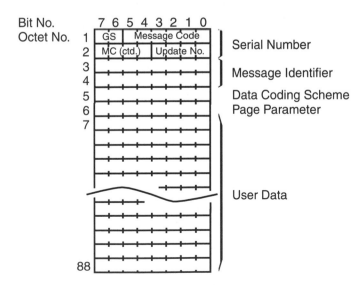

Figure 6.11 Format of a CBCH message.

Table 6.3
Geographical Scope Codes

Code	Display Mode	Geographical Scope
00	Immediate	Cell wide
01	Normal	PLMN wide
10	Normal	Location area wide
11	Normal	Cell wide

The *message identifier* is a 2-octet field that contains the channel number of the message with which one can determine the type or source of the message that is to be received. There are different identifiers for an index, news, a weather forecast, traffic information, local area identity, etc. These identifiers have even been assigned recommended values within the *Memorandum of Understanding* (MoU) for common use. For example, the zone information of Vodafone in the United Kingdom is transmitted on identifier number 50, and that of Cellnet in the United Kingdom on identifier number 200. The number is allocated by the operator. The user must enter the number into the mobile station in order to read this information. Numbers in the range of 0 through 999 will be entered in their decimal representation.

The *data coding scheme* indicates to the mobile station which scheme has been used for a message. The scheme can be either the reduced 7-bit coding or the 8-bit coding of characters. Along with the coding scheme follows the language in which this message has been written. GSM currently distinguishes between German, English, Italian, French, Spanish, Dutch, Swedish, Danish, Portuguese, Finnish, Norwegian, Greek, and Turkish.

The *page parameter* indicates whether there is a longer message and, therefore, if there is a second or even more pages to be presented. The *user data* octets contain the actual message coded according to the data coding scheme.

Figure 6.12 shows how a CB may appear in the display of a mobile station. Note that the display shows the operator name on the left side (Voda UK) along with a zone indication (0171, SW-A, W-A). In this example, "0171" indicates the local area code within which the subscriber may

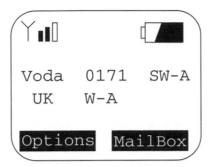

Figure 6.12 Display of a CB message for zone indication.

make calls at a reduced charge (in this case central London). SW-A and W-A indicate "home areas," which are based on the postal codes of this area. Each subscriber on a certain tariff can nominate one "home area" in which call charges are reduced. Because any cell may serve parts of more than one area, each cell broadcast can indicate up to three of them. If the subscriber can see his own area in the display, she knows her calls will be at the reduced rate.

6.2.3 Future developments for cell broadcast

There are a few things in the world that could not benefit from some improvements, and the CB is one of them.

▶ Monitoring cell broadcast channels reduces the battery life of a mobile station in standby mode. The introduction of *cell broadcast discontinuous reception* (CB-DRX) will bring relief. The proposal is to introduce an index system that allows the mobile station to power down between messages that are not assigned to the mobile user.

▶ The current capacity of cell broadcast will not be sufficient if telematic services are introduced. A second cell broadcast channel will need to be made available.

▶ In the same way and for the same reasons they will appear in the point-to-point short message service, the extended 16-bit alphabet will also appear in the cell broadcast service.

References

[1]   ETS 300 536 (GSM 03.40), "European Digital Cellular Telecommunications System (Phase 2); Technical Realization of the Short Message Service (SMS)—Point to Point (PP)."

[2]   ETS 300 559 (GSM 4.11), "European Digital Cellular Telecommunications System (Phase 2); Point-to-Point (PP) Short Message Service (SMS) Support on Mobile Radio Interface."

[3]   Redl, S. M., M. K. Weber, and M. W. Oliphant, *An Introduction to GSM*, Norwood, MA: Artech House, 1995, Chap. 5.

[4]   ETS 300 628 (GSM 03.38), "European Digital Cellular Telecommunications System (Phase 2); Alphabets and Language-Specific Information."

[5]   GSM 03.42, "Digital Cellular Telecommunication System (Phase 2+); Compression Algorithm for Text Messaging Services."

[6]   http://www.vodafone.co.uk.

[7]   ETS 300 537 (GSM 03.41), "European Digital Cellular Telecommunications System (Phase 2); Technical Realization of the Short Message Service (SMS)—Cell Broadcast (CB)."

Contents

Supplementary services

In the previous two chapters we discussed teleservices and bearer services. These are also referred to as *basic service groups*. They are the base elements for all kinds of communications that supplement the main purpose of telecommunication networks and services.

As our private time blends with our professional and public lives, we demand more control over the conditions and circumstances with which we communicate with others. For example, we may not want to miss a call simply because we are already busy on the phone with someone else, we may want to avoid the high costs of certain incoming calls when roaming, or we may want complete privacy for an hour or a day. The list is endless. The services discussed in the earlier chapters add communications services and possibilities to the networks and to our lives; they do not, however, provide many mechanisms to limit and control them. To meet the need for control, to bring order to the chaos of a user's busy day, telecommunication networks offer so-called *supplementary services* (SSs).

SSs are services that supplement teleservices and bearer services. SSs can only be offered in combination with the basic service groups. To enjoy the benefits of supplementary services, the cellular customer has to subscribe to one of the basic services. The SSs available to a subscriber depend on the offerings the network can technically handle, the terms of the subscription, and the capabilities of the subscriber's terminal.

Table 7.1 lists all available supplementary services for Phase 2 along with a short description of each service. Further developments in the GSM Phase 2+ systems are listed and described in Section 7.13.

7.1   Introduction to supplementary services

A few general statements on the supplementary services are in order before we start a more detailed discussion of them. It is up to the network

Table 7.1
Phase 2 Supplementary Services

Type of Supplementary Service	Description
Call offering SS	Call forwarding services, which also distinguish between unconditional (all calls) and certain conditions (e.g., only when busy)
Call restriction SS	Barring of outgoing and incoming calls, which also distinguishes different conditions
Number identification SS	Presentation/restriction of the calling party and the called party through their phone numbers
Call completion SS	Indication of a waiting call and then placing a call on hold
Multiparty SS	Establishing a conference call
Charging SS	Provides charging information on teleservices, bearer services, and SMSs used
Community of interest SS	Establishment of subgroups of users, such as for fleet management (e.g., taxi drivers)
Unstructured supplementary services data	Offers an open communications link for use between user and network for operator-defined services

operator to *provide* the services to an individual subscriber. Operators generally offer a few services as a standard feature of all subscriptions. One example is *call forwarding*, which has benefits for both the subscriber and the operator. The user can determine where and when to accept a call, and the operator gets more revenue from the additional forwarded communications links, which can be to another subscriber number or to an internal voice mail system. Some additional services beyond the standard packages may be offered as part of special "premium" subscriptions. Customer requests, competition, and marketing issues influence which, and in what way, supplementary services will be offered.

A service can be *withdrawn* from subscribers by the service provider or the network operator. When a subscriber wants to use specific services, for example, line identification, she would have to *register* deliberately for the service. Registration means that information related to a subscriber's desired supplementary service is stored in the network. Even though a subscriber is registered for certain services, these may still need to be *activated*. If the subscriber can perform the registration herself, then the activation goes along with the registration, which is almost always the case for call forwarding services. Services can be *deactivated* when they are not needed anymore. The catalog of available services is so large that users occasionally become confused over which service is activated, or to which number, for example, calls have been diverted. The subscriber can *interrogate* the status of a service, and request specific information about it, such as to which phone number calls will be forwarded.

The user can register supplementary services either for all teleservices or bearer services, or she can elect that a service will only be active for an individual teleservice or bearer service. For instance, a user may be traveling without the laptop computer she normally uses with her GSM phone. So, she may elect not to receive fax calls while she is away. The user would merely divert fax calls, and only fax calls, to the office fax machine, or to a fax mailbox, while she is on the road. Voice calls would still reach her terminal as long as it is switched on, and not left at home on the kitchen table. To reduce the number of subscription options, which grows as each individual supplementary service is applied to any teleservice or bearer service, the basic service groups have been grouped, and a number has been assigned to each supplementary services application. Table 7.2 shows the grouping of the basic services [1,2]. Two things should be noted in this table. Due to the nature of the cell broadcast service, no supplementary service is applicable in combination with it. Even though emergency call service is listed here, a supplementary service has yet to be defined for it.

7.1.1   Network entities

Every network operator provides a portfolio of supplementary services or groups of these services. The offering depends largely on demand and

Table 7.2
Basic Service Groups

Basic Service Group		Basic Service	
No.	Name	No.	Remark
1	Speech	TS 11	Telephony
		TS 12	Emergency call
2	Short message service	TS 21	Short message MT/PP
		TS 22	Short message MO/PP
3–5	Not allocated		
6	Facsimile service	TS 61	Alternate speech and facsimile Group 3
		TS 62	Automatic facsimile Group 3
7	All data circuit asynchronous	BS 21	Data circuit duplex async. 300 bps
		BS 22	Data circuit duplex async. 1200 bps
		BS 23	Data circuit duplex async. 1200/75 bps
		BS 24	Data circuit duplex async. 2400 bps
		BS 25	Data circuit duplex async. 4800 bps
		BS 26	Data circuit duplex async. 9600 bps
8	All data circuit synchronous	BS 31	Data circuit synchronous 1200 bps
		BS 32	Data circuit synchronous 2400 bps
		BS 33	Data circuit synchronous 4800 bps
		BS 34	Data circuit synchronous 9600 bps

| Basic Service Group | | Basic Service | |
No.	Name	No.	Remark
9	All dedicated packet assembler disassembler (PAD) access	BS 41	PAD access circuit sync. 300 bps
		BS 42	PAD access circuit sync. 1200 bps
		BS 43	PAD access circuit sync. 1200/75 bps
		BS 44	PAD access circuit sync. 2400 bps
		BS 45	PAD access circuit sync. 4800 bps
		BS 46	PAD access circuit sync. 9600 bps
10	All dedicated packet access	BS 51	Data packet duplex sync. 2400 bps
		BS 52	Data packet duplex sync. 4800 bps
		BS 53	Data packet duplex sync. 9600 bps

marketing strategies. It also depends on whether the network infrastructure supports the particular supplementary service.

The information about the status of each individual subscriber, and his related SS, is stored in the network. We will take a look at how the different network entities are involved with SS, and how the provisioning of supplementary services works. Figure 7.1 shows all the network entities involved in SS [3].

The *mobile station* (MS) holds the SIM, which identifies the subscriber and the allocated (subscribed) services in the home location register. Most supplementary services can be activated through the *man-machine interface* (MMI) of the mobile station. For some services, however, this task has to be performed by the service provider (network operator) through a customer service desk or other support function.

The *home location register* (HLR) holds the subscriber data for individual customers. A HLR is allocated to each *gateway mobile services switching center* (GMSC). The subscriber identity is stored in only one HLR. In addition to the subscriber identity, the HLR keeps information about teleservices and bearer services subscriptions, as well as restrictions and

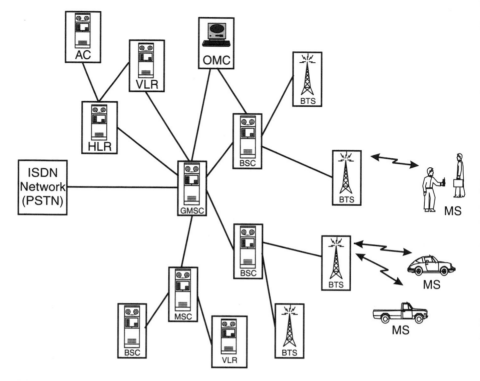

Figure 7.1 Network entities involved in the provisioning of supplementary services.

information that applies to each subscription, for example, roaming, and information about supplementary services. The information on supplementary services relates to which services are provided to an individual customer, and the status of those services: registered, activated, or both.

The *visitor location register* (VLR) keeps relevant data on mobile stations currently located in the area of the particular mobile services switching center (MSC) to which the VLR belongs. The VLR stores the identity of subscribers, the *temporary mobile subscriber identity* (TMSI), the location area where a mobile station has been registered (this information is used to page a mobile station), and the information on supplementary services that was provided by the HLR during registration (location update procedure). The details of the registration process can be found in [4].

The *mobile services switching center* (MSC) is an exchange that performs all switching and signaling functions for mobile stations located in the MSC's service area. The MSC interacts with the HLR for mobile-terminated calls, so that it can get information about the origin of the mobile station, the authentication parameters, and applicable supplementary services. When, for example, barring of incoming calls is set by the subscriber, the MSC will block the incoming calls to the indicated number. For mobile-originated calls, the MSC exchanges information with the VLR in order to get information on, among others matters, authentication and supplementary services. For instance, when all outgoing calls are barred for a mobile station, which is a popular option activated when a phone is loaned to irresponsible friends, the MSC will block the calls.

The *base station controller* (BSC) and the *base transceiver station* (BTS) play no unique role in supplementary services; their sole responsibility is the transparent movement of information between the fixed network and the MS.

7.1.2 Password handling

The barring supplementary service can be offered to a subscriber with an optional network password. Even though it is optional, the operator may decide that he offers these services only in combination with passwords, since this provides additional security for subscribers who are in the habit of loaning telephones to other people.

This password is stored in the HLR. A password is a four-digit number that is used in a manner similar to *personal identification numbers* (PINs), which are used for telephone or banking cards. The password has an initial default value at registration of "0000." Like all security and access numbers, the password has an upper limit to the number of unsuccessful accesses attempts that can be made in trying to remember or guess it. When the password is entered incorrectly three times, further access to the features requiring this password are blocked. After blocking, only the service provider can enable the service again by resetting the password to "0000." The counter that counts incorrect password entries is reset each time the password is entered correctly.

Figure 7.2 shows the traffic flow between the MS, the MSC, the VLR, and the HLR when a supplementary service with password is activated. The MS issues a request to activate an SS, which is passed through the MSC, and the VLR, to the HLR. The HLR responds to the MS with a request for the password. If the password presented by the MS is correct, then the HLR will activate the supplementary service [2].

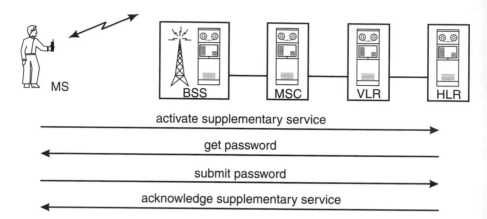

Figure 7.2   Activation of password protected supplementary service.

7.2   Call forwarding supplementary service

The call forwarding supplementary services are usually referred to as call offering services. These services have been available from the very beginning in most GSM networks. The purpose of these services is to enable the user to take the "phone line" along with him to places where a GSM phone might not have coverage, or to divert calls to a voice mailbox. The user may get some relief from his busy schedule by diverting incoming phone calls, but can still maintain some control over his life by retaining the ability to make outgoing phone calls. Call forwarding means that the phone does not need to be switched off to prevent interruptions. In general, no single phone call into a GSM network should remain "unanswered" when these services are used effectively by subscribers. When it is not possible to reach the subscriber or a diverted line number (the office secretary or the home), it should at least be possible to leave a message with a voice mail system.

The call offering SSs are made even more attractive because they allow one to distinguish different teleservices and bearer services. A user can decide that all his voice calls are to be diverted to the mail box, all fax calls are to be diverted to his office fax machine or a dedicated fax mail box, and all data calls are to be diverted to his local data network.

The call forwarding SS distinguishes different conditions when a call actually has to be forwarded:

▶ *Call forwarding unconditional (CFU):* CFU forwards all calls without regard to the condition; it does not even matter if the mobile phone is switched on or off.

▶ *Call forwarding on mobile busy subscriber (CFB):* This service diverts only those calls that meet a busy condition, or that otherwise would have returned the busy indication to the caller.

▶ *Call forwarding on no reply (CFNRy):* Calls that are not answered or not picked up by the called party within a specific time, are subsequently diverted.

▶ *Call forwarding on not reachable (CFNRc):* If the mobile subscriber is currently out of coverage of a GSM network or decides to switch off her mobile phone, this service will divert all of her calls to the indicated number stored in the HLR/VLR.

The general concept of call forwarding is the same for all four of the services, but there are some differences between the first *unconditional* service and the remaining three, which are commonly referred to as *conditional* call forwarding services [5,6].

7.2.1 General behavior of call forwarding services

Call forwarding services are initiated by the user through the MMI of the mobile station. Different implementations in a GSM mobile station, either the hard way with standard GSM commands, or the easy way with a dedicated menu structure, will be explored in Section 7.11.

Before the network can make a service available to the subscriber, the service must be *registered*. The registration includes the type of call forwarding that is selected, the number to which the calls should be forwarded, and for which basic service the call forwarding should be applicable. This information is stored in the HLR. The registration is usually done by the subscriber herself, but it can also be performed by the service provider. If call forwarding is registered by the service provider, then the user loses flexibility, because she will not be able to quickly change the phone number to which she forwards her calls. Call forwarding can be applied to all basic services except emergency calls and short message services.

The *forwarded-to number* must be a number that meets the dialing requirements of the home country. If it is a number within the home country, then the international access code and country number are not required. If it is an international number, then the home country international access code or the GSM-specific "+" character is required. It is interesting to note that the GSM network converts all forwarded-to numbers to an international GSM number (e.g., in Germany from 08912345 to +498912345). This is always done regardless of the format of the entered number. No verification is done on the forwarded-to number to check if it is a valid number that addresses an existing connection. There is only one sanity check that may be summoned; the system can verify if the indicated number is within the correct numbering range. If the subscriber enters an incorrect forwarding number, then only those callers trying to reach the subscriber will get an indication that something is wrong when an announcement from a computer is heard in the place of a useful connection.

Registration can *activate* the call forwarding service even though there is also a dedicated *activation* procedure available. The difference between registration and activation is that only the registration procedure provides the network with the necessary information about the forwarded-to number and the basic service that should be diverted. An activation is only possible if the network already knows the necessary parameters (if required) for a specific supplementary service. The parameters are established during the registration; the activation puts the service into effect. Because it includes some important parameters, call forwarding has to be registered before an activation can take place.

The *deactivation* procedure only deactivates the call forwarding service but does not erase the information about the forwarded-to number and the particular basic service from the HLR. A simple activation is sufficient to invoke the service again after a deactivation.

The *erasure* procedure erases the information from the HLR's memory that had been allocated to the call forwarding service. A complete registration is necessary before the service can be activated after an erasure procedure.

With all these procedures one might easily lose track of which services are currently active and to which numbers calls are forwarded. The *interrogation* procedure provides information about whether a service is registered, active, or idle, and to which number each service is forwarded.

After a user makes all his inputs, the network should provide the user with an indication of whether the action was executed properly, or if there was a problem executing a particular call forwarding registration or activation. In addition, the calling subscriber should get a notification that call diversions have been invoked at each termination of a mobile originated call. This indication is not made for mobile terminated calls since, in most of those cases when forwarding is on, the calls do not even arrive at the mobile station. The diversion indication is a reminder that call diversion has been set; an easy condition to forget. Some phones display an icon confirming that call forwarding is in effect.

Another choice is *endless call forwarding*. A subscriber diverts his phone, in rare cases, to another phone that is already diverted to a number that is diverted to yet another, which is diverted to. . . . This string of diversions can continue endlessly, but not in GSM or ISDN networks. The limit of consecutive call forwarding connections within these digital networks is set to a value between 1 and 5. Once the forwarded call leaves the digital networks (GSM or ISDN) into, say, an analog PSTN, the limits may no longer apply. When someone places a call into an endless trail of diversions, the network's switch (the MSC in GSM networks) will reject the phone call when the call forwarding limit is reached. The caller will get a message or an indication of an unsuccessful call setup.

One curious anecdote from the GSM community sheds light on call diversions. In the early days of GSM network service, there was an expensive period of time during which calls forwarded to another number could not be charged back to the original connection after the forwarding relays began. This was due to some inadequate software in the networks. So, someone could theoretically forward his GSM connection to a number overseas, thus making it cheaper to call the overseas number through a local GSM connection than making the long-distance call in the normal way; only the initial GSM connection was charged, theoretically of course.

7.2.2   Operation of call forwarding

All user actions on call forwarding, and the resulting information and parameters, terminate in the HLR. The HLR stores the status of all call forwarding services along with the forwarded-to number and the basic service. The HLR performs a compatibility check on whether there is a conflict with other supplementary services. Upon acceptance of an operation, the user will get a positive acknowledgment that all is well. The HLR will subsequently update the VLR about the new call forwarding status. When an operation is rejected the user gets a negative notification and the VLR is not updated.

A distinction between unconditional and conditional call forwarding has to be made for incoming calls. Because unconditional call forwarding is handled entirely from the HLR, the VLR will not get the forwarded-to number for this service. For the conditional services, the VLR not only receives the status of the call forwarding service, but also the forwarded-to number. In the special case of CFNRy, in which the called party must answer within a specified time limit, the VLR also gets the timer value.

7.2.2.1   Call forwarding unconditional

The (G)MSC first checks the HLR for the parameters of the called mobile station before connecting incoming calls, because the HLR has the latest information about the MS's location and call forwarding status. If CFU applies for the basic service in the call, then the MSC forwards the incoming phone call to the number registered with the HLR for that basic service. Figure 7.3 shows the routing of a call when call forwarding service is in effect. The call forwarding in Figure 7.3 is set to a fixed-line phone. Caller A, who might be a mobile or fixed subscriber, tries to call B. Subscriber B, who is busy trying to finish a sales report for his supervisor, has set call forwarding unconditional to C's number. Subscriber C, in this case, is a fixed-line subscriber: the supervisor anxiously awaiting B's report. The incoming call from A is routed to the MSC of B, which interrogates the HLR, only to find out that the call is to be routed to a fixed number through the *local exchange center* (LEC). The resulting connection is established between A and C. The connecting line in the top of the figure indicates that these entities communicate with each other, but they are not necessarily physically connected. The calling GSM subscriber may get a notification that the call will be forwarded on his phone's display. The SS notifications can be as elaborate as the operator chooses; even a reason for call forwarding can be shown to the caller ("call forwarded unconditional"). If there is a chain of forwarded calls, then the "reason" for the last call diversion in the chain would be indicated on the display. Instead of a call forwarding indication appearing on the display of the

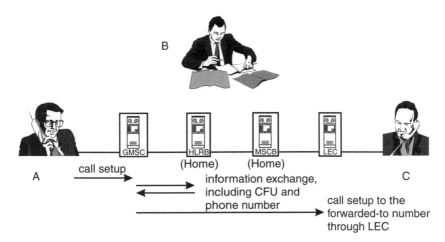

Figure 7.3 Routing of an incoming call when CFU is active.

mobile station, an operator might choose to provide an aural announce-ment indicating that the call is being forwarded to another number.

7.2.2.2 Call forwarding conditional

This situation is different from the simple unconditional case when condi-tional call forwarding is selected; some interaction among the entities within the network has to occur to establish the conditions for forward-ing. The HLR gives the MSC information about a called station's location and the status being *conditional* call forwarding. Because the MSC does not know if a call forwarding condition exists at the moment, it routes the call to the visiting MSC, which asks its associated VLR about the location of the called mobile station. The principle of conditional call forwarding is depicted in Figure 7.4. The *call forwarding* (CF) in the example of Figure 7.4 is set to a fixed-line phone (through a LEC) in a manner similar to the previous unconditional case depicted in Figure 7.3. If CF is directed to a mobile subscriber (rather than a fixed one), then the process includes the entities related to the particular mobile subscriber: MSC, HLR, and VLR. The example in Figure 7.4 includes the three participants: A, B, and C. Subscriber A wants to call B, who is still busy with his sales report, and has invoked conditional call forwarding to C only if the appropriate condi-tions are met. If B is a recalcitrant employee with a bad attitude, then he may select those call forwarding conditions that have the greatest possibility of annoying subscriber C. The connecting line at the top of Figure 7.4 indicates that the entities can communicate with each

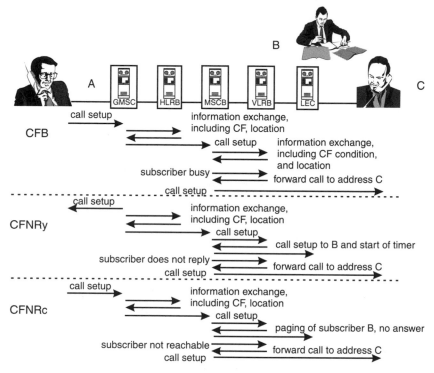

Figure 7.4   Operation of conditional call forwarding.

other even though they are not necessarily physically connected to one another. Depending on the forwarding *condition* that is set by B, the *behavior* of the network is different.

> ▶ For *call forwarding on mobile subscriber busy* (CFB) the network checks whether the mobile phone (B) is busy. When an actual "busy" condition is detected, the call is forwarded to the number (C) as indicated in B's VLR. GSM distinguishes two busy conditions: the *network-determined user busy* (NDUB) and *user-determined user busy* (UDUB). NDUB means that the mobile subscriber is, in fact, actually busy on the phone, whereas UDUB means that the subscriber actually received the call but decided that he did not want to accept it. To reject the call, the subscriber simply terminates the call by pressing the on-hook button, thus activating a busy condition. The reason for this behavior can be that the subscriber has identified the originator of the call with CLI (calling line identification), or he simply cannot accept calls and forgot to switch off the phone.

▶ For *call forwarding on no reply* (CFNRy) the network pages the mobile phone (B) and sets up a call. The phone actually rings. At the instant the ringing indication starts, the network starts a timer, which expires after the time indicated during the initiation of the particular call forwarding function (which was set in the registration procedure). After the timer expires, the network (MSC) releases the network resources allocated to the mobile phone (channels) and forwards the call to the particular number indicated (subscriber C).

▶ For *call forwarding on mobile subscriber not reachable* (CFNRc) the network pages the mobile station. If the MS does not respond to the paging, then the *not reachable* condition is met, and the call is forwarded. A mobile subscriber might not be reachable because the subscriber switched the phone off or left the coverage area. A subscriber can depart the coverage area without actually leaving it; a phone left in a steel tool box will not respond to pages form the network.

When forwarding calls to numbers outside the home country, using the "+" prefix followed by the country code makes sure that the forwarding sequences reach the proper number. The user does not have to be concerned with any other dialing rules. Omitting the prefix, in the case of foreign forwarding, diverts a call to an invalid number, or to a stranger who probably does not want to receive diverted calls. This is why networks usually add the international part of a number, in case it was entered as a national number.

7.2.3 Conflicts for call forwarding

Conflicts will arise when different supplementary services are activated. There are even conflicts between unconditional and conditional call forwarding. Here, the unconditional case would overrule the conditional case. When call forwarding conditional has been activated and the subscriber later selects unconditional call forwarding, then the conditional case would be, at least temporarily, inactive or quiescent. However, when call forwarding conditional is set for all services, and call forwarding unconditional is only set for an individual basic service, then the conditional case will be valid for the remaining services.

On the other hand, if a CFU is already set, and then the user tries to set a conditional call forwarding service, the network would reject this conditional CF with a message that this service is incompatible with an active service (CFU). Because the forwarded leg, which is the part from the MSC to the forwarded-to number, is considered as an outgoing call, the network would not accept the call forwarding when outgoing calls have already been *barred*.

Because there are many possibilities for call forwarding and call barring, there are many possible consequences. Let's have a look at a few conflicting constellations:

▶ Activation or registration of any call forwarding will be denied when *barring of outgoing calls* (BOAC) is active.

▶ Activation or registration of call forwarding to an international number will be denied when *barring of outgoing international calls* (BOIC) is active.

▶ When both conditional call forwarding (to a local number seen from the current registered PLMN) and barring of outgoing international calls are active, and the subscriber roams into another PLMN, then the call forwarding will become quiescent, because a possible call forwarding sequence would be an outgoing international call as seen from the VPLMN. After returning home or barring of international calls gets deactivated, CF will become active again.

▶ If the subscriber is currently roaming outside the home network, the activation of conditional call forwarding is denied when *barring of incoming calls when roaming outside the home PLMN* (BICroam) is active. This is because this forwarding service is activated in the visiting PLMN, and the forwarding would never become active because the incoming call would already be barred in the home country.

7.2.4   Who pays for what?

Because this question comes up frequently among users of call forwarding services, we want to prepare ourselves with a closer look. When calls are forwarded, not only is the *calling* party billed, but also the subscriber who forwarded the call. The general rule is that the calling party pays his way to the phone number he dialed, and the forwarding party (the diverter of calls) pays for the forwarded leg of the communications link. This forwarded leg can become quite expensive, but we have to distinguish between unconditional and conditional call forwarding services to figure the cost.

When *unconditional* forwarding service is active, the incoming phone call is diverted in the HLR of the HPLMN. The forwarding leg, in this case, is to the destination of the forwarded-to number. When this destination, by chance, is the mailbox (network-supplied answering machine service), the forwarding will be a very cheap service, or may even cost the GSM subscriber nothing. When the destination is a phone number outside the PLMN, or another subscriber number within the PLMN, then the forwarding leg will be treated as, and charged as, a normal phone call. Figure 7.5 illustrates charging for forwarding into a mailbox. Caller A tries to call GSM subscriber B. Subscriber B does not want to be disturbed, and sets unconditional call forwarding. Subscriber A has to pay the part from his phone into the GSM network, indicated by the MSC, whereas B has to pay the forwarded part from the GSM network (MSC) to the mailbox. The forwarded-to number could also be a fixed-line connection outside the GSM network, which would further raise the charges for B.

When conditional call forwarding service is active, the incoming phone call is diverted in the VLR of the VPLMN. This is because the HLR has no knowledge of whether the condition for call forwarding is met. The case of a nonroaming customer is identical with the one for unconditional forwarding. If the VPLMN is a foreign network, then the forwarding would become rather expensive: the calling party only pays the way to the HPLMN, and the called (and forwarding) subscriber normally pays the international leg, which is the part from the home network to the

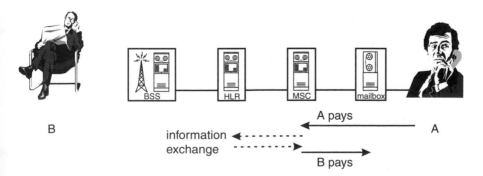

Figure 7.5   Charging for unconditional call forwarding.

currently visited network. (It is a network option to decide who is picking up the cost for the international part. Nowadays it is common practice for the mobile subscriber to pay the international part!) The same rule applies for normal roaming customers who do not forward their phones. In our example in Figure 7.6, the VLR realizes that a call forwarding condition applies for the call coming from A, and forwards the call to the indicated number. Because the number is quite often located in the HPLMN, the call is diverted back to the home network. This means that the GSM customer (B) has to pay the international leg twice! When the call is forwarded to a local number in the foreign country, the forwarding subscriber still has to pay the international leg, but only local charges for the phone call.

7.3   Call barring supplementary services

The *call barring* supplementary services are also referred to as *call restriction* services. These services allow the subscriber to restrict certain types of calls either from, or to, her phone. A GSM phone may be loaned to a friend or colleague, or it may be handed out to staff that is not authorized to place certain types of calls. In such cases the original subscriber (who pays the monthly mobile phone bill) might wish to restrict certain kinds of phone calls. To prevent unauthorized phone charges, some restrictions, such as a password in the network, can be used together with this

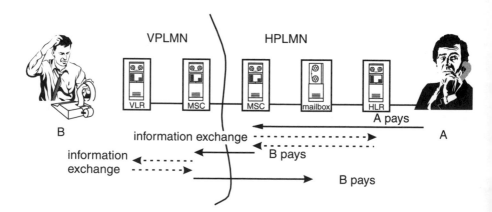

Figure 7.6   Charging for conditional call forwarding when roaming.

service. There is still only one common password for all barring services. The handling of such a password was examined in Section 7.1.2.

7.3.1 Call barring for incoming and outgoing calls

The service makes a distinction between restrictions for outgoing and incoming phone calls. The services are handled in the network after the service provider registers an individual subscriber to this service. The subscriber simply activates or deactivates the services. Just as for any SS, different conditions are distinguished:

▶ *Barring of all outgoing calls (BOAC):* All calls initiated *from* the particular mobile station will be barred; only incoming calls can be accepted.

▶ *Barring of outgoing international calls (BAOIC):* All international calls will be barred. Such calls usually imply higher call charges. When the subscriber is located in a *foreign* PLMN she would be able to make phone calls within the country, but would not be able to place calls back to her home country.

▶ *Barring of outgoing international calls except those directed to the home PLMN country (BAOIC-excHC):* All outgoing international calls will be barred, except calls to the home country. This is the more common case when roaming abroad.

▶ *Barring of all incoming calls (BAIC):* All incoming calls will be barred while outgoing calls are still possible (the calling party would hear the busy signal).

▶ *Barring of incoming calls when roaming (BIC-roam):* No incoming calls are accepted when roaming. This avoids higher roaming charges. This only applies when the called subscriber pays the international leg. There are, however, some rare cases in which the calling party would also have to pay the international leg, which would be a big surprise because he would not be aware of the cost incurred by the call until he gets his bill.

Each of these services is regarded as a single supplementary service. Outgoing and incoming barring can be combined with each other, as would be the case when BOAC is combined with BAIC. This has the same effect as switching off the phone [7,8].

7.3.2 Applicability of call barring

Call barring services are applicable to all teleservices and bearer services. This excludes emergency calls because it is commonly accepted that this important service should never be barred.

Registration of barring services is done by the network operator or the service provider. The subscriber can only activate or deactivate the service, where only one outgoing or incoming barring service may be active at a time. If another service is activated, then the previously active service will become deactivated. An activation includes the indication of the basic service group for which the call barring should be active. This allows the user to distinguish, for example, between normal voice calls and facsimile calls. Deactivation can be carried out separately for individual service groups. Assuming that call barring was activated for all service groups, the subscriber can, at some later date, individually deactivate single service groups and then make them available again.

The HLR keeps track of the barring of outgoing and incoming calls, and it also handles the password. The VLR is updated by the HLR with relevant data on the barring of outgoing calls, which allows the VLR to handle the outgoing call barring properly. The VLR keeps no information on incoming call barring, because it is not involved in this service. Incoming calls are blocked at the HLR.

The barring supplementary services must not be confused with *operator-determined barring* (ODB). ODB is a tool for the operator or service provider to control or restrict teleservices or bearer services to a subscriber. ODB applies to all supported teleservices and bearer services, including short message services. Emergency call service is excluded from ODB. ODB cannot be applied to individual services; see [9,10] for details. Save for pleading phone calls and threatening letters to the network operator, the subscriber has no influence on ODB. It is up to the operator to apply these barring services. Some examples of ODB are (1) to restrict roaming for new customers or (2) to restrict outgoing calls if the subscriber has failed to pay his subscription fee and call charges.

7.3.3 Restrictions to call barring

There are some incompatibilities with call forwarding SS that force the network to reject the activation of call barring. A difference is seen between outgoing and incoming call barring, as discussed next.

7.3.3.1 Outgoing call barring restrictions

▶ Some networks might not support BOIC-excHC, in which case it will be replaced by BOIC.

▶ The network will deny the activation of the barring of outgoing international calls when the conditional call forwarding to an international number is active. This is viewed from the local PLMN without regard to whether it is the home network or a visited network. If, for example, the subscriber is roaming and has conditional call forwarding set to a mailbox in the home network, this would be regarded as an international call and subsequent call barring will not be activated.

▶ Activation of the barring of outgoing calls will be denied when any call forwarding is active.

▶ Barring of international calls will be denied when call forwarding to an international number is active.

7.3.3.2 Incoming call barring restrictions

▶ BIC-roam remains active, but quiescent, when the mobile station is roaming in the home country. As soon as the mobile subscriber roams to a foreign network, the incoming call barring will become active again.

▶ Barring of incoming calls when roaming is denied when conditional call forwarding is active, and the subscriber is roaming outside her HPLMN. Any incoming call would follow the conditional call forwarding, which would result in a connection to the visiting network and, therefore, an incoming call when roaming.

▶ Barring of incoming calls is denied when call forwarding is active.

7.4 Line identification supplementary services

The line identification supplementary service lets a subscriber see the line identification of the calling party and/or the called party. This includes the presentation or restriction of a phone number. The phone number includes the international prefix, and the subscriber's national ISDN or MSISDN (mobile station international ISDN number) number. The identity of a subscriber may not be possible from an analog PSTN, due to lack of signaling support from analog networks. The line identification SS has four groups of services: (1) *calling line identification presentation* (CLIP), (2) *calling line identification restriction* (CLIR), (3) *connected line identification presentation* (COLP), and (4) *connected line identification restriction* (COLR).

Line identification services are activated or deactivated by the service provider upon the request of the subscriber. Depending on the legal situation in individual countries, the line identification service is, by default, switched on or off. The only exception to this is the temporary activation or deactivation of CLIR. The subscriber can only interrogate the status of the respective service.

The line identification services apply to all teleservices and bearer services, but not to short message services, which already include the presentation of the phone number of the originating party within the short message itself.

7.4.1 Calling line identification

The CLIP shows the number of a calling party to the called party. A calling party can avoid the presentation of his identity with the CLIR. GSM also provides an override function for the CLIR service. This capability can be made available, for example, to the police. The presented phone number is compared with the entries in the phone book in some mobile phone implementations. When there is a match, the name of the calling person (as stored in the phone book) will be shown either instead of, or together with, the line identity number.

How does CLIP work? The calling party's MSC provides the called party's MSC with the calling party identity. The MSC of the called party verifies whether the customer (the called party) has subscribed to CLIP. If subscription to the service is confirmed, then the line identity of the calling party will be shown to the called party. In order for CLIR to work, the MSC of the calling party also provides the line identity, but indicates that the display is restricted. The caller's identity will not be displayed to the called party, unless the called party has an override right that allows the presentation of the caller's line identity. A special case applies when CLIR is invoked on a temporary, per-call basis. The calling party can add a code (see Section 7.11) that allows the user to either display his identity if CLIR is permanently activated or disable the display of his identity in case CLIR is not permanently activated. Some mobile station models allow the user to switch CLIR on or off almost permanently. This is done by simply allowing a flag to be set in the mobile station, which determines whether this code is to be sent along with each call. Figure 7.7 shows the information flow for calling line identity. Note that the subscriber could also be located in a fixed-line ISDN environment [11,12].

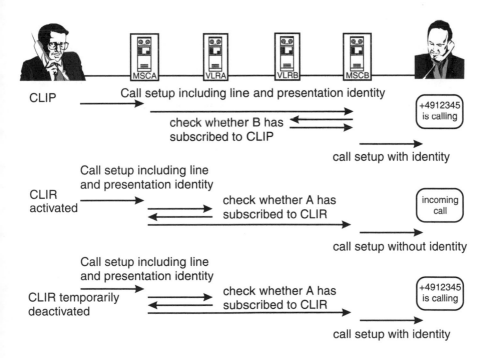

Figure 7.7 Information flow for calling line identity presentation/restriction.

Some people feel that CLIP is a great tool to use in order to decide not to accept (and even divert) phone calls from specific people or organizations.

7.4.2 Connected line identification

The COLP presents the actual number of the called party to the calling party. One might expect that since you are normally connected to the number you dialed, this is a silly and redundant service. But, this is not true when the called party has activated a call forwarding feature. COLP allows the display or presentation of the actual final destination.

A user who wishes to disguise the actual phone number a call was diverted to would invoke the COLR. Again, GSM provides an override feature for the COLR service, which could be made available, for example, to the police.

Connected line identification is handled from the called mobile station's MSC. The called MS's MSC provides the identity of the called party along with an indication of whether the identity is to be shown or, in case of connected line identity restriction, not shown. The MSC of the calling party verifies the subscription option of the calling party and subsequently presents the connected line identity if it is permitted. COLR can be overridden by the calling party if and when a certain priority exists. Figure 7.8 shows the information flow for connected line identity.

7.5 Call waiting

Call waiting (CW) belongs to the group of call completion supplementary services. CW gives, for example, a subscriber already engaged on the phone the option to decide whether he wants to accept another call. It often seems to be the case these days that while one is talking on the telephone, another, perhaps more important, call is awaiting attention. Now, the implication is that call waiting only applies to a mobile subscriber who is already busy on the phone, but this is only the best known case. CW is applicable to all basic services except emergency calls. The incoming call may be of any kind or service group; a data call may be coming in during a voice conversation.

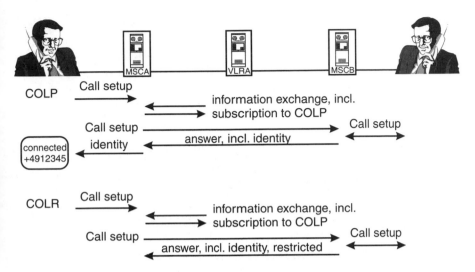

Figure 7.8 Information flow for connected line identity presentation/restriction.

The number of calls that can wait for a subscriber's response is limited by the operator. The maximum limit depends heavily on the infrastructure layout, and on the efficiency of the network. The operator considers how often it occurs that more than one call is waiting for a subscriber's attention, how many calls are likely to be waiting, and how much can be invested in the network to support CW.

Call waiting service can be activated and deactivated by the user on a basic service group basis or collectively with all basic services. When in doubt or confused, the user can interrogate the system as to whether the service has been set, and the network will report back the status of the service [13,14].

Let's look at how CW operates. As indicated before, and finally displayed in Figure 7.9, two subscribers (A and B) are engaged in a phone conversation. When one of them (A) gets a call from yet another subscriber (C), C will not get a busy indication, but a ringing indication (provided that CW is set for A). Subscriber A receives a notification, such as a low-frequency beep in the ear piece, which is not audible to B. Upon hearing the tone, A can decide to do one of three things:

1. A can put B on hold (described later), and accept the new call from C.

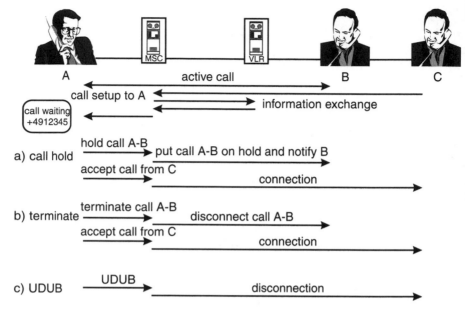

Figure 7.9 Basic procedures for call waiting.

2. A can terminate the call with B, and accept the new call from C.

3. A can continue the conversation with B, and send a *user-determined user busy* (UDUB) notification to C.

When call waiting is activated, the condition of a *network-determined user busy* (NDUB) will never occur. This is because all calls are offered to an individual CW subscriber. When call diversion on user busy is active, a rejected call will be diverted to the indicated number. The A subscriber does not have much time to decide what to do, because a network timer is running that limits the notification time of the call from C; the waiting call is disconnected after the timer expires. If a call forwarding on *no reply* was programmed into the system, then C's call is forwarded to the appropriate number in accordance with the subscriber's registration instructions.

A nice feature that works well with call waiting is CLIP, which will inform a subscriber about *who* is calling. The added provision of line identity makes the decision of whether to accept or reject a new call much easier.

Barring of incoming calls (BIC) conflicts with call waiting. It is, therefore, not possible to activate or use call waiting when BIC is active.

Information about the provisioning and the status of CW is stored in the HLR for each subscriber. The HLR also provides this information to the VLR. Figure 7.9 shows some procedures for call waiting and call hold (see Section 7.6). The initial condition of an active call between A and B, with C trying to call A, is always the same.

7.6   Call holding

Like call waiting, *call holding* (HOLD) belongs to the group of call completion supplementary services. HOLD enables a mobile phone subscriber to put a call on hold, which suspends the communication. There are many reasons for someone to HOLD a call. For example, when the subscriber wants to initiate a second call, perhaps in response to a short message received during the call. When the subscriber puts a call on hold, and no other active call is present, he has the following three options:

1. He can simply retrieve the call on hold (connect it again).

2. He can set up another call to a third party.

3. He can disconnect the call on hold.

When the subscriber is engaged in a call (which he is about to HOLD), and another call is already on hold, then there are four more options, which are depicted in Figure 7.10:

1. He can swap between the two calls.

2. He can disconnect the active call, thus making the three previous options available.

3. He can disconnect the call already on hold.

4. He can disconnect both calls.

The call hold service is only for telephony service; it is not practical to put data or fax calls on hold because the protocols used for these services are not intended to be put on hold.

The availability of call hold service will be activated by the service provider. Call hold is only active for one subscriber, which means that the

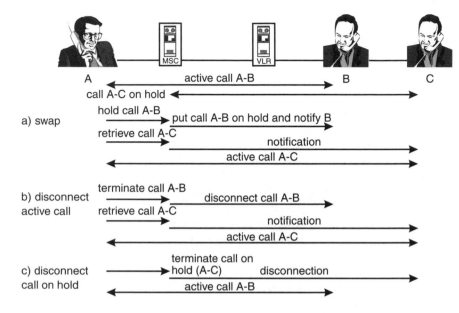

Figure 7.10 Some aspects of call hold.

party who was put on hold still has an active call, unless this subscriber also invokes a call holding service, which adds nothing of value to the connection. Figure 7.10 shows some of the aspects of call hold. The starting condition in the figure is that A has an active call to B, and another call to C is already on hold. As you study the figure, keep in mind that it is only possible to have one call on hold at a time [13,14].

7.7 Multiparty communication supplementary service

Even though *multiparty communication supplementary services* (MPTY) make up a separate group, they complement and depend on call waiting and call hold. MPTY allows a subscriber to group simultaneously from three to five other subscribers into one active multiparty phone call. The subscriber initiating a MPTY call must be involved in an active call with another call already on hold. HOLD service is a prerequisite to MPTY, and is only applicable to voice telephony. It is not necessary that all participants have subscribed to MPTY. Only the initiating subscriber has to have access to this feature [15,16].

Let's take the example from call waiting, where A and B are engaged in a conversation, and C calls A. Upon hearing the special CW tone or other indication, A can check who is calling and subsequently switch party B back into the communication. All the participants will get notification from the network that a multiparty call is established.

Every time a new member is connected to the multiparty conversation the remote parties (they were originally B and C) will get a notification. Participants in a multiparty call who are not connected to an ISDN network (perhaps they are part of an analog PSTN), will not receive the notifications.

How are new parties added? Let's say that all three phone conversationalists (A, B, and C) decide that the expertise of a fourth person (D) is required to settle an argument. Subscriber A, the originator of the multiparty call, puts the multiparty communication on hold. Subscribers B and C will still be able to communicate with each other while A (who is a registered user of MPTY) calls D. Subscriber D answers A's call who gives D a small introduction to the controversy currently on hold with B and C. Subscriber A connects the multiparty call, which was on hold, into the still active conversation with D, thus establishing a four-way conversation.

The situation is similar when someone tries to call subscriber A who has his call waiting service activated. Subscriber A receives some kind of notification from the network of an incoming call (beep and/or display) and he finally decides to put his multiparty call on hold while temporarily accepting the waiting call. Subscriber A could, if desired, add the multiparty connection to the new connection established with the intruder, or he can simply resume his original multiparty connection after disconnecting the waiting caller, who, it turns out, was a stranger trying to sell tickets to last year's football games, and could add nothing of value to the original conversation.

Once a multiparty connection is established, subscriber A can select an individual caller for a private conversation. The temporary exclusive connection with a particular caller places the remainder of the multiparty call on hold.

A multiparty call can be terminated by all members by simply hanging up their phones. The initiator of the MPTY call (A), who still controls the multiparty call, can also disconnect individual members from the multiparty call. Even as the initiating subscriber (A) has complete control of the connections, the remote parties (B, C, and D) control their own communication links. They can put their own connections on hold, they can release their calls, or they can start their own multiparty calls and swap between the two multiparty links. The MPTY abilities of all the callers is limited only by the technology (ISDN for example) of their networks and the services they decide to register.

Joining two separate multiparty calls is, in principle, also possible. This is very much dependent on additional technical implementations and provisions in the network. Whatever the connections among callers, the mobile network connection between the MSC/BSC/BTS and the MS (mobile phone) is the only physical connection (channel) actually maintained during a MPTY call. The links among all the other parties are made through bridging functions in the network (MSC). The mobile station distinguishes between the different parties in a multiparty connection by means of signaling over the air interface (U_m), which is also the mechanism used for call waiting and call hold. The management of all of these features is executed in the network, and their functions are confined largely to the switching center.

Figure 7.11 highlights some of the aspects of multiparty calls, where the initial number of members of a MPTY call is limited to three subscribers: A, B, and C. Adding more callers to the connections is simply a repetition of the three procedures listed here. In Figure 7.11(a), A and B are communicating while the communication between A and C is on hold. As the initiator, A wants to join both calls (all three parties) and requests the network to build a MPTY connection. A bridge is connected in the network (MSC) between the three parties. In part (b), A has decided to disconnect C, so he indicates his desire to the network (MSC) with whatever button-pushing procedures the network demands. The network releases the call to C without disturbing the connection between A and B. In part (c), A wants to talk with another party (D). The MPTY call is put on hold, which allows B and C to continue to communicate with each other. A is free to set up another call with D. If A decides to include D into the current MPTY call, then the condition becomes that of Figure 7.11(a), where one call is on hold as another one is active. Subscriber A can, of course, also decide to finish his business with D and retrieve the MPTY call on hold.

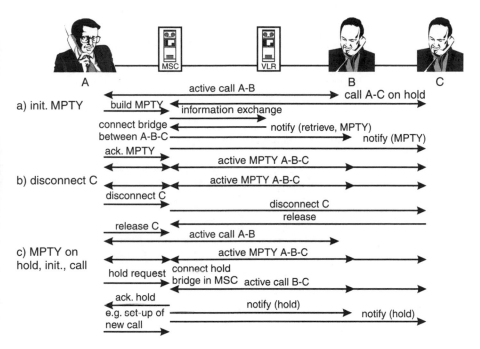

Figure 7.11 Some aspects of multiparty connections.

A hot issue for multiparty calls is charging. The general rule applies that the person who initiates a call has to pay for it. If A initiated a multiparty call to three remote parties, then A has to pay for all three connections. However, if another party calls A, and subsequently agrees to join the multiparty call, the new party pays for the call to A. This leaves us with the question about what happens when A wants to leave the multiparty call. Well, when A decides to leave, the whole multiparty call is disconnected. But when A puts the multiparty call on hold, so that the other members (B and C) of the call can still communicate, then A will have to pick up the bill for the communication links between B and C.

7.8 Advice of charge
supplementary service

The *advice of charge* (AoC) supplementary service provides the user with updated information on call charges. The charging supplementary services are divided into two different groups: (1) *advice of charge information* (AoCI) and (2) *advice of charge charging* (AoCC).

The subscriber can elect to subscribe to only one of the two services, not both. The principal information both charging services provides is identical; the major difference between them is the *application* related to AoCC. In general, the AoC services supply the mobile station with the charging advice information from which the actual call charges can be calculated [17,18].

7.8.1 Charge advice information

The *charge advice information* (CAI) is the raw data for calculating the call charges in the mobile station. The subscriber has the option to select the presentation or display of charging information. The display can be either in units, or it can be the actual charge along with a selected currency. There is seldom a perfect correspondence between calculated and displayed charges with the *actual* charges that come in the monthly bill. The small differences are caused by signaling delays between the mobile station and the network.

The mobile station is supplied with a set of seven parameters, which allow an accurate (within signaling limits) calculation of charges. This set of parameters is supplied once for a charge period and describes it completely. When the tariff changes, because of a different time of the day, for example, a new set of parameters is supplied to the mobile station. For simplicity these parameters are arranged into four groups:

1. The scaling factor, S, converts local units from the visiting PLMN to home PLMN-based units. When the subscriber is located in the HPLMN, then S is simply set to 1.

2. A constant, C, indicates a basic (minimum) charge per call for an initial time period. This is similar to a taxi charge where an initial basic rate is fixed independently of the distance.

3. A time related factor, T, consists of different parameters from which the mobile station can calculate the amount of units per time period. These also include a possible initial time period, which represents the basic charge (C) per call.

4. A data-related factor, *D*, is used to calculate charges for data calls.

With these parameters, the charge information is calculated continuously by the mobile station:

$$\text{AoC} = S \times (C + T + D)$$

The network operator would set these parameters to certain values or omit certain parameters (e.g., the *T* factor may not be used for data calls).

The whole charging process is of no use if the charging data cannot be stored somewhere in the mobile station. Two different call meters are defined in the SIM. The first is the *current call meter* (CCM). The CCM stores the call charges of the current ongoing call, which, at the end of a call, will be the amount that appears on an itemized bill for the call. CCM will store the charge of the last call until a new call is initiated and CCM is reset to "0." When multiple calls have been made during a single call, CCM will contain the sum of all these charges.

The *accumulated call meter* (ACM) holds the accumulated charges for the current call and all previous calls. The ACM is related to the *ACM maximum value* (ACMmax), which is also stored in the mobile station. When ACM reaches ACMmax, the mobile station will terminate the current call and inhibit any further chargeable calls, including outgoing or incoming international calls. ACMmax is set by the subscriber in the mobile station after entering a security code (PIN2). This ensures that no unauthorized person can change the value of ACMmax (see also Chapter 8).

In addition to these two call meter records, the SIM holds a *price per unit and currency table* (PUCT) that allows the mobile station to display not only the units stored in the CCM but also the actual price in the home currency. If a subscriber loans his phone to others, then he will probably want to cover the additional costs, which may include the monthly subscription fee and a profit. Setting the price per unit at a higher cost than actually charged by the network operator will cover these expenses. The CAI is specified in [19].

7.8.2 Advice of charge (information)

AoCI provides an accurate statement of the *total* of the bill to be expected at the end of the month. The network provides the mobile station with the CAI, where charges for non-call-related transactions and certain SSs are, sadly, not indicated. Because this is an information service, it does not matter if a visited PLMN supports AoC; calls can still be set up since this service will only provide information. This would not be acceptable for a rental phone, in which case the appropriate subscription would be to AoCC. The use of this service feature requires a subscription to it. The network provides the mobile station with a set of parameters (CAI) for each chargeable call with which the charges for each call can be calculated. The mobile station, in return, may acknowledge receipt of the CAI information. Figure 7.12 shows principal handling of advice of charge information.

7.8.3 Advice of charge (charging)

The intention for the AoCC service feature is to provide the most accurate information on all call charges that are made on a particular mobile station. Applications for the AoCC service feature are those where the actual user of a phone is not the subscriber; the user does not receive the bill from the network operator. This can be the case with phones loaned to friends, a phone given to an employee, or even a public GSM pay phone. It is important that the use of telecommunication services under these

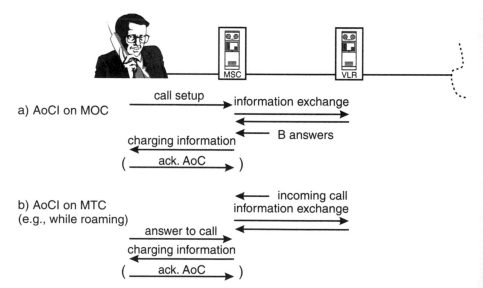

Figure 7.12 Principal handling of advice of charge information.

circumstances be restricted when AoCC cannot be stored in the mobile station, or when the visited network does not support the service. When the user is roaming into a network that does not support AoCC the HLR will reject a location update request (the MS's attempt to register in the visited network). This is a rejection of the mobile station in the VPLMN and the MS would indicate "No Service" to the user. The HLR is aware of networks that do not support AoCC, either from a database or from an interrogation of the VPLMN. When an MS that does not support AoCC is used with a subscription (information stored in the HLR) that includes and requires AoCC, the network will terminate each call setup attempt when it does not get back the acknowledgment of receipt of the AoCC element from the MS.

AoCC does not supply information on chargeable supplementary services. An operator might decide that he wants to charge a premium on individual services, for example, call forwarding or call waiting. Therefore, the network cannot inhibit the use of these services, and it is the responsibility of the subscriber not to sign up for such supplementary services.

Charges indicated through AoCC not only include mobile originated calls, but also chargeable parts of incoming calls such as the international leg of a connection when the user is roaming. The charge rates for a current call, or the last call, and all the accumulated call charges are stored securely in the SIM, which is described in Chapter 8. Figure 7.13 shows the principal handling of advice of charge charging.

7.9 Closed user group supplementary services

The *closed user group* (CUG) supplementary services allow the establishment of a dedicated user group within a GSM PLMN, which could be a group of taxi drivers in a city. The incentive for allowing this is to offer services known from *private mobile radio* (PMR) and *public access mobile radio* (PAMR) networks. GSM has the advantage, over most PMRs and PAMRs, of full nationwide coverage (plus international roaming) tempered by some additional digital cellular services. Due to the coverage of GSM networks, CUG is attractive to larger companies that operate on a national scale (transportation companies). To compete with the

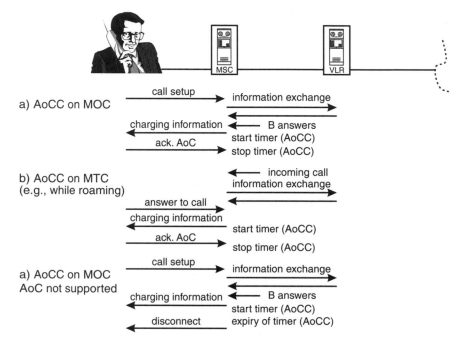

Figure 7.13 Principal handling of advice of charge charging.

PMR/PAMR systems, the pricing for calls within the CUG would certainly have to be much lower than pricing for normal cellular calls. CUG services are, by definition, available for all teleservices (except emergency calls and SMS) and all bearer services (except access to packet data networks).

As defined in GSM (through the Phase 2 specifications) a subscriber to CUG supplementary services can be a member of up to 10 CUGs. Each CUG is identified by a *CUG index*, which is to be used when calling into a dedicated CUG. Considering that a subscriber has a preferred user group to which most of his calls are directed, he can define a *preferential CUG*. Each time he sets up a call *without* a CUG index, the network will assume that the preferential CUG is the destination of the call. Members of closed user groups have several subscription options:

- ▶ Only calls within the CUG are allowed, which closely mimics existing private and trunked radio network services.

- ▶ Calls within the CUG and incoming access (cellular) are allowed. The subscriber can also receive calls from outside the CUG.

▶ Calls within the CUG and outgoing access are allowed. The subscriber can also make calls to destinations outside the CUG, such as to fixed networks or other subscribers in a PLMN.

▶ Calls within the CUG, and incoming and outgoing access, are allowed. The subscriber can receive calls from outside the CUG and can also make calls to destinations outside the CUG.

Furthermore, the subscriber may apply one of two available CUG specific *barring services* on top of his CUG definitions: (1) barring of incoming calls within a CUG and (2) barring of outgoing calls within a CUG.

A network offering CUG services will provide such services network-wide. But what would happen to roaming CUG users? If the visiting network supports CUG, then the subscriber can use her phone as if she were in her home country. If the visiting country does not support CUG, then the subscriber is treated like a normal roaming customer. However, if outgoing or incoming access is not allowed, then outgoing or incoming calls will not be possible [20,21].

7.10 Unstructured supplementary services data

Because the specification processes within GSM can be lengthy ones, it may take 1 or 2 years before the specifications for a new feature are finalized and all the different subgroups have agreed to the details. The delay deprives a network operator of the possibility of implementing a profitable feature within a short time. When they are finally available, his competitors may get access to the same features, probably at the same time, and offer services depending on the speed of implementation in the equipment from his supplier. *Unstructured supplementary services data* (USSD) provides a mechanism for an operator to design his own particular service features without waiting for the standardization process.

Before explaining the theory and mechanisms behind this service, a practical example of USSD is helpful. We begin with a look at an implementation from Vodafone in the United Kingdom that became available to all of their subscribers at the end of 1996. The implementation was around the CLIP service. The feature is called the *last calling number*, which allows a subscriber to see the phone number of the most recent person who called, or tried to call. This feature even works when the phone has been switched off. The subscriber simply has to send a USSD string to the network in order to see the number of the person who called last. The string to invoke this feature is:

$$* \# 1\,4\,7\,\#\ \text{📞}$$

The phone responds by displaying a phone number, for example, 01932816000. The little button with the handset is the off-hook key.

Even though USSD is very universal, some basic rules apply. These rules are related to a certain format of *strings*, which allow the mobile station and the network to recognize a USSD string. A USSD string is identified by the following format:

$$C_c\ [C_c]\ [C_c]\ 1\ X\ [Y]\ [*\ \text{any number of any characters}]\ \#\ \text{📞}$$

where C_c means either the "*" or "#" character, X and Y are any number between 0 and 9, and the optional parameters are in [ ] brackets.

USSD distinguishes two destinations for the strings: the HPLMN and the VPLMN. Both are distinguished by the value of X, where X [0–4] is reserved for HPLMN use, and X [5–9] for VPLMN use. For the HPLMN another string is defined:

$$7\ [Y]\ \text{📞},\ \text{with Y in the range of 0–9}$$

Service codes, which are used for supplementary services, are not used for USSD.

We only looked at mobile-originated USSD in the introduction earlier, but the opposite direction is also possible. Our example in the introduction to USSD actually showed that a string can be returned to the mobile phone for display. This is more tricky than it first appears, because the mobile station should, first of all, be able to understand the USSD, and then be able act on it. If the network requires it, the mobile station has to prompt the user for a response to the USSD strings. Each subsequent entry by the user will be regarded as a response to the network originated USSD. GSM has specified the SIM application toolkit, which has a similar mechanism as is found in USSD, but is far more universal as we will see in Chapter 9. The specifications for USSD can be found in [22,23].

7.11   Implementation of SS in a GSM mobile station

In previous sections of this chapter we got to know all the supplementary services and their uses. Now we have a look at how these are actually implemented in a GSM mobile station, and how a user can access the supplementary services.

One intention of setting up the GSM specifications was to keep the operation of the supplementary services similar to, or the same as, those found in existing telecommunications networks such as the ISDN. Operations dealing with SS are handled simply by manipulating the telephone's numeric keys and the "*" and "#" keys. The advantage of this implementation is that it is universal, but the disadvantage is that all of the control codes are difficult to remember. This is the reason that only one or two SSs are stored by default and are operational in ISDN-based PABXs. The others are usually hidden away in the dark and explained in some (unread) manuals. They get the same attention the user's guides for VCRs enjoy, which explains why so many home video recorders flash the "12:00" time regardless of what the local time may be.

Fortunately, most GSM mobile phone manufacturers took a more reasonable approach than some other makers of consumer electronics did and implemented a menu structure into the MMI. Menus are like multiple-choice examinations: the answer is visible somewhere among some unlikely possibilities—we merely have to find it. Menus allow easier access to the supplementary services: invoking them, adding their parameters, and disabling them. Even though tedious numeric control codes are not necessary with menus, it is still a basic requirement that mobile stations still support their entry. Menus lose relevance with time and increasing options for services; multiple-choice exams sometimes give way to the dreaded essay exam. In this section we provide an introduction to the codes and their meanings [24].

Supplementary services are divided into three different groups that have similar uses:

Table 7.5
Codes for Teleservices and Bearer Services

Basic Service Group Code	Teleservices / Bearer Services	Basic Service Group Number
No code required	All teleservices and bearer services	1–11
10	All teleservices	1–6
11	Telephony	1
12	All data teleservices	2–6
13	Facsimile service	6
16	Short message services	2
19	All teleservices exc. SMS	1, 3–6
20	All bearer services	7–10
21	All async. services	7, 9
22	All sync. services	8, 10
24	All data circuit sync.	8
25	All data circuit async.	7
26	All dedicated packet access	10
27	All dedicated PAD access	9

7.11.1.1 Activation of call waiting

Even though call waiting is a call-related SS, the user still has to activate or deactivate this service. This procedure informs the network if a particular mobile station should be notified of a waiting call when it is in a busy condition (already engaged in a call). After sending this string, the user will get an acknowledgment from the network.

*43#

7.11.1.2 Activation of CLIR

In the digital telephone world, people easily get accustomed to reading the phone number of the calling party on the display of their phones. This is also the case for digital PABXs. Some people may avoid conversations with a certain calling party for one or another reason, either rational or irrational. For this purpose, it is possible to switch off temporarily the CLIP function for the called party and invoke the CLIR function on a per-call basis. The same is possible in the opposite direction; that is, one can deactivate the CLIR function on a per-call basis, which is the same as the activation of CLIP for the called party. The following string will call the number 0897654321, without displaying the identity (number) of the caller at the other end. In the display of the (supporting) originating phone, the CLIR string (*31#) will not be displayed. It will only display "calling 0897654321."

*31#0897654321 📞

7.11.1.3 Registration of call forwarding unconditional for fax services

If a user wants to divert all of her incoming fax calls to the office fax machine in Munich (at number +498974880510), because there is no fax machine currently connected to her phone, the user dials:

**21*+498974880510*13# 📞

The forwarded-to number may also use the international access sign "+", which is not necessary for forwarded-to numbers in the home country, because the phone number is stored in the home HLR. The international access code of the home country applies even though the call forwarding is activated from abroad. After the call forwarding has been *registered* with a forwarded-to number, it can be *activated* or *deactivated* without the number, that is, with "*21# 📞" or "#21# 📞."

7.11.1.4 Registration of call forwarding when no reply

Sometimes it is not possible to answer phone calls, either because an important meeting is going on, or the ringer could not be heard. To make sure that an important incoming call is not lost, the *call forwarding when no reply* function should be activated. This is valid for all teleservices and bearer services in our example, and it will be activated after 20 seconds. The sequence to do this is:

61*+4989748805020# 📞

In this sequence, the second supplementary information element (the particular basic service) is not used, because the service applies to all basic services. This is why two "*" characters follow each other.

7.11.1.5 Change of password for barring SS

The password can be changed separately for the barring supplementary services. The only possible and necessary action is a registration of the password, the service code for which is "03." The user can define a password either for an individual service or for all barring services. The sequence for changing the password for barring of all outgoing calls, code 33, is:

**03*33*OLD_PW*NEW_PW*NEW_PW# 📞

When no specific barring service is specified, the new password will be applicable for all services:

03OLD_PW*NEW_PW*NEW_PW# 📞

With each incorrect entry of the *old* password the network's "wrong-password" counter is incremented, and the password access is eventually blocked when there are three failed attempts to enter the old password.

7.11.2 Implementation of call-related SS

All the call-related supplementary services require an active call except for the *activation* of call waiting. These services do not use the "*" and "#" characters; they use only the numbers and the off-hook key. Table 7.6 shows the control codes and the functions of the call-related SSs. The functions are explained in the related sections of this chapter.

7.11.3 Implementation into a menu structure of an MS

There are many ways to regulate access to and use supplementary services in a mobile terminal; the possibilities are limited only by those who design them. The problem for the typical phone user, who does not live in the same technoworld of the wizards who design phones, is that setting up access to SS is tedious without help. The user who needs to activate a particular service cannot be expected to construct and remember the codes. Hiding the SS control strings under a menu in a mobile station makes their use easier for everyone. For instance, the user simply brings up a phone's menu with a MENU key, and selects DIVERT to access

Table 7.6
Control Codes and Functions of Call-Related Supplementary Services

Control Code	Significance
0 ☎	Releases all calls on hold or sets UDUB for a waiting call
1 ☎	Releases all active calls (if existing) and accepts the call on hold or waiting call
1X ☎	Releases the specific call X
2 ☎	Places all active calls (if existing) on hold and accepts the call on hold or waiting call
2X ☎	Places all active calls (if existing) on hold, except call X with which communication will continue
3 ☎	Adds a call on hold to the conversation
NUMBER { ☎ }	Places all active calls (if existing) on hold and sets up a new call to NUMBER
ON-HOOK	Releases all calls except possible waiting calls

different call forwarding services, which appear in a submenu. The submenu shows the user some more submenus, such as one for entering the forwarded-to number, and another for selecting the basic service to be diverted. A similar menu tree may be available for all the other implemented supplementary services. Figure 7.14 gives an example of a MENU structure for call divert. The software in the mobile phone converts the particular user selections and entries into the correct strings discussed earlier.

7.12 Additional implementations in the mobile phone

Some mobile phone manufacturers have added even more additional features into the mobile phones themselves, which extend the functions based on some SS.

▸ *Call screening:* Only calls from phone numbers in the phone book are accepted and make the phone ring. This feature requires the support of CLIP from the network and in the mobile phone. The mobile phone, of course, should have some valid entries in its phone book.

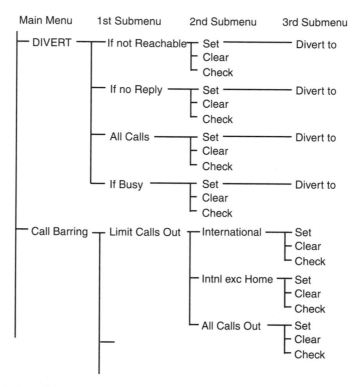

Figure 7.14   Menu structure for call divert.

▶ *Limit outgoing calls to phone book:* Only calls to phone numbers that are stored in the phone book are allowed.

▶ *Distinctive ringing:* The phone will ring differently depending on who is calling. The user has to make entries into his phone book indicating if business contacts, for example, will issue a loud bell-like sound, and private phone numbers issue a softer chirping sound from the phone's buzzer.

7.13   Future developments for Phase 2+

Many new supplementary services are being discussed and standardized through GSM Phase 2+. Let's take a look at some of these future GSM features so that we can plan our designs accordingly.

7.13.1 Call deflection

Call deflection is the transfer of a call to another third party without accepting it. Call deflection is currently supported only within an ISDN environment. The service requires that a maximum number of call diversions and forwarded calls not be exceeded. Call deflection is either invoked automatically by the mobile station, or by some kind of user interaction with the phone's MMI. The principle of call deflection can be compared to conditional call forwarding on "user-determined user busy." The advantage of call deflection is flexibility. Calls may be forwarded to different numbers depending on the number of the caller (identified by CLIP) [25].

7.13.2 Call forwarding enhancements

Currently some of the call forwarding possibilities are real *premium* services. The user is not really aware of the high costs sometimes associated with the use of such services. Consider the example of conditional call forwarding when roaming. These high costs lead to major complaints from customers who should not be expected to understand the details of international call routing and tariffs. Some enhancements are planned to clear the confusion. Users of fixed networks can already enjoy services that are not available to GSM users that could improve the situation. Some examples are (1) call forwarding depending on the time of day and (2) call forwarding depending on the day of the week. A sales executive may want to divert his calls to his office during normal business hours, to a colleague in another country outside of business hours, and to his voice mail during weekends.

Charges for call forwarding can be brought into a price range with other conditional forwarding services. This can be achieved with optimal routing, which is also planned for normal phone calls and is described in Chapter 10.

7.13.3 Call transfer

Call transfer (CT) allows a caller to transfer any call to a third party. Two scenarios are seen. In the discussion that follows, we always assume that customer A is the served one (the one who subscribed to the service), customer B is the "other party" who is engaged in a call with A, and customer C is the third party.

1. *Explicit call transfer* (ECT) applies when customer A and B are engaged in a phone call, perhaps discussing a new business plan. After they conclude their plans, they both agree it might be a good idea for B to continue the discussion with C. Subscriber A, who subscribes to ECT, puts B on hold, calls C and updates him on the latest details of the previous conversation with B, and then connects B and C. Subscribers B and C are notified that a call transfer took place [26,27].

2. *Single-step call transfer* (SCT) refers to the same situation as in item 1, but A now has the ability to transfer B, without putting him on hold, directly to C. For SCT, only B is notified that a call transfer took place; the transfer is a normal incoming call for C.

Call transfer is only applicable to telephony; this service is not possible for fax and data applications. There are three different *charging* scenarios:

1. If A calls both B and C, then A will be charged for all calls, A-B, A-C, and B-C. It might seem strange that A is also charged for the latest call, where A is not even a participant, but it may be seen as fair, because A initiated both calls to B and C.

2. If B calls A, and then A calls C, then B will be charged for the call A-B, and A will be charged for the calls A-C, and the subsequent B-C call.

3. If A is called by both B and C, then B will be charged for the A-B call, C will be charged for the A-C call, and A might be charged for the usage of the call transfer service.

7.13.4   Call completion services

There are plenty of cases where a phone call is not successfully completed, because the called party is already on the phone. The current GSM specifications (including Phase 2) only support the calling subscriber in the case where she got the busy indication. Upon hearing the busy signal, she has the option to invoke the redial option of her MS. This option makes the MS try to set up a call in periodic intervals listed in Table 7.7, in which we

Table 7.7
Redialing Attempts by a Mobile Station

Call Attempt	Minimum Duration Between Call Attempts
Original call attempt	—
1st repeat attempt	5 sec
2nd repeat attempt	1 min
3rd repeat attempt	1 min
4th repeat attempt	1 min
5th repeat attempt	3 min
:	:
nth repeat attempt	3 min

note that $n = 10$ for a busy subscriber or other network problems, and $n = 1$ when there is a problem reaching the destination, which is just to preclude a transfer problem within networks. The disadvantage of this mechanism, described in [28], is that only busy conditions, or temporary network problems during call setup, are accounted for; and that the whole infrastructure is used for each individual call setup attempt. The following enhanced call completion service avoids these disadvantages and puts more intelligence into the network to support the user.

7.13.4.1 Completion of a call to busy subscriber (CCBS)

In Phases 1 and 2 of GSM, the calling party heard the busy tone when try-ing to call someone who was already engaged in another call. The only option was to use the GSM-specified redialing scheme discussed in the previous chapter. As pointed out before, one of the disadvantages of the redialing scheme is that a complete call setup has to be accomplished for each calling attempt. This binds up network resources from the calling phone all the way out to the area in the GSM, or public network, to which the called party is connected.

Specification of CCBS brings another ISDN feature to GSM, where the infrastructure itself handles the monitoring of the busy, unreachable sub-scriber until he finishes her phone call [29]. When the busy condition finally disappears, the network routes the call to the called party as it notifies the calling party of the completed connection. Only when the calling party accepts the call from the network (the network rings the caller's phone) does the network notify the called party. If the calling party does not answer the notification (ring), then CCBS will be cleared in the network. This service applies to all telecommunication services, excluding SMS. There should be no extra charge for the use of this service, but the operator might charge a monthly fee for the subscription to this service. CCBS is much more efficient for the network than the redialing scheme specified for Phases 1 and 2, which used the whole network, including the air interface, for the duration of each attempted call setup to a busy subscriber. In the worst case, this could be 10 call setups within about 21 minutes.

7.13.4.2 An example of call completion

In the previous section, we learned about call completion to a busy subscriber. The network will monitor subscriber B until he is no longer busy on the phone. If there is some evidence that subscriber B can finally be reached, then the network will notify subscriber A. If subscriber A is still available, then the call will be connected through to B. The principle behavior is shown in Figure 7.15.

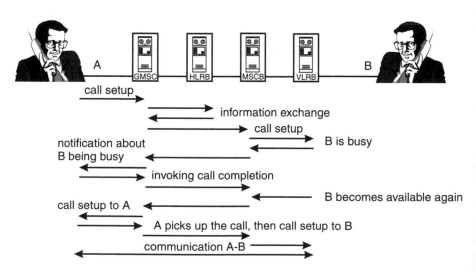

Figure 7.15 Principle operation for call completion.

7.13.5 Direct subscriber access and direct subscriber access restriction

Call forwarding and barring of incoming calls services are useful when a subscriber is in a meeting and does not want to be disturbed. Situations can arise when it is necessary to bypass call forwarding, or the barring of incoming calls services, such as when a secretary has to remind her supervisor of a change in a meeting schedule, wake-up calls, and call back service. Some of these situations might also arise because the subscriber simply forgot to disable call forwarding or incoming call barring. Two basic services are evident:

1. *Direct subscriber access* (DSA) enables a calling party to indicate that a called party should be alerted, neglecting any call forwarding or incoming call barring supplementary service which might be set.

2. *Direct subscriber access restriction* (DSAR) enables a called party to indicate that he wants to ignore DSA from specific people or groups. Because this is a very powerful service, the subscriber has to get it registered from the service provider, or the network operator; the subscriber cannot invoke this himself.

7.13.6 Malicious call identification

Phones are often used to threaten or molest people. This can normally be done without any danger of being identified, because so many ways exist to disguise one's identity to the calling party. *Malicious call identification* (MCID) is a calling line identity service that allows a threatened or molested party to mark the current call as a malicious call. Upon receiving this indication, the network operator will keep information related to the call, for example, the identity of the caller, the duration of the call, the time of the call, etc. How this information is used, or if it is even passed on to the molested subscriber, is up to local laws and the network operator. The information is usually collected for law enforcement officials. MCID can be performed after the very first instant of a call, or even after a specified time when a call has been terminated. The calling party engaged in this kind of social outrage will get no indication that his identity, through his phone number, will be stored anywhere.

7.13.7 Mobile access hunting

Service companies of all kinds find it convenient to provide a common number to their customers where they can dial in and request assistance. This might be applicable to taxi services, repair services, delivery services, or security services.

Mobile access hunting (MAH) allows a company to group individuals into a certain service group. The company subscribing to MAH may have several access numbers, each of which includes a group of several people. An incoming call will be routed to all members of a group, and all the phones in service will ring. An incoming MAH call is indicated as such to all the members of the group. When a member of the group finally accepts the call, a normal call setup to this particular member is completed as the alerting of the other group phones is suspended.

7.13.8 Support of private numbering plan

The *support of private numbering plan* (SPNP) through GSM will allow users in large companies to use their own phone numbers and extensions that are used within the company offices—the ultimate example of a user-friendly service. The numbers used on the GSM phone appear to be the same as the ones inside the PBX. Subscribers to SPNP will make their own group, and they can use the same calling procedures as those within the PBX; for example, they need to dial an escape code if they want to call someone outside the company PBX. SPNP is specified in [30].

7.13.9 Multiple subscriber profile

The *multiple subscriber profile* (MSP) is one of the examples where the DCS 1800 operators designed a feature (alternate line service) that has made its way into the GSM specifications [31]. MSP allows an individual subscriber to define up to four different subscription profiles. These profiles can be used to distinguish private calls from business calls. Parents could even allocate one profile for each of their children. The subscriber can define options, such as a subscription to fax or data services, for each of these profiles. Supplementary services can be set individually for each profile. The subscriber might set call forwarding to a mail box for the business number outside of office hours while still using the phone for private purposes.

The service provider can issue a separate bill for each individual profile. The subscriber can get a different phone number for each profile. If the service provider supports the multinumbering scheme, which allocates different phone numbers for voice, fax, and data calls, then different numbers can be allocated to each profile.

For outgoing calls, the user can select the profile to which he wants to charge the call. The phone can indicate which profile applies for each incoming call, which is how a subscriber distinguishes business calls from private calls. If the subscriber is busy on the phone, then other incoming calls to different profiles will meet normal busy conditions. If the subscriber has invoked call waiting or call forwarding, then the incoming calls will be handled in a manner appropriate to the different profiles.

The subscriber selects one profile as a default profile, which will be used in networks that do not support MSP. The other profiles are treated as if the mobile station were not reachable, in which case the subscriber can forward calls to the phone numbers of the different profiles to the phone number of the default profile.

7.13.10 Universal access to freephone numbers

Many service providers encourage their fixed network customers to dial in to a freephone number (toll-free number), with no calling charges, to ask about or order a service. This service is currently not supported by GSM, but mobile users will eventually be able to call these freephone numbers regardless of their location, even while roaming.

7.13.11 Premium rate service

Premium rate services (PRSs) are becoming popular in fixed networks. Suppliers of value-added services simply apply for a special number outside of the range of normal numbers (the "900 numbers" in the United States). The value these services add to our lives is best viewed by those who use them. Some are clearly helpful: share price quotes, current used car prices, and weather alerts. Others require patience and an open mind; one man's trash is another man's treasure. Callers to these numbers are charged at a premium rate by the operator who, in return, will pass on a percentage of the call charges to the value-adding service provider. GSM will provide the means to establish premium rate numbers and connect them to value-added services. The introduction of PRS will certainly require one addition to the barring services: barring of outgoing calls to premium rate services, which is not covered by existing call barring services.

7.13.12   Charging

Charging is currently a function that allows a simple display of present and accumulated charges. It cannot cope with future developments, especially in the value-added service area. Therefore, GSM will also tackle the charging services for improvements.

7.13.12.1   Advice of charge on special services

Today, the charging is not necessarily displayed to the user during a call. The current call meter is displayed after call clearing and on user request. If calling to special value-added services results in higher charges, then the user will probably not be aware of this unless the charges can appear during calls. Call charges during access to value-added services should immediately be displayed to the user.

7.13.12.2   Subscriber class for advice of charge

The biggest obstacles to the wider use of AoC supplementary service are the many different tariffs that apply for customers of different service providers. The introduction of a subscriber class would allow the provision of different charging levels with the same service.

7.13.12.3   Pay phone services

There is an increasing demand for GSM to provide pay phone service. The solutions, so far, are all manufacturer dependent and specific ones. The disadvantage of all the different pay phones is that there is no standard way of paying for calls. The applications range from pay phone services in international trains, to basic telephone service in rural areas, which are empty spots on the telecommunications map. In South Africa, for example, Vodacom had to set up phone booths in rural areas as part of their license requirement. Another example is offering GSM service to people with credit problems who cannot otherwise qualify for subscription service. This is not a trivial matter; there are whole economies in the world, which are so-called "gray" ones, in which credit is difficult to establish.

The most important issue for pay phone service is paying in real time. A dedicated pay phone SIM, without subscription, is useful for providers of pay phone service. A dedicated SIM would only contain limited information, since there is no need for allocating memory for a phone book or short messages.

7.13.12.4 Provision for hot billing

Hot billing refers to the on-line transfer of call records between various network entities. It might sometimes take several hours before call records are processed in the billing center. This makes the system susceptible to fraud from both normal users and prepaid applications. Hot billing also allows the identification of call charges that may be approaching certain limits in designated accounts. Getting hot billing to work is not as simple as it may seem on first glance. Network operators move billing records and other data not critical to live traffic around their networks during times when idle bandwidth is available for the task. Hot billing may work against the operator's efforts to use the latency in their networks efficiently unless it is carefully implemented.

7.13.13 User-to-user signaling

In a basic GSM network two subscribers communicating with each other have two ways to exchange simple nonvoice messages: (1) data services and (2) short message service. Both have disadvantages. Data services usually require some sort of terminal equipment and the implementation of complex protocols, and short messages are first sent to the service center before they are forwarded to the addressee. The delay can be measured in minutes. *User-to-user signaling* (UUS) provides an easy way to send simple messages, so-called *user-to-user information* (UUI), during a phone call.

The UUI can be sent and received during call setup, where the message is embedded in the call control messages, or after an active call has been established. The length of data to be transmitted can be in the range of 100 to 200 octets. UUS applies to telephony (TS11), facsimile (TS6X), circuit-switched data (BS2X and BS3X), and PAD packet access data (BS4X and BS5X). UUS is specified in [32].

References

[1] GSM 02.04, "European Digital Cellular Telecommunications System (Phase 2); General on Supplementary Services," ETSI, Sophia Antipolis.

[2] GSM 03.11, "European Digital Cellular Telecommunications System (Phase 2); Technical Realization of Supplementary Services," ETSI, Sophia Antipolis.

[3] GSM 03.02, "European Digital Cellular Telecommunications System (Phase 2); Network Architecture," ETSI, Sophia Antipolis.

[4] Redl, S. M., M. K. Weber, and M. W. Oliphant, *An Introduction to GSM*, Norwood, MA: Artech House, 1995, Chap. 3.

[5] GSM 02.82, "European Digital Cellular Telecommunications System (Phase 2); Call Forwarding (CF) Supplementary Services—Stage 1," ETSI, Sophia Antipolis.

[6] GSM 03.82, "European Digital Cellular Telecommunications System (Phase 2); Call Forwarding (CF) Supplementary Services—Stage 2," ETSI, Sophia Antipolis.

[7] GSM 02.88, "European Digital Cellular Telecommunications System (Phase 2); Call Barring (CB) Supplementary Services—Stage 1," ETSI, Sophia Antipolis.

[8] GSM 03.88, "European Digital Cellular Telecommunications System (Phase 2); Call Barring (CB) Supplementary Services—Stage 2," ETSI, Sophia Antipolis.

[9] GSM 02.41, "European Digital Cellular Telecommunications System (Phase 2); Operator Determined Barring," ETSI, Sophia Antipolis.

[10] GSM 03.15, "European Digital Cellular Telecommunications System (Phase 2); Technical Realization of Operator Determined Barring," ETSI, Sophia Antipolis.

[11] GSM 02.81, "European Digital Cellular Telecommunications System (Phase 2); Line Identification Supplementary Services—Stage 1," ETSI, Sophia Antipolis.

[12] GSM 03.81, "European Digital Cellular Telecommunications System (Phase 2); Line Identification Supplementary Services—Stage 2," ETSI, Sophia Antipolis.

[13] GSM 02.83, "European Digital Cellular Telecommunications System (Phase 2); Call Waiting (CW) and Call Holding (HOLD) Supplementary Services—Stage 1," ETSI, Sophia Antipolis.

[14] GSM 03.83, "European Digital Cellular Telecommunications System (Phase 2); Call Waiting (CW) and Call Holding (HOLD) Supplementary Services—Stage 2," ETSI, Sophia Antipolis.

[15] GSM 02.84, "European Digital Cellular Telecommunications System (Phase 2); Multiparty (MPTY) Supplementary Services—Stage 1," ETSI, Sophia Antipolis.

[16] GSM 03.84, "European Digital Cellular Telecommunications System (Phase 2); Multiparty (MPTY) Supplementary Services—Stage 2," ETSI, Sophia Antipolis.

[17] GSM 02.86, "European Digital Cellular Telecommunications System (Phase 2); Advice of Charge (AoC) Supplementary Services—Stage 1," ETSI, Sophia Antipolis.

[18] GSM 03.86, "European Digital Cellular Telecommunications System (Phase 2); Advice of Charge (AoC) Supplementary Services—Stage 2," ETSI, Sophia Antipolis.

[19] GSM 02.24, "European Digital Cellular Telecommunications System (Phase 2); Description of Charge Advice Information (CAI)," ETSI, Sophia Antipolis.

[20] GSM 02.85, "European Digital Cellular Telecommunications System (Phase 2); Closed User Group (CUG) Supplementary Services—Stage 1," ETSI, Sophia Antipolis.

[21] GSM 03.85, "European Digital Cellular Telecommunications System (Phase 2); Closed User Group (CUG) Supplementary Services—Stage 2," ETSI, Sophia Antipolis.

[22] GSM 02.90, "European Digital Cellular Telecommunications System (Phase 2); Unstructured Supplementary Service Data (USSD)—Stage 1," ETSI, Sophia Antipolis.

[23] GSM 03.90, "European Digital Cellular Telecommunications System (Phase 2); Unstructured Supplementary Service Data (USSD)—Stage 2," ETSI, Sophia Antipolis.

[24] GSM 02.30, "European Digital Cellular Telecommunications System (Phase 2); Man-Machine Interface (MMI) of the Mobile Station (MS)," ETSI, Sophia Antipolis.

[25] GSM 02.72, "Digital Cellular Telecommunications System (Phase 2+); Call Deflection; Service Description (Stage 1)," ETSI, Sophia Antipolis.

[26] GSM 02.91, "Digital Cellular Telecommunications System (Phase 2+); Explicit Call Transfer, Service Description—Stage 1," ETSI, Sophia Antipolis.

[27] GSM 03.91, "European Digital Cellular Telecommunications System (Phase 2+); Explicit Call Transfer, Service Description—Stage 2," ETSI, Sophia Antipolis.

[28] GSM 02.07, "European Digital Cellular Telecommunications System (Phase 2); Mobile Station (MS) Features," ETSI, Sophia Antipolis.

[29] GSM 02.93, "Digital Cellular Telecommunications System (Phase 2+); Completion of Calls to Busy Subscriber (CCBS), Service Description, Stage 1," ETSI, Sophia Antipolis.

[30] GSM 02.95, "Digital Cellular Telecommunications System (Phase 2+); Support of Private Numbering Plan (SPNP); Service Description (Stage 1)," ETSI, Sophia Antipolis.

[31] GSM 02.97, "Digital Cellular Telecommunications System (Phase 2+); Multiple Subscriber Profile (MSP); Service Description (Stage 1)," ETSI, Sophia Antipolis.

[32] GSM 02.87, "Digital Cellular Telecommunications System (Phase 2+); User-to-User Signaling (UUS); Service Description (Stage 1)," ETSI, Sophia Antipolis.

Contents

The subscriber identity module

The SIM, or the subscriber identity module, is a smart, plastic chip card. A chip card is a general term that refers to any of the plastic credit card-sized devices in which one or more circuits are embedded [1]. If we find a microprocessor or microcontroller among the embedded circuits, then we refer to the chip card as a smart card. A smart card that is specified for use in a GSM phone is called a SIM card, or simply a SIM.

The real identity of a GSM MS and its subscriber is actually the data stored in the SIM. It is not enough merely to identify a particular mobile phone, because the phone's use must be authorized for a particular subscriber. The subscriber's identity is confined to the SIM, which can be inserted in any phone, thus making the phone a mobile station, hence ME + SIM = MS. Without a SIM, a mobile phone cannot receive or originate calls, except for emergency calls. The requirements are that emergency calls to selected emergency services must be possible at any time without a SIM, as long as the mobile equipment is functional and within a covered service area. The SIM comes in two different physical sizes: the credit card type (ISO) and the smaller type, which is the so-called *plug-in* SIM. The little plug-in SIM is designed for use in small portable phones that cannot otherwise accommodate the larger full size SIM. Figure 8.1 shows the two different types of SIM cards. It should be noted, however, that the smaller plug-in SIM comes as an ISO type, which can be popped out to become a plug-in SIM.

The heart of the SIM card is a microcontroller that includes some memory for programs (ROM) and some additional memory (*electrical erasable and programmable read-only memory* [EEPROM]) in which to store information. The operating power supply and the clock pulses are provided by the *mobile equipment* (ME). The structure of the SIM is explained in Section 8.7.3. Several different pieces of user information are stored on the SIM, including user-specific network information and security-related functions. Because the GSM standard continues to evolve as more and more functions and service features are added, there are different phases for SIMs that follow and accommodate the enhancements. The different SIMs are named according to the different implementation phases for GSM handsets and infrastructure, namely, Phase 1, Phase 2, and Phase 2+. Still, there are some general procedures that are independent of the GSM phases. We first explore the common features and procedures that all of the SIMs support before considering the differences in each of the phases.

A SIM is structured similar to a computer. There is a microcontroller with its operating system. There is also some memory and some other standard features used for accessing the memory and functions related to security. The operating systems found on the various SIMs are not

Figure 8.1   Two different sizes of SIM cards are available (courtesy of Schlumberger).

discussed in more detail because they are different for every SIM manufacturer.

8.1 Memory structure

The SIM's memory is split up into directories in a manner similar to that of the hard drive in your computer. The main directory in this structure is the *master file* (MF). Under the MF the memory space is subdivided into different directories that are called *dedicated files* (DF). Two different DFs are specified, and it is with these differences that the *file* term becomes rather misleading. The DF should be regarded more as a location (a directory), and not as files within those locations. The files are segregated into two types by their content:

1. DF$_{GSM}$ contains specific applications for GSM, DCS 1800, or PCS 1900;

2. DF$_{TELECOM}$ contains the more common telecom service features.

One level in hierarchy below these dedicated files, we find the *elementary files* (EF), which contain the actual information for the GSM or TELECOM directories. One EF can contain only one *record*, which, for example, could be the *international mobile subscriber identity* (IMSI), which is unique for each SIM. One single EF may also contain several records of information, such as the phone book, which can hold up to 255 entries or records. A record is a small unit of information stored on a SIM. A record consists of a string of variable size, some large and some small, depending on its purpose and significance. The record sizes are measured in words, where a word is 1 byte, which is 8 bits.

The GSM directory is dedicated to the exclusive use of GSM networks and contains GSM-specific parameters and records, whereas the telecom directory can be used for other telecom applications in multipurpose cards. The phone book is an example of an EF that contains multiple records and is part of the more general telecom directory. It is possible to use the phone book for applications around a public telephone service that supports smart cards.

Figure 8.2 shows the organization of the SIM's memory. Under the master file directory there is only one EF containing the identity of the

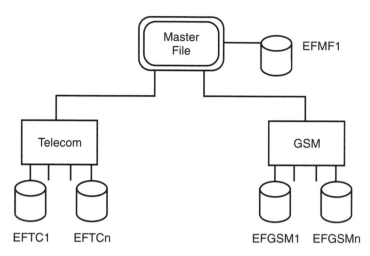

Figure 8.2   Memory structure on a GSM SIM.

SIM, which will be discussed in Section 8.4. Also under the MF we find two directories: the telecom directory and the GSM directory. The GSM directory is used for GSM, DCS 1800, or PCS 1900 SIMs. The same directory can support both the GSM and DCS 1800 systems in SIMs, which are designated (and programmed) to support both the GSM and DCS 1800 systems. Organized below these two directories are the actual elementary files, which are described in Sections 8.4 and 8.5, which are concerned with the different phases of SIMs. The additional features specified by the DCS 1800 operators call on their elementary files in the SIM, which are stored in a dedicated DCS 1800 directory.

8.2   Security

The best known GSM security feature, which is actually related to and incorporated in the SIM, is the *personal identification number* (PIN). The user is required to enter the PIN, which is only valid for a particular SIM, after switching a GSM phone to its ON condition: "ENTER PIN." The SIM also contains other security parameters, algorithms, and features related to the GSM security and authentication system:

- The *authentication result* (SRES) is computed on the SIM.
- The *ciphering key* (K_c) is generated on the SIM.

▶ The SIM has measures to protect its elementary files.

To generate the authentication result and the ciphering key, the SIM contains two algorithms:

▶ Algorithm A3 is used to generate SRES.

▶ Algorithm A8 is used to generate K_C.

It is also possible, depending on the SIM manufacturer, to have a common algorithm (A38) that computes both results at the same time. One of the inputs to both algorithms (A3 and A8) is a *random value* (RAND) that is generated by the network and transmitted to the mobile station over the radio path. The radio (ME) internally passes the RAND to the SIM. The other input to both algorithms is an *internal key* (K_i), which is stored on the SIM. The network, of course, keeps a copy of each SIM's K_i for itself in the *authentication center* (AC)[1] [2]. Figure 8.3 shows the procedure and the communications between the network, the mobile equipment, and the SIM. The network sends the RAND to the mobile equipment, which passes this number to the SIM together with a command to generate the SRES and the K_c. Both results are returned to the ME. The SRES result is transmitted back to the network, which completes the authentication procedure. The K_c result is stored in the ME where it is used to perform the ciphering function if, and when, ciphering is enabled by the network. Further details on this mechanism, the relationship between the variables, and how ciphering is invoked can be found in [2].

The information contained in each EF is protected against improper access in accordance with different security levels proportional to their significance and sensitivity. There are different protection levels for the procedures for reading, updating, invalidating, and recovering elementary files. These protection levels are distinguished from each other by the following access conditions:

1. *Always* means that the function for this field can always be executed.

---

1. The AC computes corresponding pairs of RAND and SRES values, which it provides to the HLR and VLR.

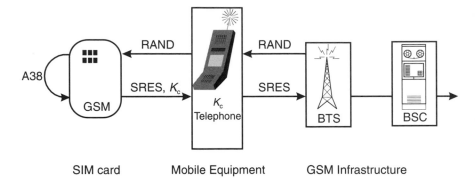

Figure 8.3 Procedure of passing RAND, SRES, and K_C through the SIM, mobile phone, and network.

2. *Card holder verification 1* (CHV1) information, also commonly known as PIN1, is usually requested when the mobile phone is turned on. If CHV1 is entered once, it will remain valid for an entire session. The use of PIN1 (CHV1) can be disabled by the user, which means it will never be requested for verification in subsequent sessions. The user has to know the *valid* PIN1 in order to disable the prompts for PIN1.

3. *Card holder verification 2* (CHV2) information, which is commonly known as PIN2, only has to be entered for access to specific procedures, for example, for changing some types of information.

4. *Administrative* (ADM) is used for actions that can only be changed or set by the network operator. This is, for instance, the case when the IMSI needs to be updated.

5. *Never* is intended for files that should never be updated. One example is the SIM identification, which can always be read but can never be changed.

In the following sections on the different EFs as they apply to each of the GSM phases, the status for reading, updating, invalidating, and recovering them is indicated later in the chapter (Section 8.5) in Table 8.2.

A common procedure that happens millions of times every day is the authentication of a user, which occurs whenever a mobile phone is switched on within the coverage area of a network. To prove that she is entitled to use the phone—that the user is, in fact, who she says she is—the user must enter a PIN. If the PIN is entered incorrectly three times, then the SIM is blocked. The SIM can be unblocked if the user flawlessly enters an eight-digit *personal unblocking key* (PUK). This key is always known by the operator and stored on the SIM. But the PUK is, in general, unknown to the user even though it is usually included in the subscriber's "welcome kit" or other user documentation. Some operators take the approach of not disclosing the PUK, but offer a dedicated service to unblock the SIM. These operators can collect a service fee from subscribers who ask for their PUKs. If the PUK is entered incorrectly 10 times, then the SIM is completely blocked, and has to be brought back to the operator for reactivation.

What are the different consequences for blocking PIN1 and PIN2? If a user uses up all three chances to guess (remember) PIN1, then the SIM is completely blocked and cannot be used to gain normal authorized access to the network, in any phone, except for emergency calls. PIN2 is only used for some much more obscure features (e.g., charging services) and is not directly related to granting access to the basic telephony service. If PIN2 is blocked, then the user has merely lost access to certain functions, but she can still can use the phone.

8.3 Phase 1 SIM

Just as the whole GSM system grew, so did the SIM grow with the demand for new features. In the beginning, the most important service was *plain old voice telephony service* (POTS) with some basic security features (authentication and ciphering). The early mobile phones did well to support the teleservices and some very basic supplementary services. The SIM only supported the mandatory features related to security. Some SIMs had space for a phone book and, eventually, some more space for storing short messages, even though short message service was not supported in the first GSM mobile phones and early GSM networks. At the time of this writing, most newly issued SIMs were Phase 2 SIMs, and many Phase 1 SIMs remained in service. Table 8.1 (see Section 8.5) provides an overview of the various elementary files supported in the different phases of the SIM card.

8.4   Phase 2 SIM

The second generation of SIMs, current at the time of this writing, supports many more features and functions than the Phase 1 SIMs. For instance, the Phase 2 SIM provides memory space for features that were initially supported only within the most expensive mobile phones. These luxury phones provided memory space for features such as the "last numbers dialed" list. This list contains the phone numbers of the most recent calls placed with the phone. Because many users tend to dial only a few numbers over and over, it is a relatively simple matter to call up the list (from a menu), and then select one of its entries (by scrolling through the list) for redialing

The reader will find a more detailed description of the fields of Phase 2 SIMs in the following paragraphs [3]. In addition to a list of the different elementary files there are descriptions of the fields' uses and, if applicable or appropriate, an example is included showing how each field (and its associated function) manifests itself in (looks like on) a typical GSM phone.

We looked at the SIM's file structure earlier in this chapter, and we can recall that the EFs can be found in different directories, which include the master file and some dedicated files. More about the file structure can be found in Section 8.7.2. The only EF located directly below the master file is the *ICC identification* (EF_{ICCID}). It contains the serial number of the SIM, which is unique to each individual SIM. It is possible to personalize the ME with this SIM number, which would only allow a specific ME to be used together with a particular SIM. This use is, however, not specified, and is left to the different manufacturers for implementation. Personalizing an ME with a SIM number removes some of the benefits those who steal phones can enjoy, because the thief will find it impossible to use the stolen phone with his own SIM. This scheme fails if another user-specific four- to eight-digit code (e.g., a dedicated PHONECODE—not to be confused with the PIN) is known by the thief. Such a PHONECODE is an invention of the manufacturers of mobile equipment. By default there is no value assigned, but as the user attempts to execute a function requiring the PHONECODE for the very first time, the phone will request entry and assignment of this code. From then on, this code has to be remembered by the user, and nobody else will know it if it is lost.

The following elementary files belong to the dedicated GSM file. The *language preference EF* (EF$_{LP}$) defines one or more languages preferred by the user of the terminal. This is used for the display of the MMI (menu) and/or for short message handling (e.g., screening of preferred languages in SMS-CB), which was explained in Chapter 6. In GSM Phase 1, many mobile phone manufacturers implemented the language selection simply by checking the *mobile country code* (MCC) part of the IMSI. If the MCC was 262 (Germany) the phone would, by default, display a German menu. If the MCC was 234 (United Kingdom), the phone would display an English menu. An advantage of having a dedicated field for the language is that an operator can specify multiple languages, which is useful in countries where two or more languages are used. Another advantage is to render the phone more user friendly: it lets the user select his own preferred languages. This value (language) would still be valid when the SIM is inserted into another phone.

The *IMSI* (EF$_{IMSI}$) contains the international mobile subscriber identity with the coding as depicted in Figure 8.4. The first three digits define the MCC, the next two digits define the *mobile network code* (MNC), and up to 10 additional digits specify the *mobile subscriber identification number* (MSIN). It is important to note that this number is not identical to the phone number with which a subscriber can be reached in a network. Actually, several phone numbers can be allocated to one IMSI, for example: (1) one number for normal voice calls, (2) another number for fax calls, and (3) a third number for data applications. The allocation of different numbers for a single IMSI depends on how the operator implements the different services in his network.

The *ciphering key* (EF$_{K_c}$) contains the key and the ciphering sequence number used during the last session with the phone. When registering into a network, the mobile station receives a new ciphering sequence from the network, together with the latest RAND value. Further information on ciphering and how it works can be found in [2].

The *PLMN selector* (EF$_{PLMNsel}$) defines the list of preferred foreign networks for an individual user. Because there is normally no national roaming, it does not make sense to list competing operators in the home country. The network operator usually presets this list with codes for preferred operators in other countries, which may be those with which the operator has set up some kind of streamlined billing arrangements or

and when this level is reached in the accumulated call meter, the mobile will not allow the user to initiate further calls. It will, however, still be possible to receive calls. Originally the ACM max value was intended for use in company phones or for loaner phones, to limit and/or control call charges. But it can also be the basis for the use of prepaid SIM cards.

The *price per unit and currency table* (EF$_{PUCT}$) defines the price per unit (time) and the currency display ($, £, DM, etc.). This information is used to compute the cost of individual calls and to display the cost in the selected currency. This field is closely related to the value of ACM, and it must be present if the EF$_{ACM}$ is allocated. Figure 8.5 shows the presentation of a price per unit and currency table in a mobile station.

The *cell broadcast message identifier selection* (EF$_{CBMI}$) contains relevant parameters for cell broadcast, such as the cell broadcast channel and, optionally, the preferred language of the cell broadcast messages to be passed to the display. When the cell broadcast is used for location area display, which is used for tariffing information, the network operator might preset this field in order to display the tariff information by default. Otherwise, it would be a difficult task for the user to find the right menu and then enter the correct numbers. Alternatively, the user might wish to display weather forecasts or traffic information, which is available through cell broadcast; the channel number would be stored in the EF$_{CBMI}$. The user can always select a preferred language; the phone would then ignore messages in other languages.

The *broadcast control channel* (EF$_{BCCH}$) stores information received from the last serving cell about the channels used for BCCH in neighboring cells. This information is used to reduce the time the phone spends scanning for a network after switching the mobile station back on. The mobile first searches on the channel(s) used for BCCH(s) by the last serving cell, and then those of the neighboring cells.

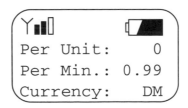

Figure 8.5 Price per unit and currency table displayed on a phone.

The *access control class* (EF_{ACC}) describes different access levels for a network. The access class is transmitted by the base station as part of the base station control channel. If the control class on the SIM and the transmitted one from the network match, then the mobile may access the network. Sixteen different classes are defined; the first 10 are randomly split and allocated to normal subscribers. Five are reserved for special users: network operator staff, emergency services (police), public utilities (gas leaks and water pipe bursts), security services, and other PLMNs (during activation or testing of cell sites). One class is reserved for emergency use, which means that if this class is allocated to the cell, but none of the first 10 is allocated, then a normal subscriber would be allowed only limited service that allows emergency calls. Mobile stations without a SIM (and without an IMSI) are allowed to make emergency calls with this emergency call access class. Emergency calls are always allowed under normal conditions, which is one of the key features of GSM. All access classes are allowed within a cell under normal conditions.

The *forbidden PLMN* (EF_{FPLMN}) list keeps track of the last four PLMNs the mobile station tried to access that ended in rejection for some reason. Some examples are (1) international roaming was not allowed or (2) access to a second, competing network was denied. If access was denied, then the mobile station would simply store the corresponding mobile country code and mobile network code on the SIM. A mobile station searches for a PLMN after power is turned on in accordance with the lists on the SIM. If it cannot find service from the home PLMN, then it searches for other GSM base channels. When the mobile phone realizes, by reading the system information of a BCCH, that the cell it wants to camp on belongs to one of the forbidden PLMNs, it does not try to register (location update), but indicates instead that access to service is not possible. This can be misleading in the rare case when international roaming was initiated since the last unsuccessful trial, but the visited network is still in the forbidden list. The only solution here is to manually select the network, in which case the mobile station is forced to try to register onto the network. If manual selection is successful, then the PLMN code is taken off the forbidden PLMN list. This list is a short cyclic list, where the oldest entry will be overwritten when the maximum number of four entries is exceeded.

The *location information* (EF_{LOCI}) stores basic information about the current location of the mobile station, such as the following:

▶ The *temporary mobile subscriber identity* (TMSI), which, for security reasons, is allocated to the MS instead of using the IMSI.

▶ The *location area information* (LAI), which describes the paging area in which the mobile was last located (including MCC and MNC). When the LAI stored in the *location information* is different from the LAI read from a tentative new serving cell during initial synchronization (read in the BCCH system information), then the mobile has to initiate a location update procedure.

▶ The *TMSI time* describes the period within which the mobile must perform a periodic location update in order to inform the network about the availability of the mobile station.

▶ The *location update status* indicates whether the mobile station is updated (after a successful location update procedure), not updated, or whether the PLMN or the present location area are not allowed.

The *administrative data* (EF_{AD}) tells the mobile equipment which mode of operation it should support. The different options here are (1) *normal operation,* which most of the available SIMs use for normal service; (2) *type approval operation,* which is used to allow certain operations (e.g., the closing of the traffic channel loop for receiver testing); and (3) certain *field test modes,* which allow anyone to test either a particular cell site before making it available to the public or to test certain features in a mobile phone before a manufacturer (e.g., hands over a phone for type approval or operator approval).

The *phase identification* (EF_{Phase}) contains information about the phase of the SIM, which is the distinction between GSM Phases 1 and Phase 2. This coding tells the mobile equipment that certain features are not available in the SIM, such as fixed dialing numbers or advice of charge in the case of a Phase 1 SIM. It is possible to code a Phase 2 SIM so that it can be used as a Phase 1 SIM.

The 16 files just discussed belong to the dedicated GSM directory; they have relevance to mobile phones in a GSM system. The following 10 elementary files (listed later in Table 8.1) belong to the dedicated telecom directory. They have more general relevance and can also be used by other applications that use these features and information, such as a phone card for public phone booths.

The *abbreviated dialing numbers* (EF_{ADN}) is simply a different expression for the phone book, which is provided by the SIM. It is possible to supply an (additional) phone book in the memory of the mobile equipment, but this is up to the manufacturer. Wherever the phone books may be—on the SIM or in the phone—the structure of both memories is identical. Like any standard phone book, the ADN contains the phone number, up to 20 digits per record, and an alphanumeric entry, which probably represents a name. The network operator may specify how long such a name tag can be. There is no restriction from the SIM itself on how long such a name tag might be, but the longer the name tags, the more memory is used for each entry. Most phones will have trouble displaying names with more than 16 characters. An average value for the size of the name tags is between 12 and 16 characters.

From Phase 2 onward, the length of the phone number is no longer restricted to 20 digits, because it is possible to append further digits into the extension field. This is especially helpful when using DTMF control strings, together with access phone numbers, which might easily exceed 20 digits. One example is dialing a secret code for some kind of access to service, or to enter a debt card number. In addition to a phone number, the phone number field can contain control characters that are available on a keypad, as discussed in the following paragraphs.

The astcrisk (*) and pound sign (#) are used for the supplementary services control strings, as discussed in Chapter 7. This shows that it is possible to store SS functions in the ADN field. It is also possible to store the SS control string for CLIR before a stored phone number, which prevents the caller's number from appearing in the display of the phone being called.

The DTMF control digit separator can be different for various types or brands of mobiles. The plus (+) character is an example. The control digit separator is used for breaks in a sequence of DTMF digits. The delay specified for the separator function is 3 seconds. Each use of the control digit separator causes the network [2] to insert a 3-second delay before going to the next digit(s). One use of this feature is to access a bank account where the user starts by dialing into a central computer, and then after being connected to the bank's server, the user dials her bank account number along with an identification number. These numbers are transmitted as

---

2. The complete DTMF string is passed on to the network, where the actual tone sequences and programmed breaks are generated.

DTMF tones (in an analog network providing only analog service) or as DTMF digits (in ISDN networks and service). In Figure 8.6, the phone number to access the account service is +49891234. After dialing this, the phone will pause for 3 seconds to allow the bank computer, or modem, to pick up the phone and request some entries from a menu structure. The digits "12" can guide someone through the menu system, where, at the end, we may find a request for entry of a PIN number (3 4 5 6). The entries are delayed for 3 seconds so that the system can distinguish among them.

A *wild* value/character, when used, will cause the mobile equipment to prompt the user for a single digit, or more digits, depending on how many wild values have been used. This is a useful feature for a user who needs to regularly dial into his office or a customer's PBX. The typical user who likes this feature is someone who knows the extensions of everyone he talks to, but cannot remember the long prefix. These forgetful users simply store the common number for everyone regularly called followed by a string of wild values. When they dial one of the numbers, the phone merely prompts for an extension to be entered. It is up to the manufacturer of the mobile equipment to specify the wild character. The "?" character is used in our example in Figure 8.7.

The *fixed dialing numbers* (EF_{FDN}) field is a type of phone book, and its content is identical to the ADN field. The difference between this and the ADN is that the user, or the network operator, can limit outgoing calls to

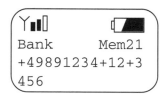

Figure 8.6   Example for DTMF tones in the ADN.

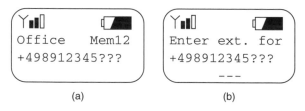

(a)   (b)

Figure 8.7   Use of wild characters: (a) entry into ADN and (b) display when dialing.

the numbers stored in the FDN field. This is helpful if a phone is loaned to someone, or if an employer gives a phone to an employee who has a bad habit of calling friends all around the world. It is possible to use wild characters, which is a helpful feature if a phone is used to call people connected to the same PBX. To make the selections easier to find in the FDN directory, a name can be attached to each entry, just as is the case with the ADN field.

The FDN prevents the SIM from being used for any calls other than those to numbers authorized by the original subscriber. A SIM with FDN capabilities has to be handled properly. When a SIM is inserted in a phone, one of the first checks is the status of the EF_{IMSI} and EF_{LOCI}. If these have not been invalidated, then the mobile equipment is operating in unrestricted mode. If they have been invalidated during an earlier session, or during initialization, then FDN is enabled. Invalidating means that the EFs are still stored on the SIM, but they cannot be used by the mobile station; they are "hidden" like system files on a computer. If the mobile equipment does not support FDN, then it is not allowed to recover these two elementary files (make them visible again) and, therefore, cannot operate with the SIM anymore. Emergency calls are always allowed. If, on the other hand, the mobile equipment supports FDN, then the phone will recover EF_{IMSI} and EF_{LOCI}, and will become operational again. Now it is up to the phone to see if the SIM has the fixed dialing number restriction activated or not. Figure 8.8 also shows the case of a Phase 1 SIM in which the phone (either Phase 1 or Phase 2) directly enters unrestricted operation because FDN has not been specified for Phase 1.

The subscriber can deactivate FDN by using CHV2 (PIN2), or the operator of the phone can do so by identifying himself with CHV1 (PIN1). If FDN is used, then the ADN field will be invalidated and not accessible. The reason for this is that all calls are restricted to the FDN phone book. Another exception in shown in Figure 8.8; if ADN is not allocated and not activated on the SIM, then it is not possible to disable FDN. This might be the case for rental SIMs where the rental company does not want anyone to store phone numbers on the rented SIM.

If FDN is disabled, then knowledge of CHV2 (PIN2) is sufficient to enable it again. Figure 8.8 shows a simple flowchart for the operation of a FDN SIM, which includes initializing, disabling, and enabling FDN.

Short-messages are stored in the *short messages* (EF_{SMS}) field. The content of this field depends on the content of a short message as described in

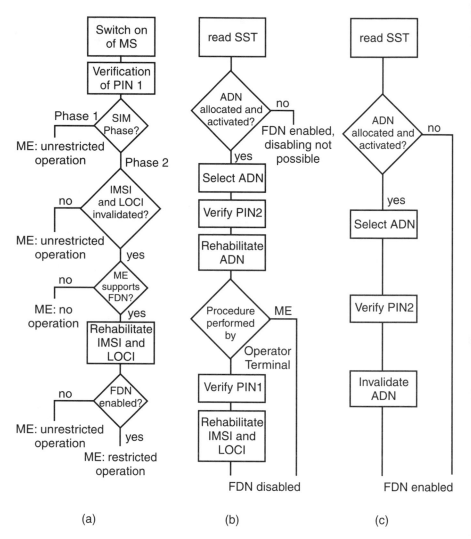

Figure 8.8 Basic operation for fixed dialing number SIM:
(a) initialization of FDN SIM card, (b) disabling of FDN, and
(c) enabling of FDN.

Chapter 6. The number of short messages that can be stored is determined
by the operator, who specifies the allocation of memory for various features of a SIM. A fixed amount of memory space is allocated for each message regardless of the actual size of a short message.

The *short message service parameter* (EF_{SMSP}) contains parameters
required for transmission of short messages. It can be preconfigured by

the network with the correct service center number, and other parameters common to each message. For Phase 1, the user of the SMS had to enter all of these parameters and had to find the various parameters and their meanings; then she had to work through the menu items and codes to store the parameters. GSM Phase 2 removes the burden of drudgery from the subscriber who wants to use short messaging. Five service parameters are stored: (1) the destination address, (2) the service center address, (3) the protocol identifier, (4), the data coding scheme, and (5) the validity period. (All of these parameters are described in Chapter 6.) It is also possible to store multiple sets of parameters, which is useful in networks with more than one SMSC. More and more service providers, including value-added service suppliers, offer SMS in connection with value-added services and have their own SMS center to support them. The user can choose a different service center for each application. The different sets of parameters are identified with a dedicated name (alpha tag).

The *SMS status* (EF_{SMSS}) holds parameters used for status messages in short message services exchanged between the mobile station and the short message service center. Two parameters have been specified for Phase 2:

1. The *message reference* is sent with each short message and enables the phone, or the SIM, to allocate status messages for certain messages coming from the service center.

2. The *memory capacity exceeded* flag allows the mobile station to assist the network in recovering from a situation where no memory is available on the SIM, or in the phone's memory, for short message storage. The user gets an indication on the phone's display of a waiting message, but is obliged to clear memory before the message can be stored. Because the network does not know when memory is cleared, the mobile phone will notify the network when this flag is cleared and memory has become available again.

The *capability configuration parameters* (EF_{CCP}) field keeps data on required network and bearer capabilities associated with a call established using ADN, FDN, MSISDN, or LND.

The *mobile station international ISDN number* (EF_{MSISDN}) field contains the phone number(s) of the subscriber within the GSM network. It is possible to specify up to 255 different phone numbers for one subscription, which is more than are likely to be needed. The operator specifies the number of MSISDNs to be stored on the SIM. The average allocation is between two and five numbers depending on the application (e.g., one phone number for voice, and another phone number for fax and data). For GSM Phase 1, the "own phone number" was stored by the user himself in either a dedicated field in the phone, or on the SIM in the ADN field. Either of these allows only one entry: a single phone number for voice service. Because many networks use different phone numbers for voice, fax, and data services, the Phase 2 SIM allows multiple MSISDNs to be stored together with a text string to identify the purpose of the number, which can be preprogrammed by the operator or service provider. The reader may question why the phone number should be stored on the SIM rather than in the phone. There are two reasons. First, we are all human and tend to forget numbers, especially if we have different numbers for voice, data, and fax services. We can recall the numbers into view at any time. The second is to allow the network operator to link the SIM card (IMSI) to a phone number when the SIM is personalized (programming of the SIM's elementary files) for a new subscriber. The old alternative was to hand the user a sheet of paper listing the numbers with their assigned services.

The *last number dialed* (EF_{LND}) contains a list of the last numbers that were dialed using the SIM, usually between 5 and 10 numbers. This feature is already known from conventional mobile phones, which supported it by using the phone's memory. Now the SIM provides dedicated memory for storing these numbers. The advantage of this new approach is that the last number dialed (your home phone number in fixed network, for example) will be displayed as digits together with a name attachment ("HOME SWEET HOME"). The name tag appears if the number was previously stored in the ADN field or the FDN field. The numbers will still be available when the subscriber uses the SIM with another phone, perhaps one rented at a train station. There can be a difference in the length of the text field reserved for the name associated with a phone number. If the field for LND is longer than those reserved for ADN and FDN, then the phone should add blank spaces to fill the difference. If the LND field is shorter, then the ME will simply truncate the names associated with the phone numbers.

The *extension 1* (EF$_{EXT1}$) field contains extension data of ADN, MSISDN, or LND. The extensions accommodate values longer than 20 digits. The *extension 2* (EF$_{EXT2}$) field contains extension data of a FDN, which is for values longer than 20 digits.

8.5 Phase 2+ SIM

GSM is an evolving standard, into which different working groups continue to specify new features. The additions have consequences for the SIM definition. Because Phase 2 is frozen, new features are included in the Phase 2+ specifications [4]. In the following pages we list and describe some new features and fields (listed later in Table 8.1) that will become available to GSM users upon introduction of this third-generation SIM.

The *SIM service table* (EF$_{SST}$), already known from Phase 1 and Phase 2, is updated with the new features that are supported by the SIM. Therefore, the SIM service table becomes larger with each new phase. Phase 2+ adds these features:

▶ *Group identifier level 1 and 2* (GID1 and GID2);

▶ *Service provider name* (SPN);

▶ *Service dialing numbers* (SDN);

▶ *Extension 3* (EXT3);

▶ *Voice group call service and group identifier list* (EF$_{VGCS}$ and EF$_{VGCSS}$);

▶ *Voice broadcast service group identifier list* (EF$_{VBS}$ and EF$_{VBSS}$);

▶ *Enhanced multilevel precedence and preemption service* (eMLPP);

▶ *Automatic answer for eMLPP* (AAeM);

▶ *Data download via SMS-CB and SMS PP* (related to SIM application toolkit);

▶ *Menu selection* (related to SIM application toolkit);

▶ *Call control* (related to SIM application toolkit);

▶ *Proactive SIM* (related to SIM application toolkit);

▶ *Cell broadcast message identifier ranges* (CBMIR);

▶ *Barred dialing numbers* (BDN);

▶ *Extension 4* (EXT4);

▶ *Depersonalization control keys* (DCK);

▶ *Cooperative network list* (CNL).

These features are explained in the following paragraphs, except for the features related to the SIM application toolkit, which are deferred to Chapter 9. The emergency call codes, which are also described here, are an additional elementary file for a Phase 2+ SIM.

The *group identifier level 1 and 2* (EF_{GID1} and EF_{GID2}) identifies different group codes in order to lock a phone to certain network operators, service providers, or individual companies. The function of the group identifier is commonly known as *SIM lock*; its purpose is subsidy protection. In countries where network operators finance the purchase of a mobile phone with a subsidy, those who furnish these phones to subscribers want to make sure that they only operate in their networks and generate revenue (air time) there. One way to do this is to allow a particular phone to work only with a particular SIM, which was issued by the operator who subsidized the phone. The GID fields can further restrict users to service providers, or even more narrowly to individual groups or companies.

The standard SIM lock feature, which existed before there were GID fields, was enabled by any manufacturer who simply stored the MCC and MNC somewhere in the phone. This early SIM lock was rather primitive because it only took the parameters of the IMSI into account in order to identify a certain operator. When a SIM from a different operator was inserted into such a phone, the user had to enter a dedicated "SIM lock" code, which was different for each individual phone.

This mechanism had the disadvantage that it was only applicable as a network operator lock, thus protecting the investment (subsidy) of the operator. If a service provider wanted to add something to the subsidy he had no way to protect his investment. Now, the GID fields make it possible to store additional identifiers in the phone and SIM, which can add additional restrictions beyond the operator identity. If the phone and SIM entries do not match, then the phone will request a specific code as mentioned in the previous paragraph.

Some manufacturers introduced an "antitheft" type of SIM lock, which allows a subscriber to lock the phone to his SIM card identified by the SIM's serial number or the IMSI. When a different SIM card is inserted into the mobile phone the user has to identify himself with a subscriber-specific "phone code," which explains why it is also referred to as antitheft SIM lock: whenever a phone is stolen the SIM card will be blocked by the operator so that the only way the thief can use the phone is to insert another SIM card. But this will not help the thief because the phone stubbornly requests the phone code when a different card is inserted. The difference between the SIM lock and antitheft schemes is that the SIM lock route protects the subsidy the operator or service provider contributed to the phone and the antitheft type renders the phone unusable in the hands of someone who steals it without also stealing the SIM card.

The *service provider name* (EF_{SPN}) field contains a text string that holds the service provider's name. In GSM Phase 1 and 2, the GSM terminals display only the names of networks on which they are camped. The service provider—the company who may issue a card and who is normally responsible for all of the customer's accounting and billing, and who may even offer popular value-added services that other competing service providers or network operators cannot supply—was never recognized as such on the phone's display. To build a stronger service provider brand image, it is now possible to add the SP name. The mechanism is that the service provider's name always appears on the phone's display even as the network names change during roaming. Figure 8.9 shows how a phone's display would appear with and without the SPN, which is "SUPERTALK."

The *voice group call service* (EF_{VGCS}) is a new service that allows railways and other closed user groups to make a group call to all the members of a group or to a subset of the members. The group call allows each member of a group to communicate with each other, similar to a multiparty connection. Unlike the normal multiparty connection, the number of participants the originator of a group call can add to the connections is not limited, and the access on the radio channel is different than for normal voice calls. (See Section 9.3 for more details on railway applications.) This field contains group identifiers for the voice group call service to which the subscriber has subscribed. The *voice group call service status* (EF_{VGCSS}) contains the activation status of the different voice group identities, which are either active (access) or not active (no access).

The *voice broadcast service* (EF_{VBS}) is a service similar to SMS cell broadcast, but based on voice. This field contains group identifiers of services to

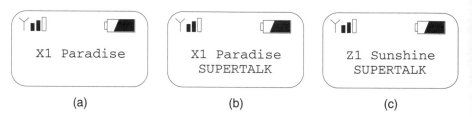

Figure 8.9 An MS display (a) without SPN, (b) with SPN ("X1 paradise" = home network), and (c) with SPN while roaming in "Z1 Sunshine."

which the subscriber has subscribed. This service is related to railway and other closed user groups, which need the ability to make broadcast calls to all the members of a group (announcements). The service mimics the open channel group calls in PMR trunked radio systems. In a manner similar to cell broadcast, voice broadcast is a single direction communication; the person setting up the call can make an announcement to everyone, but can receive no immediate responses. This may be a disadvantage to multiparty calls, but it enjoys the advantage of faster call setup and information transfer; announcements do not generally require responses anyway. The *voice broadcast service status* field (EF_{VBSS}) contains the activation status for the different groups of the voice broadcast service.

The *enhanced multilevel precedence and preemption* field (EF_{eMLPP}) allows the user to subscribe to a priority level for his outgoing calls and a preemption level for incoming calls. This field contains the priority levels a user has subscribed to, and the fast call setup parameters for the different priority levels (see Chapter 9). The *automatic answer of eMLPP service* (EF_{AAeM}) specifies the level of the eMLPP service for which the mobile station should automatically answer incoming calls.

The *emergency call codes* (EF_{ECC}) field contains emergency codes that should be used in conjunction with a SIM. Each emergency call code has up to six digits. The reason for this is that different countries use different codes for emergency calls in their fixed networks (e.g., 911 in the United States and 110 in Germany). Network operators tend to adopt the fixed network emergency call code for their GSM networks. The DCS 1800 operators in the United Kingdom already implement this in the mobile's software (emergency code 999). Providing an elementary file on the SIM card will allow a lower cost and more flexible implementation for mobile manufacturers and operators.

The *cell broadcast message identifier for data download* (EF_{CBMID}) specifies from which cell broadcast channels the content should be passed on to the SIM's memory. This field is part of the SIM application toolkit, which is described in Chapter 9.

The *cell broadcast message identifier range selection* (EFU_{CBMIR}) stores ranges of cell broadcast channels rather than individual numbers, which may be stored in the cell broadcast message identifier selection field. The advantage is that it is much easier to select a complete group of channels than to key them all in one at a time. If, for example, the movie programs of the theaters in a city are transmitted on individual channels, a user can select only one theater to get information on what's showing in the theater around the corner. If another user wishes to get an overview of the city's entire cinematic possibilities, he would have to key in the channels for all the theaters. If the subscriber is able to select a range of messages, identified by the lowest and the highest numbers, it takes less time to make the selections, and less memory space is consumed in the SIM.

The *depersonalization control keys* (EF_{DCK}) contain the keys related to depersonalization of a phone. As we already described in the paragraphs on the GID fields, a mobile phone may be personalized to SIM cards issued by a network operator, a service provider, or a company (group). These DCK fields hold the depersonalization keys for all of these locks. The purpose of storing the keys on the SIM is to relieve the phone manufacturers of generating them in the phone's software, and give those who issue SIMs the freedom to alter the keys. A detailed description of the personalization process can be found in [5].

The *cooperative network list* (EF_{CNL}) contains the list of networks that cooperate in the provision of services and who prefer to lock each other's subscribers to their networks. The field allows the MCC and the MNC of cooperating networks to be stored in a SIM, thus making a phone operational with SIM cards from multiple networks. This field is used during personalization of mobile equipment in order to lock it to all the networks stored in the list.

The *service dialing numbers* field (EF_{SDN}) is a type of a phone book, and has the same structure and contents as described in the earlier section on the ADN field. The service dialing numbers are stored by the operator, and often include useful phone numbers such as (1) phone number inquiry (directory) service, (2) help desk service, (3) a travel service, (4) traffic advisories, and (5) a weather alert service. In contrast to ADN and FDN, the user cannot update or change the numbers in this field. For Phase 1 and 2 SIMs, these service numbers are simply stored in the ADN area of the SIM, leaving it up to the user to keep the numbers stored in the phone book; the user can simply erase these numbers and use the space for other numbers. The use of SDN in a Phase 2+ SIM adds permanence to the numbers stored in the field. It is possible for manufacturers to create special operator-branded phones that have dedicated access keys for the SDN. When the special keys are pressed, a complete list of service dialing numbers may appear. The network operator can provide real value-added services for his customers, or he can even create special supplementary services through access numbers stored in this SDN field. The user will appreciate meaningful names attached to each number and service.

Extension 3 contains extension data for a SDN, which is provided for values longer than 20 digits. *Extension 4* contains extension data for a BDN to accommodate values longer than 20 digits.

The *barred dialing numbers* (EF_{BDN}) field follows the phone book structure of EF_{ADN} and allows the subscriber to bar certain numbers from being dialed with a SIM. This is the reverse function of FDN, where the user is only allowed to dial numbers from the list stored in the FDN file. The structure of the BDN elementary file and its control mechanism are similar to FDN.

Table 8.1 compares the content of the SIM for different phases of implementation. Phase 1 and Phase 2 are fixed, whereas items and features are still being added to Phase 2+. The large number of optional fields reveals the freedom of choice individual operators enjoy in personalizing a SIM for a specific use within their networks.

Table 8.1
Content of a SIM for Various Phases of Implementation

SIM Elementary Files (name of EF in parentheses)	Phase 1	Phase 2	Phase 2+
ICC identification (ICCID)	M	M	M
Language preference (LP)	—	M	M
IMSI (IMSI)	M	M	M
Ciphering key (K_c)	M	M	M
PLMN selector (PLMNsel)	O	O	O
HPLMN search (HPLMN)	—	M	M

SIM Elementary Files (name of EF in parentheses)	Phase 1	Phase 2	Phase 2+
SIM service table (SST)	M	M	M
Accumulated call meter (ACM)	—	O	O
Accumulated call meter maximum value (ACMmax)	—	O	O
Price per unit and currency table (PUCT)	—	O	O
Cell broadcast message identifier selection (CBMI)	—	O	O
Broadcast control channels (BCCH)	M	M	M
Access control class (ACC)	M	M	M
Forbidden PLMNs (FPLMN)	M	M	M
Location information (LOCI)	M	M	M
Administrative data (AD)	M	M	M
Phase identification (Phase)	—	M	M
Group identifier level 1 (GID1)	—	O	O
Group identifier level 2 (GID2)	—	O	O
Service provider name (SPN)	—	—	O
Voice group call service (VGCS)	—	—	O
Voice group call service status (VGCSS)	—	—	O
Voice broadcast service (VBS)	—	—	O
Voice broadcast service status (VBSS)	—	—	O
Enhanced multilevel preemption and priority (eMLPP)	—	—	O
Automatic answer for eMLPP service (AAeM)	—	—	O
Emergency call codes (ECC)	—	—	O
Cell broadcast message identifier for data download (CBMID)	—	—	O
Cell broadcast message identifier range selection (CBMIR)	—	—	O
Depersonalization control keys (DCK)	—	—	O
Cooperative network list (CNL)	—	—	O
Abbreviated dialing numbers (ADN)	O	O	O
Fixed dialing numbers (FDN)	—	O	O
Short messages (SMS)	O	O	O
Short message service parameters (SMSP)	—	O	O
SMS status (SMSS)	—	O	O
Capability configuration parameter (CCP)	O	O	O
MSISDN (MSISDN)	—	O	O
Last number dialed (LND)	—	O	O
Extension 1 (EXT1)	—	O	O

Table 8.1 (continued)

SIM Elementary Files (name of EF in parentheses)	Phase 1	Phase 2	Phase 2+
Extension 2 (EXT2)	—	O	O
Service dialing numbers (SDN)	—	—	O
Extension 3 (EXT3)	—	—	O
Barred dialing numbers (BDN)	—	—	O
Extension 4 (EXT4)	—	—	O
M: mandatory; O: optional; —: not present			

Table 8.2 gives an overview of how elementary files contained on the SIM are accessed. Some files can always be accessed, and some only after PIN1 or PIN2 has been entered. The normal PIN entry is valid throughout the time period the phone is switched on. Similar restrictions apply to updating, invalidating, and recovering (rehabilitating) fields. Many fields are left to network or administrative control.

Table 8.2
Access to Different Fields on the SIM

SIM Elementary Files (name of EF in parentheses)	Protection of Elementary File			
	Read	Update	Invalidate	Rehabilitate
ICC identification (ICCID)	ALW	NEVER	ADM	ADM
Language preference (LP)	ALW	CHV1	ADM	ADM
IMSI (IMSI)	CHV1	ADM	ADM	CHV1
Ciphering key (K_c)	CHV1	CHV1	ADM	ADM
PLMN selector (PLMNsel)	CHV1	CHV1	ADM	ADM
HPLMN search (HPLMN)	CHV1	ADM	ADM	ADM
SIM service table (SST)	CHV1	ADM	ADM	ADM
Accumulated call meter (ACM)	CHV1	CHV1/ CHV2 [1]	ADM	ADM
Accumulated call meter maximum value (ACMmax)	CHV1	CHV1/ CHV2 [1]	ADM	ADM
Price per unit and currency table (PUCT)	CHV1	CHV1/ CHV2 [1]	ADM	ADM
Cell broadcast message identifier selection (CBMI)	CHV1	CHV1	ADM	ADM
Broadcast control channels (BCCH)	CHV1	CHV1	ADM	ADM

SIM Elementary Files (name of EF in parentheses)	Protection of Elementary File			
	Read	Update	Invalidate	Rehabilitate
Access control class (ACC)	CHV1	ADM	ADM	ADM
Forbidden PLMNs (FPLMN)	CHV1	CHV1	ADM	ADM
Location information (LOCI)	CHV1	CHV1	ADM	CHV1
Administrative data (AD)	ALW	ADM	ADM	ADM
Phase identification (Phase)	ALW	ADM	ADM	ADM
Group identifier level 1 (GID1)	CHV1	ADM	ADM	ADM
Group identifier level 2 (GID2)	CHV1	ADM	ADM	ADM
Service provider name (SPN)	ALW	ADM	ADM	ADM
Voice group call service (VGCS)	CHV1	ADM	ADM	ADM
Voice group call service status (VGCSS)	CHV1	ADM	ADM	ADM
Voice broadcast service (VBS)	CHV1	ADM	ADM	ADM
Voice broadcast service status (VBSS)	CHV1	ADM	ADM	ADM
Enhanced multilevel preemption and priority (eMLPP)	CHV1	ADM	ADM	ADM
Automatic answer for eMLPP service (AAeM)	CHV1	CHV1	ADM	ADM
Emergency call codes (ECC)	ALW	ADM	ADM	ADM
Cell broadcast message identifier for data download (CBMID)	CHV1	ADM	ADM	ADM
Cell broadcast message identifier range selection (CBMIR)	CHV1	CHV1	ADM	ADM
Depersonalization control keys (DCK)	CHV1	ADM	ADM	ADM
Cooperative network list (CNL)	CHV1	ADM	ADM	ADM
Abbreviated dialing numbers (ADN)	CHV1	CHV1	CHV2	CHV2
Fixed dialing numbers (FDN)	CHV1	CHV2	ADM	ADM
Short messages (SMS)	CHV1	CHV1	ADM	ADM
Short message service parameters (SMSP)	CHV1	CHV1	ADM	ADM
SMS status (SMSS)	CHV1	CHV1	ADM	ADM
Capability configuration parameter (CCP)	CHV1	CHV1	ADM	ADM
MSISDN (MSISDN)	CHV1	CHV1	ADM	ADM
Last number dialed (LND)	CHV1	CHV1	ADM	ADM
Extension 1 (EXT1)	CHV1	CHV1	ADM	ADM

Table 8.2 (continued)

SIM Elementary Files (name of EF in parentheses)	Protection of Elementary File			
	Read	Update	Invalidate	Rehabilitate
Extension 2 (EXT2)	CHV1	CHV2	ADM	ADM
Service dialing numbers (SDN)	CHV1	ADM	ADM	ADM
Extension 3 (EXT3)	CHV1	ADM	ADM	ADM
Barred dialing numbers (BDN)	CHV1	CHV2	CHV2	CHV2
Extension 4 (EXT4)	CHV1	CHV2	ADM	ADM
ADM = administrative; ALW = always; CHV1/2 = PIN1/2				
1 This is fixed during SIM personalization				

8.6   The SIM initialization process

By the time this book is published, most new SIMs and ME will conform to GSM Phase 2. We will, therefore, confine our discussions to the SIM initialization for GSM Phase 2. It is clear that the difference between GSM Phase 1 and GSM Phase 2 SIMs is in the number of elementary files. Anyone initializing a Phase 1 SIM ignores the files that only appear in Phase 2.

Once the ME is switched on, it supplies voltage and a clock signal to the SIM, which is activated when these inputs are present. The ME selects (accesses) the GSM dedicated files (EF_{GSM}) and reads the language preference (EF_{LP}). If this elementary file is not present, then the ME assumes the SIM is a Phase 1 SIM. If the ME does not support the language indicated in the elementary file (EF_{LP}) it will select a default language, for example, English.

If the card holder verification 1 (CHV1, PIN1) procedure has not been switched off in a previous session, then the ME prompts the request for PIN1 and continues only when the correct PIN is finally entered. If the PIN was blocked in an earlier session, then the ME demands the PUK. Ten wrong entries of the PUK will completely block the SIM.

When a Phase 2 SIM supports FDN or BDN, and these have been activated, the ME should follow the FDN/BDN restrictions. If the ME does not support FDN and it has been activated on the SIM, then the ME will fall into a state that only allows emergency calls. Finally, if the ME does not support FDN and FDN has not been activated, then the ME will go into normal operation. The ME requests the following information from the SIM during normal operation:

- Administrative information;

- SIM service table;

- IMSI request;

- Access control;

- HPLMN search period;

- Location information;

- Cipher key;

- BCCH information;

- Forbidden PLMN.

Loaded with this information, the mobile station is ready for full GSM operation. It is useful, but an option left to the manufacturers of mobile equipment to read the phone books (ADN, FDN, SDN, BDN) into the ME's memory (RAM) right after the initialization process. Because SIM memory access is much slower than ME memory access, this method spares the user the long wait direct access to the SIM's phone books would entail.

8.7   Electrical characteristics of the SIM

Having explored the software aspects of the SIM, we turn our attention to some electrical and hardware characteristics.

8.7.1   SIM Power Supply

Beyond second- and third-generation GSM terminals there is a lot of talk about 3V phones. What does this mean for the SIM? The latest generation silicon technologies enable most functions in a GSM terminal and the SIM, and they are based on lower operating voltages. The steady fall in the required operating voltages follows a continuous decrease in the size of the component cells (transistors) on the wafers. The benefits of shrinking the cell (transistor) size, and the operating voltage, are widely accepted: smaller area (more transistors and functions per square millimeter), less power consumption, and lower cost.

Normal Phase 1 SIMs and first-generation mobile equipment are based on 5V silicon technology, whereas Phase 2 SIMs first appeared with 3V technology. A 3V specification normally includes a 10% tolerance so that the operating range is actually somewhere between 2.7V and 3.3V. The new 3V SIMs can, however, also work with higher supplies up to 5V, thus making them compatible with ME working from a nominal 5V supply; the industry will not accept a 3V SIM that dies when it is installed in a 5V terminal. The typical user cannot be expected to know the technical and operating restrictions of SIMs and mobile stations.

This compatibility issue has some impact for mobile phone manufacturers. Although both types of SIMs are still on the market, a phone manufacturer might choose to support only 5V SIMs. Restricting support to 5V SIMs avoids the need for pull-up resistors, which use space and consume power. When using 3V silicon technology in the ME together with a 5V SIM, the ME's chips have to be able to accept the higher input voltage without damage, and accommodate pull-up resistors in order to interface the ME's output to the 5V SIM.

8.7.2 SIM memory

In the early days of GSM it was sufficient to have 2 kilobytes (KB) of memory[3] (EEPROM) on the SIM, but with all the new elementary files (and especially the different phone books) this size has long since failed to be sufficient. In 1996 8-KB SIMs were the industry standard, whereas in 1997 more and more chip card manufacturers moved towards 16-KB SIMs.

To give the reader some idea of the SIM's memory requirements, the typical sizes, in bytes, for some elementary files are listed in Table 8.3. Note that the table only covers fields that have been used in GSM for Phase 2 and that, therefore, there are no entries for, say, service dialing numbers. The table includes the maximum and typical sizes for most files. The operator can specify her own field sizes within the constraints listed

---

3. Unless otherwise stated, *memory* in this chapter refers to "user" memory. The operating system is stored in another place.

in Table 8.3. Fields with fixed sizes are listed with only one value. The memory size indicated in the "Memory Required" column takes into account the space for the elementary file's header. Table 8.3 should only be used as an indication of the SIM's memory needs, because the dedicated files also require some bytes of memory.

Table 8.3
Typical Sizes of SIM Memory Fields (in bytes)

SIM Field	Maximum Size	Typical Size	Memory Required
PIN1/PIN2	1	1	168
ICCID	1	1	32
LP	255	4	20
IMSI	255	1	28
K_c	1	1	40
PLMNsel	255	8	40
HPLMNsearch	1	1	20
ACMmax	1	1	20
SST	1	1	20
ACM	1	1	24
PUCT	1	1	24
CBMI	255	4	24
BCCH	1	1	32
ACC	1	1	20
FPLMN	1	1	28
LOCI	1	1	28
AD	1	1	20
Phase	1	1	20
ADN	255	100	2,416
FDN	255	20	496
SMS	255	10	1,776
CCP	255	3	60
MSISDN	255	2	64
SMSP	255	3	132
SMSS	1	1	20
LND	255	10	272
EXT1	255	5	84
EXT2	255	2	44
Total			5,972 bytes

We have seen that there are three directories in the SIM, and that each of them contains elementary files, which are fields. The purposes of the different fields were reviewed, and their relevance in the various phases of GSM were listed in this chapter. As a review, the complete file structure is summarized in Figure 8.10, in which we see the complete organization of the Phase 2+ SIM.

8.7.3 SIM architecture

The SIM holds some other functions on its silicon outside the user memory discussed in all the earlier sections of this chapter. In this section we take a look at the nonuser utility functions of the SIM. Let's start with Figure 8.11, which gives an overview of the entities on a SIM's chip, and which some readers will recognize as the block diagram of a typical computer.

The *central processing unit* (CPU) is a small microcomputer that performs the myriad routine processes typical of any computer, as well as

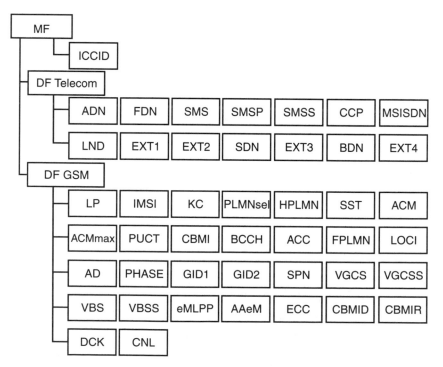

Figure 8.10   Organization of SIM directories.

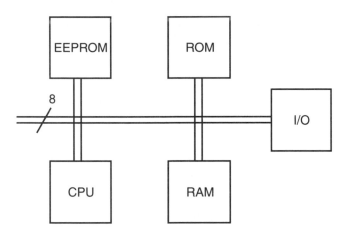

Figure 8.11 Internal structure of a SIM chip.

some specific tasks more visible to the world outside the SIM, such as cal-culating the authentication keys. In future applications, such as the SIM application toolkit, which is presented in Chapter 9, the CPU will have even more work to do.

The *read-only memory* (ROM) contains the basic operating system for the SIM, which is different for each manufacturer of SIM cards. The size of the ROM depends on the chip supplier but was, in 1996, between 16 and 24 KB and is expected to grow to around 64 KB by the year 2000. There is no such thing as a standard SIM operating system, such as the MS-DOS or Windows 95 systems we see in the world of the PC. The operating system performs the tasks the SIM demands: communications with the mobile equipment and manipulating data read from the elementary files (fields) in the SIM's electrically erasable and programmable read-only memory.

The EEPROM contains the actual user data, which includes all the elementary files that we have looked at in previous sections. In 1997 the SIM's EEPROM size was about 16 KB, and by the year 2000 we expect to see user memory sizes of around 64 KB.

The *random access memory* (RAM) is the actual working memory for the CPU. Upon initialization, the CPU loads the operating system into the RAM and starts working on its assigned tasks. In 1997 the typical RAM size was between 256 and 512 bytes, but it can be expected to grow up to 1 or 2 KB by the year 2000.

Besides these typical entities we find a communications bus over which data are exchanged. There is also a type of *input/output* (I/O) circuit that handles the communications between the SIM and the mobile equipment. The internal bus is typically 8 bits wide, and the communication to the outside world is confined to a serial bus, thus limiting the number of contacts on the SIM.

The SIM, itself, has between six and eight pins, which are always visible as a small group of metallic pads in the SIM's surface. Only five pins are required and defined as shown in Figure 8.12. We can identify the voltage supply and its return (VCC and GND), the clock (CLK), which drives the CPU, the serial I/O port, and the reset (RST) pin.

8.8 Outlook for future applications

It should be evident by now that the SIM is a more powerful device than its tiny size would indicate; it is more than some memory with an I/O port. In Chapter 9 (on the SIM application toolkit) we will see how the GSM community can make rather extreme use of intelligence on the SIM. Because these advanced applications will take some time to implement, some SIM manufacturers and network operators are offering their own interim solutions, which are already being marketed. The advantage of all these temporary solutions is that they are compatible with standard mobile phones as long as they support some basic GSM features such as SMS-MT. These solutions are, however, only intermediate step toward

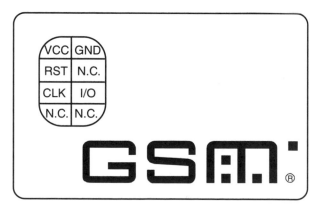

Figure 8.12   Contacts on a SIM.

the SIM application toolkit. All of the features described in the remaining sections are going to be incorporated in the application toolkit.

We have selected two examples rather than a review of the complete list of possible advanced applications. The first is an example of how the SIM can handle enhancements to value-added services, and the second is the prepaid SIM. Other developments not mentioned here are sure to change before they finally mature; there are plenty of market forces behind GSM. When we look at the SIM application toolkit, we will see that it is an elaborate but logical extension of the temporary enhancements our two examples illustrate.

8.8.1 NATELsicap by Swisscom

NATELsicap (SIM card application platform) [6] is the Swisscom approach to introducing more value-added services to the GSM system and the mobile telephone. It only requires changes to two system entities, both of which are controlled by the operator:

1. The *short message services–mobile terminated* are modified in order to support an additional header, which identifies a SMS-MT as a Sicap message. This not only has an influence on the message structure, but also on the short message service center that has to prepare the Sicap messages.

2. The SIM, which normally stores short messages, must be a "Sicap" SIM, because it has to check each individual incoming short message: Is it a "normal" short message, which can be treated as such and stored in the SIM memory in the usual way? Or is it a Sicap short message that requires some alternative action?

Figure 8.13 shows the difference between a normal short message and a Sicap short message.

Different applications and value-added services are possible on this platform:

▶ Modification of the subscriber options, which is done by writing to the SIM service table—the actual modification is done through a SMS-MT;

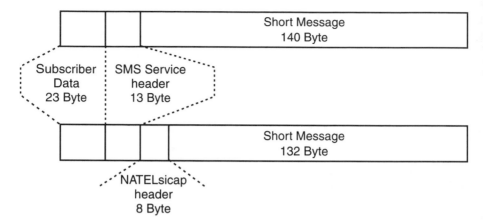

Figure 8.13   Difference between normal and Sicap short message.

▶ Location dependent services such as traffic information;

▶ Teleshopping, where the Sicap SIM provides a secure means to transfer information about bank accounts;

▶ Customer-tailored services for airline or train ticketing;

▶ Prepaid card with the possibility to recharge the card through special messaging. A Sicap short message can modify the ACM field on the SIM, and the user can continue to place phone calls once the field change takes effect.

8.8.2   Prepaid SIM

Today, more operators are approaching their markets with the new prepaid SIM card. This concept is already established in many European GSM markets, such as Portugal, Spain, the United Kingdom, Italy, Switzerland, Germany, and France. Judging from its early success, we can assume that by the time this book is published many more operators will offer this kind of service.

The principle of this service is that a customer buys a SIM card with a fixed amount of air time already stored on the card. When the initial air time is used up, the customer is notified with a SMS, or a voice call, from the operator that his account is used up and it should be refilled. Depending on the operator's implementation, the subscriber will be able to recharge his account and continue making phone calls. This concept is very attractive for both the operator and the subscriber. The operator has the advantage that he only sells air time for which the subscriber has already paid. The advantage for the customer is that prepaid SIM cards do not normally have a monthly subscription fee, and he only pays for the air time he actually uses. Even if the air time is used up, a prepaid subscriber is usually entitled to receive phone calls for a limited period of time. This kind of package is very attractive to customers who are very price conscious and need cellular phones more as an emergency tool than for normal communication, or who cannot otherwise enjoy cellular phone service because of credit problems. The operator hopes that prepaid customers will become accustomed to mobile telephony, and that he will eventually use an upgrade path to permanently subscribe to cellular service. In the short term, prepaid users are the *best* users: For all prepaid services available in the beginning of 1997, there were no international roaming and no international calls; all the revenue went to the mobile network. With increased use of prepaid SIM cards, and the necessary network support, these service restrictions will probably be overcome. Then it is only a question of time before this temporary advantage for the operator disappears. There are two aspects to prepaid cards that we should explore: (1) Where is the charging information handled? (2) How can the customer recharge his card?

8.8.2.1 Handling of charging information

As the name implies, the customer prepays or deposits money into an account with the operator with which he pays for his phone calls. The account must be handled somewhere in the network or in the phone. The GSM system allows two options.

The first option is to have credit information stored on the SIM in the specific charging files. This makes full use of the SIM and the AoC supplementary services. This method limits the service to phones that support AoC. The network would, at each call setup, provide the phone with the charging information, which the phone would have to confirm. When the accumulated charge reaches the maximum value, the phone would not allow any further outgoing calls that are subject to a charge. The customer would still be able to make a phone call to a service that can recharge the card (see Section 8.8.2.2). When the network receives payment, or proof of payment, it issues a dedicated SMS to the SIM telling it to change the accumulated call meter by the amount that has been paid by the user.

In the second option, credit information can be stored in the network's billing center. The advantage is that this does not require specific fields on the SIM, and no specific functionality in the phone. However, the effort in the network is much greater because it requires real-time billing of these customers. This option does not give the user instant access to the remaining air time or charges in his account. The network will, therefore, provide a means for the customer to investigate his current charge level. This could be done by either sending a specific SMS to the network, which interrogates the billing center, or by working through a voice server. If his account is used up, then the customer would get notification from the network to settle his account. When the customer deposits additional money into his account the operator is obliged to update immediately the information in the billing center.

Summarizing the two possibilities, we should mention that the method of storing charging information on the SIM is becoming out of date. Nowadays more and more intelligence is put into the network, where it is under the sole influence and access of the operator. Storing charging information on the SIM card is too susceptible to fraud, even though a SIM card is supposed to be secure. On the other hand, this also limits prepaid applications to phones supporting AoC, which is also not in the interest of the operator.

8.8.2.2   Recharging of prepaid SIM cards

The recharging mechanism gives the operator a great deal of freedom and lots of options. The appropriate way to recharge prepaid SIMs depends on the lifestyle of the typical subscriber, and is tempered by the culture and social practice in the country. Because there are many ways to transfer money we present just a few methods and leave the reader to her own creativity for more possibilities:

1. *Bank deposit:* When the customer realizes that his account is used up he will transfer money to the operator indicating his phone number or his customer number.

2. *Automatic teller machine (ATM) recharging:* If the information is stored on the SIM, then the customer could insert his SIM into a

dedicated teller machine, perhaps located in banks or post offices. He would then insert money into the machine, which would increase his account on the SIM.

3. *Phone card:* Some operators introduced the concept of selling "secret numbers" to their customers. A customer would buy a phone card that has a rubber field stamped on it. When the rubber is peeled away a secret number appears, which has the cash value coded into it. The customer is able to pass this number to the network operator with a short message or through a voice server.

4. *Deduction from bank account:* The first three options are not particularly elegant for the customer, but they allow him to remain unknown to the operator. If the user's bank account is known to the operator, then the customer could issue a request to the operator via a short message, or through a voice server, to credit his prepaid mobile account directly from his bank.

8.8.3   Future parameters

Before we end our chapter on the SIM we need to remind ourselves that, just like all the other features of GSM, the SIM card is still going through new modifications. We want to introduce these newest changes briefly in this chapter. Two general areas can be distinguished.

The first is the addition of new elementary files required to support new GSM features, where the parameters are stored on the SIM rather than in the phone. Adding fields to the SIM rather than to the phone is a consistent theme in GSM enhancements. The SIM method allows portability from one phone to another, which is useful for international travelers and phone exchanges; it leaves a considerable measure of control with the network operator or the service provider. All of the known improvements have already been listed in Section 8.2.5, but we can expect even more elements will be proposed and made available on the SIM.

Because the amount of data transferred between the mobile equipment and SIM is increasing in Phase 2+, there is great interest in increasing the speed of the serial interface between SIM and mobile equipment. The work on this is referred to as the *SIM interface data transfer speed enhancement.*

The second area of change is the exhaustive search for new applications around the SIM card's ability to enable value-added services. One big example is described for the SIM application toolkit. Another example is in the area of *universal personal telecommunication* (UPT), which puts emphasis on using a SIM for other card applications outside of GSM, such as a normal pay phone card, where the user can have alternative access to his phone book stored on the SIM [7].

References

[1]   Zoreda, J. L., and J. M. Otón, "Old Cards and New Cards," in *Smart Cards*, Norwood, MA: Artech House, 1994.

[2]   Redl, S. M., M. K. Weber, and M. W. Oliphant, *An Introduction to GSM*, Norwood, MA: Artech House, 1995, Chap 3.

[3]   GSM 11.11, "European Digital Cellular Telecommunications System (Phase 2); Specification of the Subscriber Identity Module–Mobile Equipment (SIM-ME) Interface," ETSI, Sophia Antipolis.

[4]   GSM 11.11, "Digital Cellular Telecommunications System (Phase 2+); Specification of the Subscriber Identity Module–Mobile Equipment (SIM-ME) Interface," ETSI, Sophia Antipolis.

[5]   GSM 02.22, "Digital Cellular Telecommunications System (Phase 2+); Personalisation of GSM Mobile Equipment (ME), Mobile Functionality Specification," ETSI, Sophia Antipolis.

[6]   Aebi, P., "Value-Added Services via NATELsicap," *Mobile Communication International*, No. 27, December 1995/January 1996, pp. 83–86.

[7]   Universal Personal Telecommunication (UPT), "UPT Phase 2; Functional Specification of the Interface of a UPT Integrated Circuit Card (ICC) and Card Accepting Devices (CADs); UPT Card Acceptation Dual Tone Multiple Frequency (DTMF) Device."

Contents

New Phase 2+ functions

In previous chapters we discussed new Phase 2+ services for service groups (e.g., HSCSD for bearer services). In this chapter we take a look at completely new services that are not mere additions to service groups. Instead, they will make GSM more versatile and mark the transition to a third generation of mobile telephony. We commence with the following three Phase 2+ features:

1.   The *SIM application toolkit* allows the operators to establish their own applications and value-added services;

2.   *CAMEL* is a significant move toward *intelligent networks* (INs) that allows a network operator to make personal applications and unique value-added services available to roaming subscribers even when the visiting network does not support them;

3. *Railway applications* show how other interested user groups can make use of GSM's coverage and integrate their needs into the standards.

9.1 SIM application toolkit

With more than 100 operational networks all over the world, the demand for new and different features grows to meet the needs of diverse users. The specification process cannot always keep up with the demand. ETSI, therefore, tries to provide the tools and vehicles for new features rather than specifying the features themselves. As was done for the *unstructured supplementary services data* (USSD), the SIM application toolkit [1] provides operators with a mechanism to create new services; the applications behind this tool or platform are only limited by the imagination of the industry and will take widely different forms in different regions of the world.

9.1.1 Overview of the SIM application toolkit

GSM Phases 1 and 2 saw the SIM as a kind of portable memory that could separate the subscriber from his phone: all the user-related data would be stored in the SIM rather than on the phone. Because the small chip on the SIM actually contains a microcontroller, the SIM can be called on to do much more than simply exchange data between the SIM's memory and the mobile equipment; the communication process can become more than a simple master/slave one. The SIM application toolkit exploits the SIM's full potential. The SIM can assume some responsibility as it notifies the mobile equipment that it has completed a task and that some data are ready for the mobile's use. The kit allows much more than this rudimentary example, but those given in this chapter are not necessarily available as actual services because the kit's specification was just frozen at the time of this writing, and it is entirely up to the industry to create the services and applications. The examples given here are meant to show what is possible with the SIM application toolkit and to spread the germs of ideas for even more features.

9.1.2 Profile download

The profile download is a function of the SIM initialization process that is carried out by the ME to recover the SIM's subscription information. The ME reads the SIM's phase indicator early in the procedure so that it knows if the SIM is, in fact, a Phase 2+ SIM and if the ME should, therefore, initialize the profile download procedure that informs the SIM about the ME's capability concerning the SIM application toolkit. This prevents the SIM from sending information that is of no use to older MEs.

The application is split into three different classes that define the features the ME has to support as part of one of these classes. Table 9.1 in Section 9.1.4.6 shows the classifications; the SIM must know the ME's limitations. The simplest case is the one in which the ME does not issue a profile download procedure at all, which means there is no SIM application functionality available in the ME and no further action should be taken.

9.1.3 Proactive SIM

The proactive SIM part of the toolkit describes how the SIM actually takes action. Because the interface between ME and the SIM was initially intended for the ME as the permanent master, a small modification allows the SIM to take the initiative, thus making the ME the slave.

In principle, the SIM provides the ME with answers for each of its requests for information such as a request to calculate the cipher key or to return specific entries from the phone book. The proactive SIM still returns these kinds of results to the ME but adds a kind of message to the response indicating that the SIM also wants to issue or request some information depending on the relevant feature of the proactive SIM. Ten of the features are discussed next.

The *display text* feature allows the SIM to force a text message of up to 160 characters, similar to short messages, to appear on the mobile's screen. There are two delivery priorities. One is the *normal* priority, which allows text to appear only when the display is not otherwise being used, which applies to the phone's idle mode. As the user browses through the phone's menu, for example, the text message remains hidden until he exits the menu. The other mode is the *high* priority one, which brings the text onto the display regardless of the state of the user interface, which may be its idle menu display mode. After the user reads the information the ME returns its display to the state current at the instant the text message appeared (e.g., the menu structure or entering a short message).

The SIM can initiate the *transmission of a short message* after getting some data from the user (see the initiating dialogue feature discussed later in this section). This user data are packed into a short message that is passed transparently through the ME, or the data are packed by the ME itself. An example of this feature's utility is when money transactions are completed with the GSM telephone: the user enters an amount and an optional PIN number. The data are sent to the network or, more precise, to an entity within the network for further processing.

The SIM can *set up a voice call to a number held by the SIM*. The operator service center number is a typical possibility of this kind for the SIM. An application for this feature appears when the call meter of a prepaid SIM reaches the maximum allowed value, at which time the SIM places a call to the service center in order to negotiate additional prepaid funds.

Depending on the urgency, three different types of call setups are defined. We can also imagine applications that automatically provide a phone number to which the SIM card initiates a call setup.

▶ *Type 1:* Set up a call only when no other call is connected (the phone is in its idle state).

▶ *Type 2:* Put an ongoing call on hold then place the SIM-initiated call.

▶ *Type 3:* Disconnect all calls then place the SIM-initiated call.

The SIM's call setup procedure is identical to normal call setups; the SIM may even initiate the automatic redial mechanism if the dialed line is busy, but before dialing the ME will alert the user in a manner similar to what happens on an incoming call. The ME may also display a message, provided by the SIM, describing what is going on. The user can accept the SIM's call setup action by pressing a menu key or by *not* pressing a cancel key. When the user decides not to allow the call to be set up, the SIM is informed of this and the procedure aborted.

Another feature is the ability to *set up a data call to a number held by the SIM*. This feature is not further specified at this time.

The SIM can initiate the *transmission of a supplementary service control string* to the ME (e.g., for barring calls). The resulting *outgoing calls barred* message may be displayed on the screen of the mobile equipment. The use of USSD for operator-specific supplementary services is reserved for further study.

The SIM can *play a tone in the ear piece*. The SIM can inform the user of an event (running out of prepaid cash is an example), and one way to get the user's attention is to make a noise. A tone or signal can be generated in the phone's loudspeaker without interrupting a call setup or a call. The tone can also be played through the ringer or through the external speaker of a car kit when the ME is in idle mode. The tone can be similar to the notification of a received SMS: a short beep.

The SIM can *initiate a dialogue with the user*. For many features it is important for the user to enter commands or text strings as answers to requests from the SIM or from the network. When the SIM asks a question of the user, it does so through the ME, which displays a text message explaining the request and soliciting a response. Upon entry of the requested data (a PIN, or some other number or general response) the input from the user is transferred back to the SIM by the ME. The information might be a single character or digit, or it could be a string of several characters or digits.

For example, let's assume that Oscar, a GSM mobile user, would like to go to the movies in the evening. He calls a particular service number and, via a short message, a selection of movie titles is listed on the display of his mobile phone. Oscar selects a movie from the list and, because it is late and he is short some cash and the banks are closed, he elects to pay for his ticket from his billing account with the GSM operator. For this transaction he has to key in his personal PIN number, which is passed on to the SIM for verification, which finally passes the information via the short message service to the network.

This kind of dialogue with the user can be enabled with modifications to the menu structure initiated by the SIM. This can occur, for example, when the user wants to add a feature to his subscription such as call barring that was not previously available to him. The network operator issues a command to the SIM via a SMS-MT, and the SIM issues a command to the ME that causes the new menu items to be added in the appropriate locations in the menu tree. A menu item can be operator specific such as when the operator wants to invoke a new and specially named service.

The SIM even has *refresh* capability, the ability to tell the ME that the SIM-related telephone settings and data have been changed. This notification is important because many MEs load copies of the SIM's file contents into internal memory to speed the selection process. Take as an example the phone book with more than 100 entries, which takes a long time to read from the SIM. Depending on the type of changes the SIM will request, the ME performs the SIM initialization process again, only reads files that have been changed, or completely resets the SIM. The first two cases, however, do not include a deactivation, reactivation, or electrical resetting of the SIM; the user will not be asked to enter his PIN because the ME will read all information from the SIM again, including sub-scription-related information. This technique lets the operator change SIM settings via a dedicated SMS-MT, like the SIM service table, thus acti-vating or deactivating features in the background. The reason for such action can be a change in subscription (upgrade or downgrade in fea-tures). Resetting the SIM allows the operator to activate or deactivate features immediately during the session. The user is not aware of the refresh procedure; its effect will be seen when the user enters the menu and finds that new items have been added or old items have been erased. It is also possible that after reset and during the recharacterization of the fixed dialing numbers (see Chapter 8), the user will only be able to call these specific numbers. The user will realize that his SIM has been com-pletely reset when a new SIM session is initiated with the PIN request.

The ability to *provide local information* is another feature. Upon receipt of this command the ME has to pass local information to the SIM. The local information includes the *mobile country code* (MCC), the *mobile net-work code* (MNC), and the *location area code* (LAC) and the cell identity. It can also include information on the ME—the *international mobile equip-ment identity* (IMEI), which is the unique electronic serial number.

There are also features that make sure that normal GSM mode is not inhibited; just because the mobile equipment is busy waiting for an input from the toolkit is no reason to deny the subscriber the use of his phone. In this case the SIM simply tells the ME that it needs more time to process a certain function. The SIM can set a polling time after which the ME has to check on the SIM's progress on pending matters.

9.1.4   Data download to SIM

The data download to the SIM is a tool to provide the SIM with informa-tion or commands directly from the network. There are two different means of transporting the data. The first is a dedicated one via SMS-PP and the other is a more general one via the SMS-CB.

9.1.4.1 Data download via SMS-PP

The network can communicate with a particular SIM via a SMS-PP message. The user should not be aware that his mobile phone just received a command or information, thus the protocol identifier for this SMS-PP is set to data download and it is indicated as a class 2 message (see Chapter 6), which goes directly to the SIM. A dedicated *envelope* command is used for this purpose. Even as the ME should give no indication to the user that it has received the message, the ME should acknowledge the SIM's confirmation that the command or information has been received.

An application for data download could be the modification of the charge file on the SIM of a successful recharging of a prepaid SIM, or a change in subscription options that modify the SIM service table. When the SIM is not a toolkit SIM, the ME should treat the SMS as a normal message, which would be stored in the normal message storage area on the SIM even though the content might appear quite strange to the user!

9.1.4.2 Data download via SMS-CB

The SIM can be prepared to process data from the SMS-cell broadcast for which purposes the *cell broadcast message identifier* (CBMID) is stored in the SIM. The CBMID contains the channel numbers on which the phone has to listen for information; these numbers will usually not be made public. Upon reception of a cell broadcast message, the ME checks with the SIM to see if the message identifier is stored on the SIM. If it is, then the ME passes the message on to the SIM without displaying it. The information from the CB is processed in the SIM and the results are displayed to the user with a display message (SIM-to-ME).

An example of this application might be traffic information, which is displayed in an encoded style with numbers representing certain routes or types of routes (highways or small streets) and the density of traffic which is represented with numbers. In this case the SIM processes these data to make them more useful to the subscriber. The highways would be displayed along with their numbers and the direction and the density of the traffic in plain words. Another use is described in Section 9.1.5, the application section of this chapter.

9.1.4.3 Menu selection

The ME's menu system can be modified through the SIM toolkit as described earlier. Because the added menu item is likely to be operator specific, the ME might not know the exact meaning of the new selection. In this case the ME passes the selected item from the SIM-specific part of the menu to the SIM itself, which has to decide what to do. There can be some kind of supplementary service implemented along with the new feature, or the SIM may need some additional information (a phone number) from the user in order to respond completely to the menu selection.

An example of this could be mobile banking. The subscriber is enabled by his bank, in cooperation with his network operator, to check his account or perform transactions via his mobile phone. The operator issuing the SIM card might wish to generate a standard menu for checking the account, performing a transaction, or reviewing the latest transactions. The advantage of this is that the application would look the same on each mobile phone regardless of the mobile's manufacturer.

9.1.4.4 Call control by SIM

When the SIM takes responsibility for call control, each attempt from the user to set up a call is first passed to the SIM, which has to approve the attempt. For mobile originated calls, the SIM checks whether the dialed number is valid or authorized before it will allow the call setup procedure to continue. If the called line is busy and a redial procedure is set up, the subsequent attempts are not subject to the SIM's verification process. When the SIM denies the call setup, the call will simply not take place or it will be completed with different parameters, perhaps a different phone number. The ME may indicate the changes to the user; that the call was barred or that the dialed number was changed.

This feature applies when there is a destination number included for supplementary services, which would be true for a call divert SS. Now the SIM has to approve the number to which calls are to be diverted, and there are two options: Either the supplementary service is processed as intended by the user, or the SIM disapproves and either stops the procedure or changes the destination number. An application for this kind of feature can be a loaner phone or a pool of phones in a big company that wants to confine calls to business uses. As we saw earlier, the FDN feature prevents this kind of misuse, but the call control by the SIM feature provides even more flexibility. Fixed dialing numbers and barred dialing numbers can be applied simultaneously. The ME first checks the FDN/BDN list before setting up a call. If the requested number is not included in the FDN list or contained in the BDN list, then the call setup is terminated. Turning matters over to the SIM, if the number is restricted by FDN or BDN it still has to pass the SIM call control check before the same procedure applies.

9.1.4.5 Security

From the descriptions given here we can see that the SIM application toolkit is a very powerful feature that allows useful applications not limited except by one's imagination. These kinds of applications often work with personal data such as account numbers and PIN numbers. Security is an important issue that, at the time of writing, was still the topic of discussions within relevant ETSI groups.

9.1.4.6 Classes of support

So that the SIM application toolkit can be introduced in a smooth, easy, and orderly way, the features have been divided into different classes. This helps the terminal manufacturers and the operators because the applications require a lot of verification and testing.

- ▶ *Class 1* is a first step that expands the services available to the subscriber and then activates them with the refresh/reset function.

- ▶ *Class 2* makes the whole range of capabilities available to subscribers with the exception of the USSD.

- ▶ *Class 3* adds the USSD to the range of features in the SIM application toolkit.

Table 9.1 lists the features in the SIM application toolkit, including to which of the three classes each feature is assigned, and the source and destination of the commands involved with each feature.

9.1.5 Applications using the SIM application toolkit

Having explored a few features of the SIM application toolkit, we turn our attention to some practical examples that incorporate the combined use

Table 9.1
Features and Classes Within the SIM Application Toolkit

Feature	Classes			Source	Destination
	1	2	3		
Call control		X	X	ME	SIM
Cell broadcast download		X	X	Network	SIM
Display text		X	X	SIM	Display
Get inkey		X	X	SIM	ME
Get input		X	X	SIM	ME
Menu selection		X	X	Keypad	SIM
More time		X	X	SIM	ME
Play tone		X	X	SIM	Ear piece[1]
Polling		X	X	SIM	ME
Refresh	X	X	X	SIM	ME
Select item		X	X	SIM	ME
Send short message		X	X	SIM	Network
Send SS		X	X	SIM	Network
Send USSD			X	SIM	Network
Set up call		X	X	SIM	Network
Set up menu		X	X	SIM	ME
SMS-PP download	X	X	X	Network	SIM
Provide local information	X	X		SIM	ME

1 ME might route this to other loudspeakers or to the ringer if more appropriate.

of different aspects of the toolkit in ways that are more likely to be seen in actual networks.

9.1.5.1 Activation or modification of a subscription

The SIM application toolkit provides an ideal way to perform *over-the-air activation*. A new customer simply buys a telephone along with a SIM in some convenient place. The new SIM may be initially programmed with only basic telephony service, or our new subscriber may have selected all the services she wants from a catalog on display in the shop where she purchased her new mobile phone. Some service providers and network operators sell services in classes or bundles (gold, silver, bronze; or basic, business, elite) each with different levels of subscription service and fees. The mobile phone outlet (the shop) passes the subscription data to the network operator (through SMS, intranet, or the Internet), and the operator activates the subscription and its services with a SMS-MT. The SMS-MT writes the relevant data to the SIM so that the subscriber's desires are set up correctly. After a successful setup the operator can send another SMS-MT welcoming the new user and listing how each of her selected options is set up. Another option is to put the responsibility of displaying welcome messages on the SIM itself, which is better able to determine when all the features are actually activated.

9.1.5.2   Zonal indicator

The cell broadcast service is currently used in networks to indicate area codes that provide the user with some information about where she is located and whether calls from the location will incur local or long-distance charges. This type of information has a problem: it makes the mobile phone appear to be *different* from a landline phone; the user has to remember her own area code and may become confused when so many numbers appear on the mobile phone's display. In some cases the cell broadcast messages use up the entire display. People are reluctant to use things that are confusing or mysterious; unused phones do not bring much revenue to operators.

The SIM toolkit allows the display of only the important keywords such as *local* or *long distance*, which is easier to understand than the standard displays (see Figure 9.1). How does this work? First, the message identifier for a type of cell broadcast message is stored on the SIM. After the SIM examines the message identifier, the ME detects that the SIM wants to work on the message and passes the contents of the CB to the SIM. The home zone is stored in the SIM and by simply comparing the home zone with the cell broadcast message the SIM is able to determine if a call will be a local or a long-distance call. The SIM passes the appropriate keyword to the ME with the *display text* command. One advantage of this approach is that it is downward compatible. If either the SIM or the ME does not support the application toolkit, the ME simply displays the zonal information in the traditional way as shown on the left side of Figure 9.1.

9.1.5.3   Flight booking and confirmation

Many mobile phone users are very busy businesspeople who are always pressed for time. Even though this is changing as mobile telephony

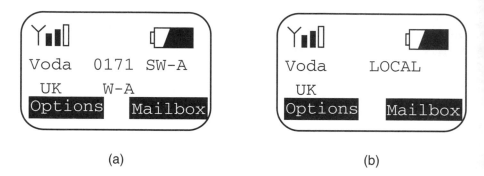

(a)   (b)

Figure 9.1   (a) Standard and (b) toolkit-modified display for zonal indicator.

becomes more and more consumer oriented, the busy users remain, and they still appreciate anything that can save time in handling routine matters of all kinds. Making travel arrangements is a common task among the most heavy users of mobile phones. With little time on their hands and their plans always changing, they tend to book and change their flights often and at the very last minute. And since they are usually away from their offices, they are on their own with nothing to help them except the SIM application toolkit.

A businessperson can use a normal SMS to send his travel office a departure city, a destination, and a preferred time of travel. Those who routinely use this kind of help with their travel arrangements will store a few standard SMS messages on the SIM that allow easy programming and modification of the queries. The travel agency receives the SMS and responds with a number of options (different airlines, flights, and tariffs) within several minutes by a special SMS-MT message back to the busy subscriber. The response can also originate from an e-mail tool. The response will be interpreted as a special SIM application toolkit message that is processed in the SIM and presented to the ME with the *select item* command, which allows the user to make a selection from one of several choices. The selection is sent back to the travel agency. Further exchanges of data could follow: payment options and which mileage account to credit. The travel agency would, ideally, provide the user with an electronic boarding pass indicating the flight number, the date, the check-in time, and a check-in number. See Figure 9.2 for example displays.

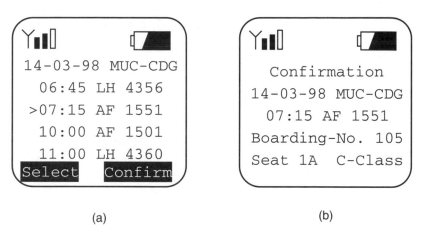

(a)   (b)

Figure 9.2   (a) Flight selection and (b) electronic boarding pass.

9.1.6   Conclusion

In this section on the SIM application toolkit we demonstrated that this GSM feature offers an abundance of new services and applications to the GSM subscriber that are particularly attractive to users. It is an easy exercise for the operators to define and introduce new services with the SIM application toolkit. Differentiation of network services among operators is supported. The advantage for the operator is that she can define and implement features independently from the mobile phones' manufacturers. Once the manufacturers support the SIM application toolkit the operators can use off-the-shelf products to realize their value-added services. Specific applications might ask for such differentiation in different countries and for different user groups. The SIM application toolkit is a box of tools with which different combinations of different features can be constructed to offer many new services and benefits.

9.2   Customized applications for mobile network enhanced logic (CAMEL)

In increasingly competitive GSM market environments, operators want to develop their own new value-added services, so-called *operator-specific services* (OSS). But OSS only work satisfactorily if customers have access to them even when roaming. Because subscribers get used to special features, they will not understand why networks have different implementations or even different legal requirements such as for calling line identification. *Customized applications for mobile network enhanced logic* or *CAMEL* was defined to support OSS while roaming. CAMEL lets a GSM network offer IN features. This is made possible through the exchange of relevant information between, for example, the HLR and VLR. The logic providing the services and the call processing functions are separated from each other, which makes provisioning new services easier since the switch-related signaling functions remain transparent. CAMEL-related standards will be introduced through GSM Phase 2+ in two steps (see Chapter 4).

9.2.1 Functional description of CAMEL

CAMEL makes dedicated network features available to roaming subscribers even though a visited PLMN may not support them. A popular example of this is dialing into the voice mail. Most networks nowadays use a short network internal number (e.g., "3311" mentioned in Chapter 6) rather than a complete national or international number (e.g., +49511XXXXXXX) through which the subscriber gets voice mail. The only problem with this approach arises with roaming subscribers, since these numbers may not exist in the visiting network or they may have another significance. CAMEL would extend the reach of the home network's unique dialing sequences to roaming customers. How can this work? The *CAMEL service environment* (CSE) controls all outgoing and incoming calls to a CAMEL subscriber regardless of her location. Whenever a roaming customer registers in a foreign PLMN, the HPLMN provides the visited PLMN with relevant information on the subscriber, most of which should be familiar to us: authentication parameters, call forwarding parameters, etc. The *CAMEL subscription information* (CSI) is added to the CSE information for a CAMEL subscriber. The CSI notifies the visited network about specific features that have to be supported for the visiting subscriber. CSI contains an identifier for OSS that is transparent to the visited network. CAMEL applies to all circuit-switched basic services except for emergency calls, which can be placed by anyone in any network.

The HPLMN has to be notified for each call setup with the CAMEL subscriber regardless of whether it is a *mobile-originated* (MO) or *mobile-terminated* (MT) call. When the VPLMN realizes a call setup involves a CAMEL subscriber, it suspends the call and informs the HPLMN (with the CSE) about the event, the parties involved in the setup, and the location of the calling subscriber. The CSE at the subscriber's HPLMN provides the VPLMN with information on how to proceed with the call: to continue with the call unchanged, to bar the call, or to continue with modified information (e.g., a different called party as indicated in our example on the voice mail). All this has to happen with minimum delay.

Together with some operator-specific services, the CSE must investigate the whereabouts of a subscriber from time to time. The HPLMN will inform the CSE about the current subscriber status (idle, busy, detached) and the location information. As with all advanced features, so it will also be with CAMEL that some networks provide certain services before others do, but CAMEL also requires that some signaling links dedicated to CAMEL be supported even if nothing else is implemented. When the HPLMN realizes that a CAMEL subscriber wants to register in a network that does not support it, the HPLMN has the ability to apply operator-determined barring or deny location updating if necessary. Which option is finally exercised depends on the CAMEL subscriber's subscription profile. The CAMEL subscriber will, in most cases, at least be able to use his phone without his OSS in the visited network. The specifications for CAMEL are detailed in [2,3].

9.2.2 Network architecture

Now we look at how CAMEL is implemented in the home (HPLMN) and visiting (VPLMN) networks. Figure 9.3 depicts the different entities and functions that are required. The main signaling link between the HPLMN and the VPLMN is through the two GMSCs of both networks. The *home location register* (HLR) stores the CSI, including both mobile-originated and mobile-terminated calls, for each of the individual customers who subscribe to the HPLMN's service. The HLR provides these data to the *visitor location register* (VLR) when a mobile station starts a registration procedure or when the CSI is updated. The GMSC will also get the data when routing information is required during a call setup to or from the mobile station. The *GSM service control function* (gsmSCF) is a support function for CAMEL allocated to the HLR. It keeps track of the current CAMEL status

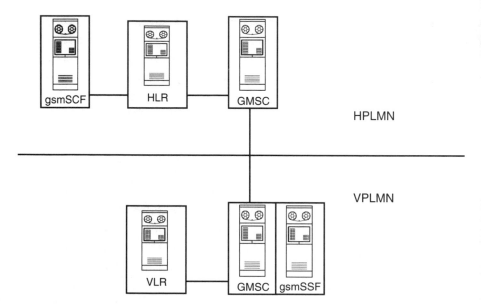

Figure 9.3   Network architecture required for CAMEL support.

of a subscriber and tracks the location data. The *GSM service switching function* (gsmSSF) is an interface inside the GMSC that detects requirements for CAMEL support, for example, a roaming CAMEL subscriber initiates an outgoing call. The VPLMN's VLR gets a copy of the CAMEL data needed for outgoing calls.

9.2.3   A CAMEL example

We now illustrate the uses of CAMEL with a trivial example. Many network operators add service numbers to the SIM cards they supply to their subscribers. Some examples of service numbers are (1) the voice mail number, (2) the network's operator, (3) the network's help desk or customer care center, (4) a travel service, (5) a pizza delivery service, and (6) a flower delivery service. The service numbers do not follow the numbering rules for standard phone numbers; they may be four-digit numbers that only have significance within the home network. If a customer wants to use one of the service numbers stored on the SIM while roaming, he will not be able to get through to his home system because the meanings of the service numbers are totally unknown to the visiting operator. When CAMEL is applied to the situation the call control is located in the home country, which knows all about each of the service numbers; the roaming subscriber is properly connected to the service he seeks, which would probably be the travel service rather than the pizza delivery service in most roaming examples. This simple example should clarify how CAMEL works. The same mechanism applies to all operator-specific features that will be offered in the future.

9.3 Railway applications

The features and applications described in this section are not limited to railway use even though they meet the specific needs of the *Union Internationale de Chemin de Fer* (UIC). The chief requirements of the European railways dealt with priorities and preemption for call setups, the speed of call setup procedures, and simple procedures for group call services and voice broadcast services (see Chapter 3). These requirements ensure that GSM can respond to their internal demands for a safe and secure railway communications system based on a European-wide standard that specifies the same communications gear in all countries.

Even though UIC features will be available in future GSM-based systems, we should note that the railway organizations will implement their networks themselves on their own frequencies set aside for the purpose, and will not simply add something to an existing GSM network. An additional 4 MHz of spectrum below the extended GSM frequency band is reserved for the railways.

9.3.1 Enhanced multilevel precedence and preemption

The *enhanced multilevel precedence and preemption service* (eMLPP) actually consists of two services: The *precedence* service gives certain calls priority, and the *preemption* service makes sure that calls of high priority get the required resources. The eMLPP service belongs to the group of supplementary services to which users subscribe as they need them. The service is applicable to speech and fax teleservices as well as to all bearer services [4].

9.3.1.1 Priority levels and preemption

Seven priority levels are defined as indicated in Table 9.2. Levels A and B are used only within the network and we cannot subscribe to them. Both

Table 9.2
Priority Levels Defined for eMLPP

Priority Level	Purpose
A	Highest, for network internal use
B	For network internal use
0	For subscription
1	For subscription
2	For subscription
3	For subscription
4	For subscription

the A and the B levels are mapped to level 0 outside of the network area. The other five levels are used not only within the network but also apply globally, which means they are also applicable in an ISDN, which provides the *multilevel precedence and preemption service* (MLPP); see [5–7] for further details.

The maximum priority level for an individual subscriber is stored in the SIM. The user may select the level she wants to use for a specific call or she may set a default value for all calls. The user cannot select a priority that exceeds the maximum level specified by the priority level.

The advantage for a user with high priority is that when the system is congested or there are limited radio resources in a part of the network, sufficient resources are freed up so that the priority user can proceed with his call. The resources are taken away from users without an eMLPP subscription or from eMLPP subscribers with a lower priority level. Lower priority users who have their traffic channel resources taken away from them simply hear a congestion tone. The decision on which resource should be preempted is made by the MSC.

The reason for introducing such a service is that certain customers, such as a railway, must ensure that service or a traffic channel is available whenever needed. An engine driver cannot wait to report a problem; human lives or the general public safety may be in jeopardy.

9.3.1.2 Call setup time improvements

A logical extension of the priority requirement is the demand for improved call setup times. Normal call setup times range from a few seconds up to 15 or 20 seconds, which is the time between the instant the send button is pressed and the instant a ringing indication is heard from the other side of the connection; this may be too long in critical or emergency situations. Additional classes for call setup times have been specified as listed in Table 9.3, which reflects the allocation for UIC requirements. The classes are allocated to the priority level by the operator.

It should be obvious that call setup times of less than 1 second are very difficult to achieve in GSM. These times will be achieved for 95% of all calls in case of railway networks; the remaining 5% of the calls may suffer setup times 1.5 times longer. The setup times for class 1 calls depend on the required service. Broadcast or group calls, which are described later, demand setup times of less than 2 seconds. These fast setup times are achieved by omitting authentication and ciphering procedures. Figure 9.4 shows the message exchange requirements for normal GSM calls and Figure 9.5 shows the procedures for high-priority calls. The reader will note the diminished number of transactions in Table 9.4 relative to those in Table 9.3. Because messages are not only transmitted over the air interface but have their origins somewhere in the network, it is obvious that fewer messages reduces the time required to set up a call.

9.3.1.3 Automatic answer for eMLPP

The eMLPP service allows a mobile station to automatically answer incoming calls of certain levels. The parameter for the level of calls to be answered thus is stored in the SIM. If the called user is busy on the phone, the eMLPP service can preempt the current call and put it on hold in order to connect a call with a higher priority. A subscription to eMLPP should always include a subscription to call waiting and call hold. The only call that does not allow preemption is an emergency call.

Table 9.3
Call Setup Performance

Class	Setup Time	Priority Level
1	< 1 sec	A
2	< 5 sec	B, 0
3	< 10 sec	1, 2, 3, 4

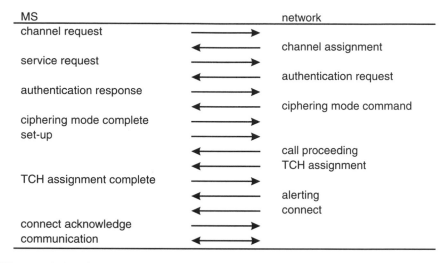

Figure 9.4 Call setup for normal calls.

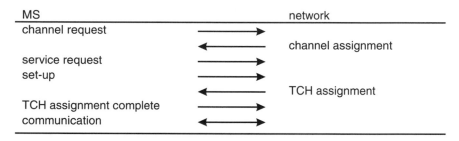

Figure 9.5 Call setup for high-priority calls.

9.3.1.4 Railway (UIC) applications

To summarize the eMLPP service and the setup times allocated to the respective levels, we will examine some typical UIC applications, the details of which can be found in [8]. Table 9.4 not only gives examples, but also shows that the preemption applies only to the first four priority levels, which, looking at the examples, should get very high priorities. The allocations given in these example apply to UIC requirements; it is the operator's choice to allocate the setup time class or allow preemption for individual priority levels. This kind of service is useful for any large group that has dedicated tasks in our society and will make use of the good coverage of GSM networks.

Table 9.4
Examples of eMLPP Service

Priority Level	Setup Time	Preemption	UIC Example
A	Class 1	Yes	Railway emergency application, broadcast/group call or even a data call (e.g., to stop a train)
B	Class 2	Yes	Automatic train control
0	Class 2	Yes	Emergency services
1	Class 3	Yes	Operational train calls
2	Class 3	No	Train data services
3	Class 3	No	Railway information services
4	Class 3	No	Passenger services (e.g., phone booth in train)

9.3.2 Voice group call service

GSM already offers the multiparty supplementary service that mimics a kind of voice group call service. The *voice group call service* (VGCS) introduced by UIC reflects their different requirements. The building of a multiparty communications link takes much more time than can be tolerated in an emergency and there is a limit to the number of participants. VGCS allows call establishment times of a few seconds to all members of a group and offers half-duplex service to limit the network resources used for the service [9,10].

9.3.2.1 Service description

The members of a voice group are fixed by the network operator or the service provider upon the request of the subscriber. Members of a group are identified by the group identification or group ID. Individual subscribers can be members of several groups. Where and how the group ID is stored in the network is detailed in the following section.

VGCS distinguishes different participants of a group call. A *dispatcher* is not necessarily a mobile subscriber but can also be located in the fixed network; he will always be addressed when a group call is placed to his group. The dispatcher has several privileges that are discussed later. The *calling subscriber* is the one who initiates a group call; individual subscribers need to have authorization to initiate group calls. The *destination subscriber* is anyone who receives a group call.

A voice group call can be initiated by a dispatcher or by any authorized subscriber. The call setup is performed via a normal traffic channel. The members of the group are notified (paged) via a dedicated *notification channel* (NCH). When all the members of the group are connected, or after a predefined time if some members are missing, the initiator of the group call will be advised that the group call is established. All members of the group are now listening to a dedicated voice group call channel on the downlink. This channel is the same for all members within a cell, which drastically reduces the required resources. After the initiator has finished talking he will lose the dedicated link and join with the others in the voice group channel. From now on only one member can talk on the uplink. An uplink channel is allocated upon request. When several members start talking at the same time, the network will select one on a first come, first served basis. The voice group call service is a half-duplex teleservice.

VGCS also offers security against eavesdropping. In this case the uplink and the downlink channel apply ciphering with a group key that is identical for each member of the group.

Dispatchers are the only exception to use of the voice group call channel; they always have a full-duplex channel and can, therefore, always talk. The network will forward the dispatcher's voice without regard to a group member who may already be talking. If there are several dispatchers connected to a group all of them can talk and all of them will be heard by the group members at the same time. This is possible because such calls are linked within switches (MSCs) with one dedicated link for each of the individual members.

Figure 9.6 illustrates the voice group channel and the associated half-duplex links. In our example a central location (the switch tower in a railway yard) starts a voice group call. The two trains in our example are paged and connected to a voice group channel on which they would be able to listen to the message coming from the switch tower. When the announcement is finished the dedicated link to the switch tower is terminated and replaced by the same half-duplex resource allocated to the trains. Now everybody gets an uplink channel upon request from which individual acknowledgments of the message are received from the trains. The only exceptions are the dispatchers who might talk at any time.

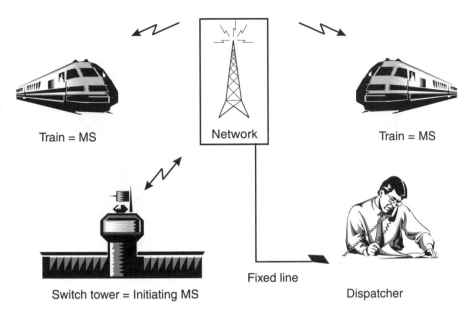

Figure 9.6 Voice group call.

9.3.2.2 Network architecture

The only addition to the standard GSM network for the support of VGCS is the *group call register* (GCR). The GCR is logically collocated in the MSC. The interface between the GCR and MSC has not been standardized and depends on individual manufacturer implementations. The GCR may, therefore, be an integrated part of a MSC. The GCR stores all data related to the VGCS (see Figure 9.7), such as the following:

- A list of cells that belong to the service area. For UIC applications it is obvious that only cells along the train tracks will belong to the service area.

- A list of relay MSCs required to cover the complete service area.

- A list of members for individual group IDs.

- Parameters for group calls, for example, a time-out value when calls will be terminated after no activity, information on the cipher algorithm, and the group key to be used for a group call.

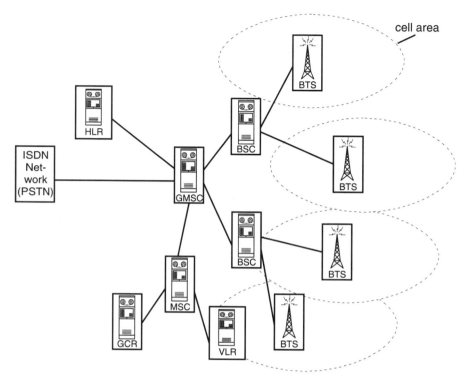

Figure 9.7   VGCS infrastructure.

Members are only identified via the group ID; the *temporary mobile subscriber identity* (TMSI) or *international mobile subscriber identity* (IMSI) have no meaning. The group ID is used to page members of the group. The group ID is broadcast for the duration of the voice group call to allow members entering the service area to join the group call.

9.3.3   Voice broadcast service

The *voice broadcast service* (VBS) is very similar to the VGCS except that VBS only allows announcements from one party to the remaining members of the group; there are no responses from the group. The network entities involved in VBS are identical to those used for VGCS. The GCR contains a separate set of parameters identifying the VBS. The SIM contains separate data to identify different options for VBS. Further details can be found in [11,12].

References

[1] GSM 11.14, "European Digital Cellular Telecommunications System (Phase 2+); Specification of the SIM Application Toolkit for the Subscriber Identity Module–Mobile Equipment (SIM-ME) Interface," ETSI, Sophia Antipolis.

[2] GSM 02.78, "Digital Cellular Telecommunications System (Phase 2+); Customized Applications for Mobile Network Enhanced Logic (CAMEL); Service Definition (Stage 1)," ETSI, Sophia Antipolis.

[3] GSM 03.78, "Digital Cellular Telecommunications System (Phase 2+); Customized Applications for Mobile Network Enhanced Logic (CAMEL) —Stage 2," ETSI, Sophia Antipolis.

[4] GSM 02.67, "Digital Cellular Telecommunications System (Phase 2+); Enhanced MultiLevel Precedence and PreEmption Service (eMLPP) —Stage 1," ETSI, Sophia Antipolis.

[5] ITU-T Recommendation I.255.3, "ISDN MultiLevel Precedence and PreEmption (MLPP) Stage 1."

[6] ITU-T Recommendation Q.85, "Stage 2 Description for Community of Interest Supplementary Services (Clause 3: Multi-Level Precedence and Pre-Emption MLPP)."

[7] ITU-T Recommendation Q.735, "Stage 3 Description for Community of Interest Supplementary Services Using SS No. 7 (Clause 3: Multi-Level Precedence and Pre-Emption (MLPP)."

[8] GSM 03.67, "Digital Cellular Telecommunications System (Phase 2+); Enhanced MultiLevel Precedence and PreEmption Service (eMLPP) —Stage 2," ETSI, Sophia Antipolis.

[9] GSM 02.68, "Digital Cellular Telecommunication System (Phase 2+); Voice Group Call Service (VGCS)—Stage 1," ETSI, Sophia Antipolis.

[10] GSM 03.68, "Digital Cellular Telecommunications System (Phase 2+); Voice Group Call Service (VGCS)—Stage 2," ETSI, Sophia Antipolis.

[11] GSM 02.69, "Digital Cellular Telecommunication System (Phase 2+); Voice Broadcast Service (VBS)—Stage 1," ETSI, Sophia Antipolis.

[12] GSM 03.69, "Digital Cellular Telecommunications System (Phase 2+); Voice Broadcast Service (VBS)—Stage 2," ETSI, Sophia Antipolis.

Roaming and call routing

GSM is a mature system accepted throughout the world. Most of us who travel with our GSM phones take it for granted that when we cross borders, or switch on our phones after leaving the plane, we have reliable access to mobile telephone service. With all the GSM roaming agreements in place and many manufacturers making dual- and triple-band phones, service coverage will no longer be limited to Europe but will extend to Asia, the Pacific Islands, Africa, the Middle East, and even to North America. It does not seem to matter where we switch on our GSM telephone—we are connected and people can find us; we get service and we can make calls. Today's reality sounded like dreams or the ravings of unbalanced people just 10 years ago. Some of us have grown to depend on the global nature of GSM, and others in the industry base their wealth and their ability to meet their monthly rent payments on the miracle. How does all of this work? What are the mechanisms behind roaming?

In this chapter we explore how calls are routed to our mobile phones with seemingly little regard for our location, and we briefly consider how we are billed for this convenience. Because such costs cannot be expressed in absolute figures, we will give an indication of which part of a call is billed to the calling party and which part is billed to the called party.

10.1   Routing in GSM PLMNs

Because mobile users are, by definition, always on the move, the mechanisms the networks use to locate each of us within the GSM world are important. We will confine our attention to how calls are forwarded to our mobile stations, which are also referred to as *mobile-terminated* (MT) call setups. We will check different cases as we see how the networks route calls to specific mobile stations [1].

10.1.1   Location registration

The most important tool used for finding phones within GSM networks is location updating. The location of a mobile station is uniquely identified by the *mobile country code* (MCC), the *mobile network code* (MNC), and the *location area identity* (LAI). The MCC is a three-digit value that identifies the country where the network is located. The MNC is a two-digit value (three digits in North American PCS 1900 systems) that identifies different (competing) networks within one country. The LAI identifies the physical area in which a mobile station is located. A location area may consist of one or more physical cells; it is also referred to as a *paging area*, which a network designates as an MS's location for paging tasks.

With each location updating procedure, the mobile station reads its location from the control channel transmitted by its serving *base transceiver station* (BTS) and reports it back to the network. In a GSM network, as shown in Figure 10.1, two registers store the location-related data of a mobile station: the *home location register* (HLR) and the *visitor location register* (VLR). The HLR keeps data that are permanently associated with individual mobile stations as well as the current location. The permanent data include subscription details such as the teleservices, bearer services, and supplementary services allocated to the subscriber. The billing information and the subscriber's home address are not in the HLR; they

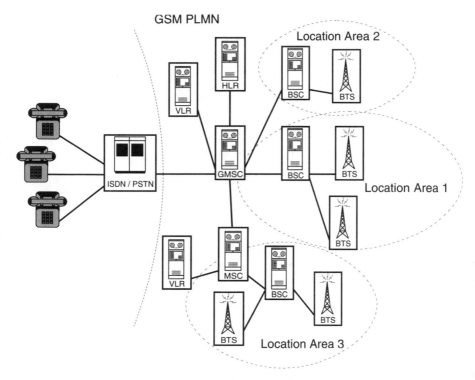

Figure 10.1 Network infrastructure of a GSM network.

are located elsewhere in the *billing center*. The VLR keeps temporary data on a subscriber for only as long as the subscriber is located in the area belonging to a particular VLR. The temporary data contains the subscription-related data (obtained from the subscriber's HLR) as well as the MS's exact location in the VLR's area. Figure 10.1 shows two VLR areas. Location area 1 and location area 2 belong to the first VLR, and location area 3 belongs to the second VLR. Note that the resolution of the location areas is the same as the *base station controllers* (BSCs).

As an aid to understanding the hierarchy of locations (MCC, MNC, and LAI), we briefly explain the situation in North America's PCS 1900 systems by noting the large size of the North American continent and the likely possibility that a single North American PCS 1900 operator may have multiple HLRs. In the United States (MCC = 310), there are several operators, two of which are VoiceStream [2] and Omnipoint [3]. VoiceStream operates systems all over the continental United States including one system thousands of kilometers away on the Hawaiian Islands. It seems logical, therefore, that VoiceStream maintains several MNCs (200, 210, 220, 230, 240, 250, and 260). Omnipoint runs a large system in the New York City region (MNC = 160) and operates a much smaller system in Wichita, Kansas. The subscription details for the Wichita subscribers resides in the HLR located in New York such that all of Omnipoint's subscribers, as of this writing, have the same MNC.

A network may consist of several *mobile services switching centers* (MSCs), but only those connected to an ISDN or a PSTN are used as *gateway MSCs* (GMSC), with an attached HLR, since this is the switch through which incoming calls enter the system. The incoming calls for mobiles in the network cannot be routed properly until the system's HLR is checked for the location of the target mobile and the subscriber's authorized services. The GMSC and the MSC each have their own VLRs. Upon performing a location update, the message from the mobile station that marks its location is first sent through the MSC to its attached VLR. The VLR determines if it already has a record of the MS and, if so, only the location area is changed if it has, in fact, changed from an earlier one. If the MS is not known to the VLR (it is a new visitor to the system), then the VLR will apprise the mobile's HLR of the MS's location and routing information as it sends along a request for the subscriber's home data, which the visited VLR will need before it can complete a mobile originated call from the mobile. After all of this is done, the HLR knows which VLR should receive incoming calls for the roaming mobile and the VLR knows the exact location of the mobile and which BTS should send paging messages that attract the mobile's attention. If, later, the mobile wants to originate a call, the VLR has a record of the subscriber's subscription data, that is, authentication variables and service options.

There are three different types of location updating procedures [4]:

1. The *registration* takes place whenever a mobile station is switched on. After an internal initialization, including necessary SIM procedures, the MS checks for an available network. When it finds one the MS is able to read important information such as the location information. Two cases are distinguished here: (1) *IMSI attach required* or (2) *IMSI attach not required*. In the first case, the *IMSI attach required* case, the mobile station always initiates a location update procedure. A location update procedure is also completed in the second case (the *not required* case) but only when the

location area identity the MS reads from the downlink control channel has changed since it was last switched off. In the first case, the mobile also starts an *IMSI detach* procedure when it is switched off, thus telling the network that the MS is no longer idle, that it cannot respond to incoming calls. This prevents the squandering of radio resources when mobiles that cannot possibly respond are paged. In the *not required* case, the detach procedure is not required. The first case should be the rule; the second case the exception.

2. *Periodic location updating* is performed after a period of time predefined by the network and constantly sent to all active mobile stations monitoring the control channel. If the mobile station does not register after this time, perhaps because the subscriber drove into an underground parking lot, then the network will assume that the mobile station is no longer available and will mark it as "not reachable" in the HLR and VLR. This means that incoming calls for this MS will not be routed to the location area but will be blocked at the home GMSC. If the subscriber invoked *call forwarding on subscriber not reachable*, the network will forward the call to the indicated number.

3. When the MS detects a *location area change* it will notify the network that it is now located in a different area. As shown in Figure 10.1, an MS could move around within location area 1, from one cell to another, without the need for a location update. If it moves from location area 1 to location area 2 the VLR must be notified. If the MS moves from location area 1 or 2 into location area 3, then the HLR must be notified of the change and the VLR in area 3 will store the mobile's new LAI.

10.1.2   Routing within a PLMN

Figure 10.2 shows the routing of a call from mobile subscriber A to mobile subscriber B. When A calls B's number the call will be routed to the GMSC of the network (1). If the network has more than one GMSC, then the call will be routed to the GMSC to which subscriber B belongs, that is, to the GMSC the attached HLR of which hold's subscriber B's permanent data (2). The distinction between the HLRs and the attached GMSCs is made through the phone number—the first digit(s) after the network access

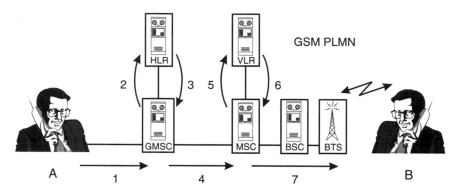

Figure 10.2   Routing of a call within a GSM PLMN.

code. The GMSC will discover the location of subscriber B as it determines which HLR responds to its location request. The responding HLR tells the inquiring GMSC which MSC area or location area the subscriber is currently visiting (3). With this information the GMSC can route the incoming call to the particular MSC that can complete a proper call to subscriber B (4). Since the MSC area is covered by many BSCs and BTSs, it has to check (5) the exact location of B within the VLR, which passes the latest subscriber information, including the location, back to the MSC (6). With this information the MSC is able to forward the call to subscriber B (7) through the proper BTS.

Routing of a call from the fixed network follows the same principle. The only difference is that the GMSC will be accessed from an ISDN or a PSTN.

10.1.3   Call routing when a mobile station is roaming

When the mobile station is roaming, the process includes a foreign network to which calls have to be properly routed. Figure 10.3 shows an example of a call from the fixed network to a roaming mobile subscriber. The originator of the call does not usually know where the called mobile is actually located and may not even know the called number is a mobile phone that could be in a different country roaming in a foreign network.

The ISDN/PSTN subscriber A sets up a call to GSM subscriber B by dialing his normal PLMN access number. The call is routed through the

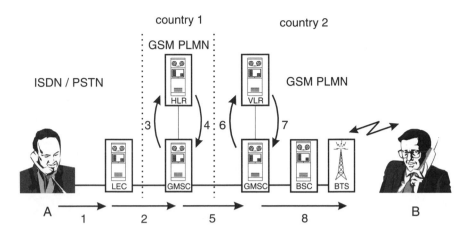

Figure 10.3 Routing a call to a roaming subscriber.

local exchange center (LEC) to the HPLMN (1 and 2). The GMSC asks its HLR where the subscriber is currently located (3 and 4) and routes the call to the (G)MSC that serviced B's last known location (5). After determining the exact location (6 and 7), the call is eventually routed to subscriber B (8). The interesting and important fact is, as stated earlier, that A will not notice that he is actually calling into a *foreign* network.

A special case applies when B (normally in a country foreign to A) is roaming in the home country of A, which is depicted in Figure 10.4. This time A calls B, which is an international call because B is permanently registered in a foreign country. The call is routed to B's home PLMN. From there the call is routed back to the originating country because B's HLR knows that B is actually located there. Here we have call routing with two international links.

10.2 Charging principles

The standardization body for GSM agreed on charging principles on which networks should base their charges to their customers [5]. The home network operator will collect all charges an individual subscriber incurs, including those from foreign networks. A general rule is that the calling party pays. This does not only hold true for a GSM subscriber, but

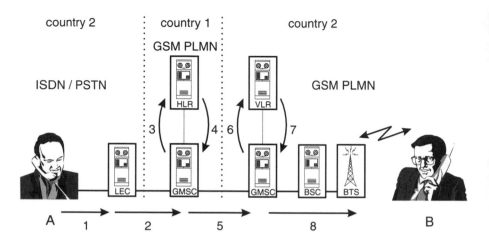

Figure 10.4 Special case for routing a call to a roaming subscriber.

also applies to callers in the fixed network calling into a GSM PLMN. The rate that is fixed for calling out of or into a GSM PLMN should cover the investment for the infrastructure that allows the call to be completed in the first place. Thus calling a mobile subscriber from a fixed phone is usually more expensive than completing a long-distance call to a fixed-line phone from the same fixed phone. But calls made from a mobile phone are usually more expensive.

Just like all rules in this world there are exceptions to general practice; rules are made to be broken. The exceptions are discussed in the following sections as we make distinctions according to the location of a mobile subscriber and the mechanisms the networks use (e.g., call forwarding) to connect callers to mobile subscribers. In the examples that follow we always assume a fixed-line subscriber A, and a mobile subscriber B.

10.2.1 National call charges

When both parties, the calling and the called party, are located in the same country, the calling party always pays the charges incurred. Figure 10.5 distinguishes both cases; when subscriber A calls subscriber B and the other way around. Both subscribers get their respective monthly bills from their individual operators. In some countries the same operator

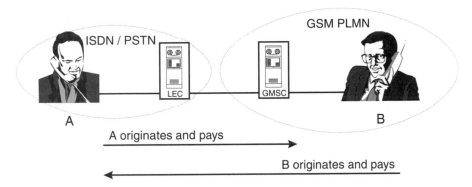

Figure 10.5 National call charges.

might run both the mobile and the fixed networks so that he might receive a combined bill.

10.2.2 Call charges when roaming

When the GSM subscriber, B, is roaming he also has to pick up the additional international part of the costs, the so-called roaming leg. The roaming leg is the part from the home PLMN to the visiting PLMN. The reason for allocating the call charges to the called party is that there is no way the caller could be aware of B's location in another country. Subscriber A still has to pay for the part from the ISDN/PSTN network to the GSM PLMN. The situation in which subscriber B calls subscriber A is handled as a normal international call where B has to pay the international rate of the visiting country plus a roaming surcharge, which will be collected by his home operator as a reward for properly handling the roaming customer. Figure 10.6 illustrates the situation.

The worst case for charging occurs when subscriber A tries to call roaming subscriber B, who lives and pays his monthly GSM bill abroad, but is currently located in A's home country, as shown in Figure 10.4. In this case A incurs international call charges (A should know where B lives) and B has to pay the international leg back to A's country. A cheaper connection appears when B calls A. Subscriber B has to pay only a local call charge for the noninternational call through the local GSM network plus a roaming surcharge to his home operator.

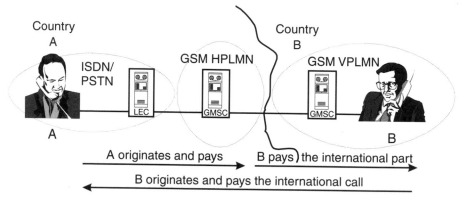

Figure 10.6   Call charges when roaming.

10.2.3   Call forwarding

The third case applies when subscriber B activates the *call forwarding* supplementary service that was explored in Chapter 7. When A tries to call B he will be charged for the call into the GSM network. The remaining part from the GSM network to the forwarded-to-number will be charged to subscriber B.

10.2.4   More exceptions to the rule

The three examples just given should serve only as guidance. There are, of course, many more cases where a subscriber incurs charges that are at variance with the "caller always pays" rule. We are reminded that a monthly subscription fee is collected from subscribers by mobile network operators that covers the expenses associated with making the network available day and night, making sure that monthly bills are accurate, and paying for improvements to the network that make it even more valuable.

That there are also exceptions to the rule becomes more evident with the example of India where the mobile subscriber always pays [6]. It is considered normal practice among a growing number of PCS 1900 operators in North America to charge nothing for the first minute of incoming calls to mobiles. Subscribers are spared the expense of answering "wrong numbers" or unwanted calls, and are thus not discouraged from giving out their numbers, which increases traffic. Many of us turn to complaining at bill paying time; we complain that the charges for mobile phone services are too expensive. It is at such times that we can consider the alternative: how would we have to alter our daily lives if the mobile network we have grown to depend on were to suddenly and completely fail or disappear. What is its value to us then?

10.3   Phase 2+: support of optimal routing (SOR)

In previous sections we saw that call routing does not always follow the most direct route, particularly in the case of roaming subscribers (see Section 10.2.2). In addition, the call charges for nonoptimal routing are quite high, and the traffic unnecessarily occupies and ties up international lines. The international link is paid twice in some cases. There must be a better way. One small example should help as an introduction to this section about optimal routing.

Two employees (A and B) of the same company are on a business trip in a foreign country, say, Italy. Both A and B have their mobile phones with them and both are registered in their home country, say, Germany. If A, who is attending a meeting in a hotel in Rome, wants to communicate with B, who is finishing a report at another hotel in Rome, A has to call B through his home number (in Germany), which results in an expensive international call, and B has to pay, again, for the incoming roaming leg from A (through Germany). This is a rather expensive call if both are located in the same city!

GSM Phase 2+ brings a solution to this kind of situation that will be implemented through *support of optimal routing* (SOR). The intention of SOR is to reduce unnecessary international links. The initial phase of SOR will focus on two situations:

1. Optimal routing of calls within one country, which applies to the extreme but not uncommon example presented here;

2. Optimal routing of calls to the country where the call normally would have been routed, and which especially applies to conditional call forwarding.

Optimal routing applies to all teleservices and bearer services, but not to dedicated PAD and dedicated packet access. Callers from fixed networks will not benefit from optimal routing since their networks do not have the ability to investigate a GSM PLMN and to apply optimal routing. Only GSM mobile subscribers, who typically pick up lots of extra charges, will benefit from SOR. If one of the networks involved in a connection does not support SOR the normal GSM call routing will apply.

The advantage of SOR is reduced unnecessary international call traffic and a resulting reduction in call charges. Each call record for *optimal routed* (OR) calls or call forwarding will contain the connection charge (for the reduced distance) and an OR flag indicating that optimal routing was used. Further information is found in [7,8].

The impact of optimal routing is illustrated by some examples where we use mobile subscriber A as the calling party, mobile subscriber B as the called party, and subscriber C as the forwarded-to-party or voice mail server, which could be a mobile or fixed subscriber line.

10.3.1   Roaming mobile subscriber

A and B are mobile subscribers from different countries. A calls B, who happens to be located in A's home country. Under normal circumstances both subscribers would have to bear the high international call charges associated with these kinds of calls. With optimal routing applied, A's HPLMN would try to route the call to B's HPLMN, but B's HPLMN reports back to A's HPLMN that B is currently located in country A. So, A's HPLMN withdraws the call and routes it directly to subscriber B (in country A) rather than through country B, as shown in Figure 10.7. If A is a fixed-line subscriber rather than a mobile phone subscriber, SOR would not be applied and both would pay the higher call charges.

10.3.2   Call forwarding to home country

B is currently traveling in a foreign country C, and B has activated conditional call forwarding to his voice mail located in his home country. Another mobile subscriber, A tries to call B. The call is routed to the visited country C, where a call forwarding condition is met. Because the call is

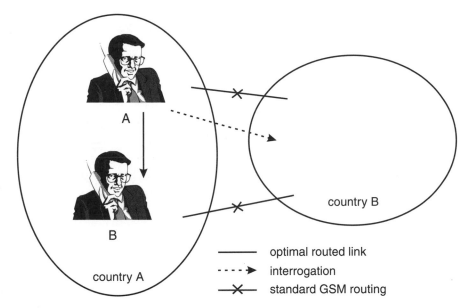

Figure 10.7 Roaming subscriber B located in A's home country.

redirected back to the home PLMN, optimal routing applies and B covers only the local call charges. This situation is depicted in Figure 10.8.

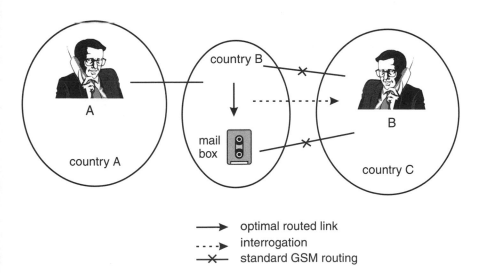

Figure 10.8 Roaming subscriber B with conditional call forwarding.

10.3.3 Call forwarding to visited country

Subscriber B, who lives in country B, is currently visiting a subsidiary of his company in a foreign country A, and has activated conditional call forwarding to a secretary C, located in the visited country (A). Subscriber A, who lives in country A, tries to call B (in country B). The call is routed back to the visited country (country A) where a call forwarding condition is met and the call is forwarded to C (Figure 10.9). Because the call is redirected to the home PLMN of subscriber A, optimal routing applies directly to secretary C, and subscriber A has to cover only local call charges.

10.4 Conclusion

Here ends our look into the networks, at call routing, at some call charging issues, and at SOR. In the next and last chapter we pull the mobile phone from the subscriber's hands and, over the subscriber's persistent objections, remove the phone's battery and pry off the covers to discover how it works and begin to see how we can make one for ourselves.

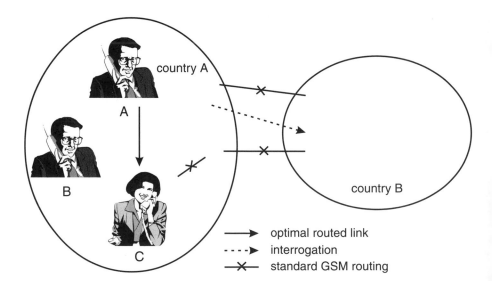

Figure 10.9 Roaming subscriber located in the same country as the calling subscriber with call forwarding to a local number.

References

[1]   GSM 03.04, "European Digital Cellular Telecommunication System (Phase 2); Signaling Requirements Relating to Routing of Calls to Mobile Subscribers," ETSI, Sophia Antipolis.

[2]   http://www.voicestream.com.

[3]   http://www.omnipoint.com.

[4]   GSM 03.12, "Digital Cellular Telecommunication System; Location Registration Procedures," ETSI, Sophia Antipolis.

[5]   GSM 02.20, "European Digital Cellular Telecommunication System (Phase 1); Collection Charges," ETSI, Sophia Antipolis.

[6]   Shankar, N. K., "Regional Focus: India," *Mobile Communication International,* Issue 33, July/August 1996, p. 17 ff.

[7]   GSM 2.79, "Digital Cellular Telecommunication System (Phase 2+); Support of Optimal Routing (SOR); Service Definition (Stage 1)," ETSI, Sophia Antipolis.

[8]   GSM 3.79, "Digital Cellular Telecommunication System (Phase 2+); Support of Optimal Routing (SOR); Technical Realization," ETSI, Sophia Antipolis.

PART

III

GSM technology
and
implementation

Introduction to GSM technology and implementation

Even though the TDMA concept was not new when GSM was introduced to the world along with some other second-generation "digital" wireless systems, the industry had to take some big steps into new territories. The first halting and plodding steps were in the direction of getting everything to work. A mountain of details had to be sorted out: the radio techniques had to be implemented, speech codecs needed to be tested, and systems had to be designed in the light of actual commercial practice. The steps became a brisk walk as the industry transformed the technology so that it could be reproduced under different manufacturing circumstances, and it broke into a run as it slashed the cost of the technology so that millions of people all over the world could afford it.

The different implementations of digital mobile telephones made tremendous progress in only a few years. Just compare the size, weight, and cost of the early models of only a few years ago with those available today. A glance into the trade publications proves that progress is far from over. The network components, base stations, switches, location registers, antennas, and supporting systems have sustained equal transformations even as they are hidden from the gaze of those who buy digital phones. No single chapter in any book can do justice to the fascinating technology and extraordinary accomplishments of the industry, but we can show the way. What are the issues we should consider when we design a modern digital cellular product? What are we getting into? All GSM phones have specific jobs to do. What are their functions, and how do we implement them? What are the typical building blocks that can make up the functions, and how do they work?

There are about as many ways to build a digital cellular phone as there are people who endeavor to do so. Different implementations are carried out through different combinations of methods, technologies, and architectures; each approach has its own advantages. Rather than give advice on how to design certain functions, we treat the salient aspects of the technology, and point out valuable references. Though we explore the whole GSM system, we concentrate on the phone itself: the most visible part of GSM, which, as far as the subscriber is concerned, *is* the GSM system. We offer detailed explanations of important and uniquely digital functions in GSM phones that are often left to specialized literature. One of the functions is the *speech codec*, which, though it is well known and the object of many informal discussions and marketing hype, is surrounded by myths. Another one is the forgotten and mysterious *equalizer*, which, since it uses so many important digital radio techniques, we cover in some detail.

With these two "hard parts" exposed and explained, the remainder of the phone is reduced to (1) variations on the functions and techniques used in the speech codec and equalizer and (2) traditional radio functions and circuits (mixers, amplifiers, oscillators, etc.). The aim is not to offer up all the details required to build a GSM phone, but to achieve a general understanding of the building blocks, architectures, and implementations; to point the way into the technology rather than explain it all Benchmark figures are given where necessary. Only a basic understanding of communications theory and related system matters is expected of the reader. The last section of this chapter is a guide to the references and further reading.

11.1 Breaking GSM down

Unlike some introductory books that consider only the broad system aspects of GSM, just as this book does, we reserve some space in this final chapter to probe into the building blocks, functions, devices, and techniques of digital radio.

11.1.1 Physical and logical blocks of a GSM mobile station

As in any communications systems, a digital cellular system's purpose is to provide reliable, cost-effective transmission and reception of the user's information. Digital cellular systems carry user information, including speech, computer or message data, and even facsimile information in digital form. Even though we talk broadly about a digital communications system, the physical transmission from a digital radio transmitter to a digital receiver is analog, and so is the audio interface to the users on both sides of the channel. The digital processes are confined *within* the phone.

If we look inside a GSM MS we can identify a number of building blocks that process audio signals and user information before transmitting it. Some other blocks work in the opposite direction because they convert the received signals back into the form in which they entered the transmitter at the other end of the link. A measure of a phone's performance can be how "invisible" the phone appears to the user: the voice from the receiver's speaker should sound exactly as it did at the transmitter's microphone, and computer data should arrive at the receiver free of errors. Even as we strive to make the phone invisible, we need some kind of control and interface functions to do so. Figure 11.1 shows a logical block diagram of a GSM mobile station. This figure concentrates on the physical and logical *signal processing* flow. The blocks in Figure 11.1 represent functions, not necessarily actual devices. As we will see, because many of the signal processing functions share similar processes and techniques, only a few physical devices perform the tasks.

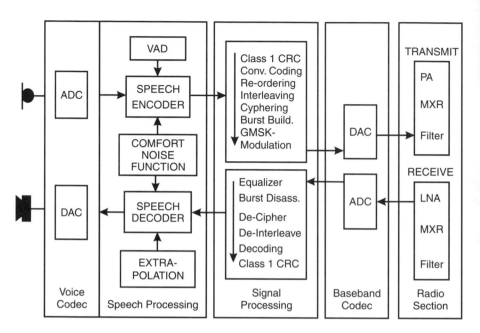

Figure 11.1   Block diagram of a GSM mobile station signal processing.

11.1.1.1   The voice codec

The *voice codec* performs the conversions between the radio's speech processing functions, which are digital processes, and the analog audio domain outside the phone. We find two converters in the voice codec: (1) the *analog-to-digital converter* (ADC) for the analog input signal from microphone and (2) the *digital-to-analog converter* (DAC) for the recovered analog output signal to speaker. The output of the voice codec's transmit path is the digital input to the GSM speech codec. The input to the speech codec is a 13-bit representation (13-bit resolution) of the voice signal sampled at 8 kHz. This 13-bit linear *pulse code modulation* (PCM) is represented by a 104-Kbps data stream ($13 \times 8,000 = 104,000$), which means there are 8,000 samples per second, where each sample is represented by 13 bits.

The voice codec contains band-limiting low-pass filters for anti-aliasing in the transmit path, and waveform reconstruction in the receive path. There are also some gain amplifiers for the signals coming from the microphone and the signals that drive the speaker. Multiplexers can be used in order to switch acoustic signal paths, for example, from a built-in handset microphone or speaker to a hands-free microphone or speaker.

11.1.1.2 GSM speech processing

The GSM *full-rate* (FR) speech codec (speech encoder and decoder) compresses the transmit side's 104-Kbps voice signal into only 13 Kbps by drastically reducing the amount of redundant information from the voice codec [1,2]. The 13-Kbps speech data rate is more suited for transmission over the limited radio resource, which is a sharp contrast to the typical ISDN (fixed-line) PCM signal, which represents speech at a relatively high rate of 64 Kbps.

For GSM speech encoding/decoding on the fixed network side, a conversion between the fixed-line 64-Kbps (8-kHz sampling rate × 8-bit resolution) and the GSM 104-Kbps speech codec (8-kHz × 13-bit resolution) is done by companding and expanding the digital representation of the audio in accordance with the A-law or μ-law scales. A compander converts the higher resolution 13-bit values to 8-bit values by quantizing low-amplitude signals more precisely than high-amplitude values. Low-amplitude samples are given a higher resolution with smaller step sizes, and high-amplitude values get a lower resolution with larger step sizes. The reverse process of compressing is called expanding, hence, *companding*.

Further significant reductions in redundant speech information and data rates call for a much more intricate processes beyond the simple quantization process; the final reduction in payload data occurs in the speech coder.

The counterparts of the mobile terminal's speech codec can be found in the networks in different physical places, for example, with the BTS, BSC, or MSC, although it is logically associated with the BSS. The placement depends on network topology and line traffic considerations. The transport of speech data on the network takes place either in encoded logical 16-Kbps subchannels (13 Kbps + synchronization data, submultiplexed from 64-Kbps channels), which is the most economical way in terms of line traffic, or in standard 64-Kbps channels.

The GSM full-rate speech coder is a type of hybrid coder (see Section 11.5.4) called a *regular pulse excited long-term prediction codec* (RPE-LTP). The output of the GSM speech coder occurs in frames of 260 bits at a rate of one every 20 ms (260 bits/20 ms = 13,000 bps). This speech data (separated into class I and class II bits) is passed on to the channel coding process, which prepares the coded speech bits for transmission on the radio channel [2,3]. The reverse side of the process in the receiver sees 260-bit frames coming from the channel decoder, which are used by the speech decoder to recreate the speech sounds that specified the bits the coder generated. Auxiliary functions such as *voice activity detection* (VAD), extrapolation, and a comfort noise function support features such as *discontinuous transmission* (DTX) and substitutions for lost frames. DTX is an effective way to save battery capacity by minimizing transmit cycles when no voice activity is detected. This feature also reduces interference in adjacent cells.

Two more speech coders are specified for use in GSM networks. One is the *half-rate* (HR) coder for use in half-rate channels [4]. GSM half-rate frees up the use of every other time slot in a single full-rate traffic channel, thus doubling the channel capacity. The *enhanced full-rate* (EFR) coder uses the same rate as normal full rate [5]. It was required by the North American personal communications industry for use in PCS 1900 networks. The EFR algorithm that was chosen (known as the US-1 codec), and also adopted by ETSI (through GSM Phase 2), offers much better perceived voice quality.

General speech coding techniques, and the GSM speech coders in particular, are explained in considerable detail in Section 11.5.

11.1.1.3 Signal processing

The signal processing function works with the user data (speech or data traffic), as well as signaling and control information. Payload data must be well protected before transmission over the radio channel, and the protective measures must be "undone" on the received data. The GSM *channel coding* block protects speech coder information with several processes: *cyclic redundancy check* (CRC) for the most important class I bits, followed by convolutional coding, and interleaving. User data (other than voice data) are transmitted over GSM data channels, which are convolutionally coded and interleaved with their own unique schemes, depending on the data rate, that are different from those used on voice data. Signaling data are block coded, convolutionally coded, and block interleaved according to their own rules. A common ciphering scheme can be applied to the channel coded data. The *channel decoding* block reverses the transmitter's coding for the received data. The burst building and multiplexing process adds a midamble (training sequence code), tail bits, and guard bits to the encoded bits after the interleaving, and makes sure that the burst data (radio bits) are delivered to the modem and the radio at the right time. The radio bits are differentially encoded and modulated through a Gaussian filter function as part of the modem function [6].

The reverse side of the modem function is dominated by the channel equalizer, which adaptively compensates for the adverse effects (intersymbol interference) applied to the original signal through time-variant multipath interference, delay spread, and Doppler shifts. This function is explained in Section 11.6. After deciphering and deinterleaving the data from the equalizer, a decoding algorithm recovers the original payload data.

11.1.1.4 Baseband codec

The baseband codec converts the transmit and receive data into analog or digital signals, respectively. The transmit process delivers analog baseband I- and Q-signals (from two DACs), which modulate the radio carrier [2]. The receive process converts the analog I- and Q-signals from the radio back to a digital data stream, which is filtered, sampled, and quantized (again, two ADCs) before presentation to the equalizer. A typical value for the converter's resolution is 10 bits for both directions.

11.1.1.5 The radio section

The radio section, which includes the IF stages, is one of the interfaces to the outside world; the other is the audio interface. Functions such as frequency synthesis, a local oscillator, up-and-down conversion with mixers (MXR in Figure 11.1), analog RF and IF filtering, transmitter power amplification (PA in the transmit path of Figure 11.1), preamplification in the receive path (LNA in Figure 11.1), RF pulse shaping, automatic gain control, and frequency correction and control are handled in the radio section. A general view of the radio section is offered in Section 11.2 of this chapter.

11.1.1.6 Other functions

Apart from routine signal routing and conditioning tasks between audio, baseband, and radio blocks, there are some other functions that complete the GSM mobile station. Some control units are required to organize and schedule logical and physical entities. On top of the physical layer functions, which should be viewed as the physical conditioning and processing of payload data (coding and decoding), we need to accommodate some higher level tasks such as (1) synchronization, (2) frequency and time acquisition, and tracking, (3) monitoring (serving cell and adjacent cells), (4) received signal strength measurements, and (5) radio control. Above these radio tasks we need to add the higher protocol layers (layers 2 and 3) with signaling functions, which allow the phone to operate in an orderly way within a network. Other control functions within a GSM mobile station handle the user interfaces (keypad, display, and the user menu), the SIM interface, and other auxiliary (data) interfaces.

11.1.2   Physical and logical blocks of a GSM base station

A GSM base station is the mobile's counterpart, but not in all its aspects. The base station does not provide for most of the layer 3 signaling functions, and none of the higher layer functions, because it is transparent to those functions coming from the network. Because the base station converts the network's wire environment to an air interface, base stations exercise their own control over the layer 1 and layer 2 processes. The signal processing functions in a GSM base station are similar to those in a mobile station. Additional logical channel structures, not found in the mobile station's uplink, must be supported by a base station: (1) *common control channels* (CCCH), which comprise the *paging channel* (PCH) and the *access grant channel* (AGCH); (2) the *broadcast control channel* (BCCH); (3) the *frequency correction channel* (FCCH); and (4) the *synchronization channel* (SCH). All of these are unidirectional channels; a mobile does not support these channels beyond being able to read and understand them. The interface to the fixed network requires some wireline physical and logical entities that provide transmission and reception capabilities for the user data, signaling data, and control information.

One of these entities is the *transcoder rate adaptation unit* (TRAU), which supports the multiplexing of speech data to and from the Abis interface [7]. Just as there are many configurations possible within a GSM network, so there are many architectures of GSM base stations. There are different applications, different numbers of channels, different configurations for standby channels, and different network interface configurations. Figure 11.2 shows only one example, which assumes that the speech codec is placed remotely in the network (e.g., within the mobile

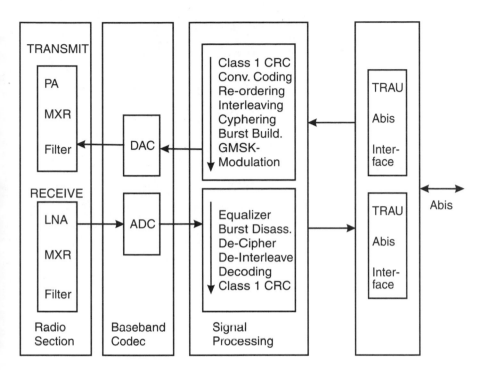

Figure 11.2 Sample block diagram of a GSM base station signal path.

switching center), and shows the architecture for a single transceiver. The appropriate signal processing and radio sections would exist in parallel in multiple-transceiver architectures.

11.2 Transmitters and receivers

The radio sections in cellular phones and base stations require an enormous design effort. Circuits need to meet specifications even as they are manufactured in large quantities at low cost. In particular, this means that analog circuitry must be produced with an absolute minimum of individual "tweaking" (tuning and calibration) exercises such that millions of "equal" radios are the result. Radio design, which is strictly separated from the digital design task, is sometimes regarded as "black magic" or "alchemy" rather than a disciplined and orderly engineering task. These kinds of references arise from myths surrounding RF work that are given prolonged life as "lab legend" by those unfamiliar with the techniques and skills of RF practice. Radio design is a balancing act in which the successful practitioners find the optimum combination of technology, technique, cost, and performance. The skill and experience of the attending engineer is the key to meeting design goals.

Figure 11.3 shows one sample block diagram of a digital radio structure that may be used in a GSM transceiver. Here we find a single conversion transmitter and a dual conversion, superheterodyne receiver with quadrature (I/Q) modulation and demodulation, respectively. Many other architectures and techniques are possible, and the reader is referred to existing literature on the subject. Section 11.12 is a starting point for those new to radio work. The most important aspects, parameters, functions, components, and requirements are discussed in the following sections.

11.2.1 Transmitters

There are two kinds of radio transmitters in GSM: (1) the ones in the BTS, which create the physical forward side of the RF link to the mobile station (downlink), and (2) those we find in the MS, which create the other—the

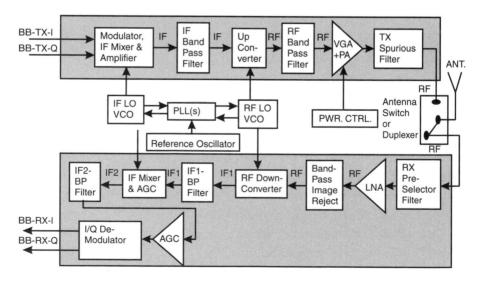

Figure 11.3 A GSM radio transceiver architecture example.

reverse—side of the RF link back to the base site (uplink). The chief difference between these are (1) the much greater RF power, relative to the MS, generated in the BTS, and (2) the tiny size of the MS with the proximity of the transmitter to the MS's receiver. The similarities outnumber the differences. Those who design transmitters for GSM service are obliged to consider the following parameters:

1. The *power output* is seldom more than 2W for an MS, but it can be more than an order of magnitude greater in a BTS. The amplification job usually needs to be distributed over several stages in the high-power applications in base stations [8]. The transmitter's output power needs to be managed over temperature as well as with changes in the source impedance of the power source for the RF amplifier stages, and the RF output components (including the antenna).

2. *Power consumption* (efficiency of amplifiers) is critical in the MS.

3. *Load pull* is the transmitter's tendency to change its output frequency with impedance changes at the antenna, or in the power source of the transceiver [9].

4. *Modulation accuracy* (linearity) is compromised by attempts to increase the efficiency of the transmitter [10].

5. *SNR, hum,* and *noise* are general terms for those factors that limit the best *signal-to-noise ratio* (SNR) that we can expect from a transmitter. Linearity problems in the modulation conspire with noise sources and unwanted RF coupling to put a limit on our best efforts.

6. *Spurious outputs* can be either harmonic ones or all of those undesired outputs that are not harmonics of the desired output [9]. The former usually come from linearity problems in the RF amplification stages of the transmitter, and the latter find their origin in mixers and frequency synthesizers.

7. Excessive *adjacent channel power* comes from too much *single sideband* (SSB) phase noise from the transmitter's oscillator, or poor fractional frequency stability performance of the frequency synthesizer [9].

8. *Intermodulation distortion* appears when a strong signal from outside the transceiver is coupled into the transmitter, through the antenna, and mixing products are created in one of the many nonlinear mixing sites anywhere in the transmitter's signal path. This is a particularly troublesome problem in base sites, because base stations often share site space with other high-powered transmitters such as TV transmitters.

Seven of the transmitter's functional components that are not a part of the digital baseband functions are listed next in the order in which they appear as the transmit path approaches the antenna:

1. An *oscillator* is one of the reference inputs to a *phase-locked loop* (PLL) synthesizer.

2. The *frequency synthesizer* functions as *an intermediate/radio-frequency voltage-controlled oscillator* (IF/RF VCO), which is a means to add frequency agility to the fixed reference oscillator. The relatively high RF levels of the transmitter's output reduce the required noise performance of the synthesizer compared to what is required of the same function in the receiver's *local oscillator* (LO).

3. The *modulator* imparts intelligence to the carrier. GSM's GMSK modulation affords many possibilities for modulators, but some kind of I/Q process is the general method [10]. I/Q modulators are sometimes referred to as "universal modulators." We could, for example, use a more direct technique than an I/Q method to generate MSK, such as a pair of oscillators operating 67 kHz on either side of the assigned channel frequency, and then use the modulator's data stream input to enable one or the other oscillator. Chief among the many reasons such direct methods are not used in digital radio is that the frequency shifts are sloppy and abrupt. The more complicated I/Q method can be made to generate a wide variety of digital modulation schemes including GMSK. They work by enlisting two double-balanced mixers into which we apply alternate halves of the symbol data stream; first into one mixer, and the next into the other mixer. Each mixer has a reference input of the transmitter's carrier frequency shifting from

each other by 90 degrees. The outputs are summed and passed into the up-converter. The utility of this *quadrature* technique is that the phase and the amplitude of both parts of the data input can be adjusted and filtered through look-up tables before they are applied to the mixers.

4. An *up-converter* (mixer) brings the transmitter's output up to the assigned frequency. There are many kinds of mixers, and all of them produce odd order products that have to be filtered from the transmitter's output [10]. The double balanced variety is often used, because it isolates the frequency synthesizer from the variable load of the RF power amplifier.

5. *Amplification* raises the transmitter's output to the required RF level specified for the task. The job is made particularly difficult in the GSM mobile station, because the power has to be adjusted over a wide range and intermittently keyed within strict burst power-time templates without the AM splash that usually accompanies intermittent duty transmitters. All of this has to be accomplished in a small volume with a limited power source (battery).

6. *Filters* attenuate most kinds of spurious outputs before they reach the transmitter's antenna, but they have none of the effects of unwanted outputs caused by linearity problems in the modulator. Filter design has become a very specialized skill, and the possibilities are enormous [8].

7. Though the half-duplex operating mode of GSM may seem to eliminate the need for a *duplexer*, it is often included in mobile stations. The duplexer is a filter that is common to the transmitter and the receiver, and it is included to (1) add receiver front-end filtering to reduce the harmful influences of nearby transmitters in other phones and (2) increase the isolation between the transmitter and the receiver. (See the same subject in Section 11.2.2.)

The transmitter's performance characteristics, which we listed in this section, are mostly a matter of regulatory compliance. Designs must comply with the requirements, but there is no compelling reason to increase costs by improving on them. When our transmitter designs fail to meet the required minimum performance we cannot expect type approval. The requirements are, however, not capricious and arbitrary barriers designed merely to increase the costs of radios offered into GSM service; they are designed to make the best use of the spectrum, and yield a balance between system capacity, costs, and the best quality of service the technology can deliver.

11.2.2   Receivers

The wide use of traditional radio technology and techniques in the RF functions contrasted with the use of integrated functions on silicon for the baseband functions that we examined in the transmitter is also evident in receivers. The receiver's job is to extract the signal from a distant transmitter and recover the modulation. Once the modulation is recovered, advanced techniques (1) equalize and decode the channel, (2) separate signaling from traffic information, (3) recreate the voice sounds that entered the transmitter, and (4) bring audio or user data out of the transceiver.

GSM's minimum requirement is to be able to recover a transmitter's symbol stream with less than only 1 bit error in 100 for signals of less than -102 dBm, which is only 6.3×10^{-14} W. The noise present at the input of a GSM receiver will be, at best, about -121 dBm. This *bit error ratio* (BER) requirement as a criteria for receiver sensitivity is rather modest when compared to wireline systems where one error in one million symbols is considered intolerable. Because channels in wireline systems are relatively stable and orderly compared to those in wireless systems, the protection coding in wireline systems would be too weak to be able to respond to the error rates typical in radio channels. Current GSM receivers perform far better than the minimum -102-dBm requirement. Typical implementations achieve the required BER performance at nominal levels of between -105 and -110 dBm. Outside system influences, including the presence of strong blocking signals, or co-channel and adjacent channel interferers, Doppler and deep fades, quickly reduce margins in the designs. To recover information with low BER from a signal that is suffering interference from a co-channel signal (from a distant base station or mobile station using the same physical channel), the available signal bandwidth should be high. The opposite is the case for adjacent channel interference, which typically occurs at higher levels than co-channel interference. Here we want to have an as narrow as possible filter characteristics in the signal recovery path in order to separate the good signal from the bad ones. (See Figure 11.5 in the following pages for the effect of IF filters that perform this task.)

Surface acoustic wave (SAW) filters are the device of choice for IF filters. They are piezoelectric transducers on which interfering mechanical waves are set up such that the devices become bandpass filters with sharp cutoff characteristics. SAW devices have high insertion losses (about 20 to 30 dB). The most narrow bandwidth characteristic is selected such that it can still pass the received signal without distorting the symbols that eventually appear from the demodulator; too narrow a filter will hurt the BER performance of the receiver.

In general, improved system performance in wireless networks is limited to optimizing receivers, not the transmitters, and all kinds of tricks and techniques are brought to receiver designs to improve BER performance. The improvements tend to be confined to the baseband processes (DSPs or ASICs) after the demodulator or detector. All air interface standards go through a maturing process, such that we can see, as a rough rule, a steady improvement in the average BER performance of all the receivers deployed in a particular kind of air interface technology of about 1 dB each year. Eventually, we see receiver improvements slow down such that each 1-dB improvement takes much longer than a year to achieve, and each improvement quickly becomes more and more costly.

This slowing in receiver performance improvements occurs as they approach fundamental thermal noise limits. When we finally observe this slowing in receiver improvements, then we declare the host air interface technology to be "mature," and we can start to look for spurts of innovation elsewhere in the systems that wring additional capacity, performance, and quality from the system by further lowering average BERs throughout the systems. We also see increased interest in new air interface proposals as system improvements slow and become more expensive.

A receiver fulfills its duties by processing the small amounts of energy at the antenna such that the original information is recovered with minimum distortion and errors. The specifications that are used to judge a receiver's particular abilities to do its work are discussed in the following paragraphs.

The *sensitivity* of a receiver refers to its ability to react properly to a weak signal. Digital receivers use the maximum BER at some low RF level as a measure of their performance. This is analogous to the SNR technique used in analog receivers [9,11]. The SNR technique is not useful in digital receivers because they have some kind of decision circuit early in their signal recovery and analysis stages that makes constant determinations on what the actual transmitted symbols probably were. If the BER = 1%, then the decision circuit makes one bad decision for each 100 symbols it is called on to judge (see Section 11.6).

The *selectivity* is a receiver's ability to reject unwanted signals that are very close to the frequency of the desired signal [9]. These interfering signals are usually spurious emissions from nearby transmitters. The duplexer and the IF stage's SAW filter are the devices of interest here. The average price of phones can be held down with strict adherence to outside system matters such as low transmitter adjacent channel power emissions, strict frequency planning, and proper system design.

Intermodulation rejection is the receiver's ability to overcome its own natural tendency to generate an internal on-channel signal from off-channel signals present at the antenna. The receiver's spur-free dynamic range is a measure of its immunity to intermodulation, which is primarily determined by the receiver's input stages. Figure 11.4 illustrates the derivation of the IP3, or third-order intercept point, and the spur-free dynamic range of a receiver.

A receiver's nonlinearity can be measured by injecting two closely spaced signals (f_1 and f_2) of increasing but equal amplitude into the RF input (antenna port) while observing the rise in third-order intermodulation products at both $2(f_1) - f_2$ and $2(f_2) - f_1$. Intermodulation products appear in the receiver when the input test signals exceed whatever level drives the receiver beyond its spur-free dynamic range. Figure 11.4 shows that the level of the third-order intermodulation products (the dark curve with the greatest slope in the figure) rises at a rate three times greater than the rise in the input signals (the dark curve to the left of the steep curve).

We confine our attention to the third-order products because all of the other intermodulation products are far outside the receiver's own passband. As the amplitude of the input signals increases, the level of the receiver's output, both the fundamental signals and intermodulation products, increases until the input stages start to saturate (the curved parts of the dark traces in the figure). If we extrapolate the straight parts of both curves in Figure 11.4 beyond the saturation levels, then the point at which both lines intercept is called the *third-order intercept* (IP3).

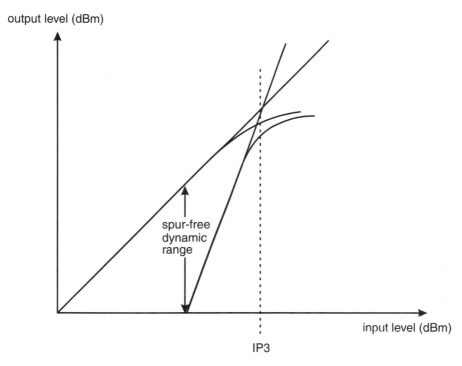

Figure 11.4   Receiver dynamic range.

The further to the right IP3 is in Figure 11.4, the better the receiver's intermodulation rejection performance.

Receivers exhibiting poor intermodulation performance will loose sensitivity; we say the receivers suffer from *desensitization,* as they pass close to transmitters in a different network from that whose services the subscriber is using.

Image rejection is a measure of the receiver's ability to attenuate signals that appear in the receiver's IF stages, or output, as a result of an off-channel signal with the same offset from the LO frequency as the desired input signal. Figure 11.5 illustrates how images are rejected from a receiver's output. The top part of the figure depicts a spectrum analyzer display on which we see our desired signal (made fatter then the other signals for ease of identification) together with others that are not particularly interesting to us and need to be filtered out. The frequency of our LO is plotted in the center of the scale. The second part of the figure has the passband of the RF preselection filter, our duplexer, plotted on the scale; signals

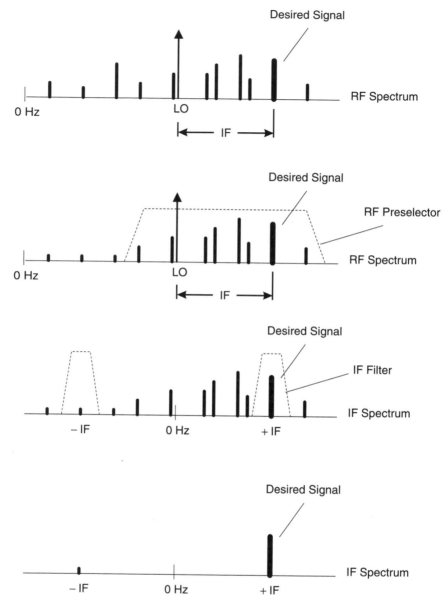

Figure 11.5 Image rejection.

outside the passband are attenuated. The third part of the figure plots the downconverted output of the receiver's IF where the LO frequency translates itself to 0 Hz. The subsequent IF filter's passband is plotted on the right side of the "IF spectrum" scale together with another image channel the mixer translates to –IF. The bottom part of Figure 11.5 depicts the result out of the IF stages. Since the receiver's detector cannot distinguish between the desired signal and the image, it accepts the superposition of both signals. If the lower side of the preselector filter is less then 2 × IF away from the desired signal (the condition shown in the second part of the figure), then the image will be attenuated. Raising the IF will allow more freedom in preselector specifications but will demand IF filters with a smaller fractional passband specification.

With the exception of the demodulator or the detector, the devices and functions in the receiver have the analogous duties they had in the transmitter, but in the reverse order. The chief difference is that since the signal levels are millions of times smaller than they are in the transmitter, the required performance of the amplifiers, mixers, and oscillators (including the synthesizer) is higher. Just as we did in the transmitter, we will follow the signal from the antenna back toward the user [9,10]:

1. A *duplexer* is a three-port filter. The antenna is attached to a common port, and the receiver is attached to another port where we find low loss at the receiver's operating frequencies between the receiver port and the antenna. The transmitter is connected to the remaining port where another low-loss path, at the transmitter's operating frequencies, is found to the antenna. The filter is constructed in such a way as to isolate the transmitter from the receiver. Even though the TDMA/TDD timing in a GSM mobile station calls for a minimum idle time of two time slots between transmit and receive instances, the isolation the duplexer affords between the transmitter and the receiver is valuable.

2. A *preselector filter* limits the bandwidth of the receiver, which reduces *intermodulation* (IM) distortion. Additional preselector filtering is sometimes required to reduce the cost of the duplexer.

3. The *RF amplifier*—also referred to as a *low-noise amplifier* (LNA)— determines the IP3 of the receiver, and isolates the preselector filter from the image filter that follows the amplifier. As we discovered in Figure 11.4, the location of IP3 is a way we can judge a receiver's ability to limit any harmful effects off-channel signals may have on the receiver's performance. Intermodulation distortion is generated in active components in the receiver's input

circuits, which act as mixers for the input signal and another off-channel signal that yields an output within the receiver's pass band. The RF amplifier raises the level of the input signal together with the noise. Improvements in intermodulation performance of any RF amplifier are usually accomplished at the expense of additional current requirements for the amplifier, thus reducing battery capacity. The influence of the noise a device can contribute in a circuit also fights our efforts to optimize a receiver's treatment of weak signals. Space restrains us from including a treatment of these important influences, but the reader is invited to refer to [9] and also Chapter 4 in the same reference for guidance.

4. An *image filter* can follow the RF amplifier to attenuate spurious signals and images. It also attenuates harmonic distortion generated in the RF amplifier.

5. The LO is the mix-down reference for the input signal, and it is generated in the same frequency synthesizer that creates the transmitter's carrier, but not on the same frequency. An injection filter between the LO and the mixer (see next item) will attenuate noise from the LO.

6. The *mixer* accepts the LO's input together with the amplified and filtered input signal to generate a lower frequency copy of the input signal. The output is called the *intermediate frequency* (IF). There are many kinds of mixers, each with their own advantages and disadvantages [9].

7. The *IF amplifier* and appropriate control logic are responsible for most of the analog gain of the receiver, which is constrained with an *automatic gain control* (AGC) function.

8. The *detector* or *demodulator* is the last analog function in the receiver. It reverses the transmitter's modulation process in order to recover the original I- and Q-signals, which are sent to the baseband processes as shown in the lower left corner of Figure 11.3.

Designing a receiver is an elaborate balancing act that starts with allocating gains, losses, and signal levels over all eight blocks in the receiver's

RF and IF paths. The process is reduced to an orderly one in [9], which is highly recommended for the reader new to this kind of work. As the design proceeds, the designer will find herself clearing spurious signals generated in the last mixer stages, which starts the balancing act over again.

Interest is growing in using digital techniques in front of the baseband chips in receivers. One of these new techniques is the *digital IF*, which replaces the last mixer in a typical analog superheterodyne receiver with an *analog-to-digital* (A/D) converter. If the first IF is, say, 10.7 MHz and we are interested in designing a wideband receiver that can be used for a variety of services in addition to cellular and PCS ones, then the A/D converter must perform 21.4 million samples per second. If we limit the IF bandwidth, as we would if we are designing an alternative handset for PCS services, we can reduce the sampling rate. All further processing duplicates the analog mixer, but performs it in the digital domain. The I and Q baseband components can be extracted from the digitized 10.7-MHz IF by performing all the tasks an actual analog I/Q demodulator accomplishes within a virtual digital copy of the process. Such a virtual machine requires digitized sine and cosine versions of the reference signal from an oscillator or frequency synthesizer, and can recover the I and Q components by performing multiplying operations between the digitized IF and sine/cosine references. Digital filters can notch-out unwanted signals, and then, once the result is decimated in proportion to the reduction in IF bandwidth, DSP processes can be applied as they are today with current techniques.

Readers new to digital radio are usually surprised by the wide use of old and traditional radio technology and components used in the RF sections of base stations and phones [12]. This practice has two origins. The first is that optimizing any of the circuits and functions listed above is not a particularly straightforward and orderly task: impedances change for a variety of reasons, circuits tend to oscillate even when they are not designed to do so, and ground loops and shielding problems are cleared through experimentation. Once the RF functions finally work, they tend to work very well. It is often easier and cheaper to "borrow" an old circuit from another radio than to design a new one. Second, most of the components in the RF parts of phones are passive devices that are very small, cost very little, and have excellent quality. There is currently no cost advantage in employing integrated RF functions.

As the *power consumption* of a portable phone becomes an increasingly important differentiating factor in the market, transmitter performance is examined more closely. Standby times in mobile phones are limited by the receiver and the baseband section's performance. Exploiting functions like *discontinuous receive* (DRX), in which the receive path in an idle state needs to be activated only intermittently (for reading paging channels or monitoring serving and adjacent cells), yields long standby times. Talk time is greatly influenced by the efficiency of the RF transmitter. Taking advantage of the fact that average speech activity, which can be detected with *voice activity detection* (VAD), is typically around 45% during a normal speech conversation helps save battery power. Enlisting VAD in GSM is called *discontinuous transmission* (DTX). We are still left with the power amplifier's efficiency (loss vs. actually radiated power) as the remaining parameter to improve as best we can. The efficiency of a transmitter's RF amplifier remains the chief limiting factor in battery capacity today; but when DSPs were first employed in radios, they contributed a significant contribution to battery drain. The efficiency of RF amplifiers is a function of numerous factors, including the modulation scheme; a design that works well with GMSK may not be optimum for DQPSK. Efficiencies in the range of 45% to 50% are normal today, and the industry is currently striving to improve this figure beyond 70%. Trade-offs will abound as the 70% efficiency goal is reached. The balance between cost and the ideal efficiency will be difficult to reach without the use of advanced technology.

11.3 MS and BTS—new roads to the ultimate radio

Although they perform similar tasks, there are some differences between the radios in base stations and mobiles, and the differences are likely to increase in the future. As we noted in the previous two sections, there is little incentive these days to consider integrated RF functions in GSM radios. This does not mean that new ideas in RF implementations are doomed to be ignored by the industry, for the growing interest in digitized IFs shows that designers are always looking for something better. The feasibility of monolithic GaAs circuits, for example, offers the possibility of increased sensitivity with better control over SNR compromises and the balancing act that is typical in receiver design [13]. Since today's analog processes are adequate, new digital techniques must substantially enhance performance and reduce costs. Mobile stations are designed to tough cost goals, and network operators shop for smaller base stations with decreased costs per channel.

The number of base stations will always be a small fraction of the mobile population. Base stations are designed to pick up the slack the mobiles leave in the air interface: (1) mobiles have tiny, low-power transmitters that require the base station to have very sensitive receivers, and (2) base station transmitters generate much more power than mobiles to compensate for the suboptimal receiving conditions in which users place their handsets. Because base stations do not need to be carried in pockets, all kinds of measures can be employed to make their receivers more sensitive than those in mobiles. Because base sites are expensive and their proliferation over the landscape is less acceptable to the public than in the past, innovations that strive to make GSM systems support smaller and smaller mobiles with fewer base sites are always welcomed by the industry. Increased base station transmitter power can decrease the density of base sites in a region only if there is a corresponding way to increase the base station's receiver sensitivity. The more sensitive a base station's receiver becomes, the more likely it is to suffer the effects of interference. Inserting very high Q (highly selective) filters in the receiving path can mitigate this situation, but filters add loss to the path. Cryogenically cooled filters (filters cooled with liquid helium) of very high Q have recently appeared in the market, thus offering filters of extraordinary selectivity without the high loss. Overall system performance can be further improved on the receiver side with smart antennas, which are adaptive arrays that can notch-out interfering signals, or improve the base site's receiver directivity toward distant mobiles.

Another salient example of new innovations in radio technology is the so-called *software radio*, which is a concept that describes a generalized wideband radio that can be configured, with software, to work in a large number of systems employing diverse modulation and access techniques. The digital IF mentioned in the previous section is an example of this kind of approach. The techniques demand a great deal of performance from baseband functions, primarily in ADCs, which need to drive AGC circuits over a wide range. The increased costs and battery drain that accompany this kind of highly adaptive technique will force the new technologies into the base stations rather than the mobile terminals. There are some advantages in software receivers: (1) any desired bandwidth can be selected, (2) the receiver's center frequency and phase noise can be controlled, and (3) the I and Q data can be recovered in exact quadrature. Given the increased public intolerance for the proliferation of base sites and antennas, interest will increase in RF multiplexing schemes that allow different base stations to share a single antenna site. A logical extension of this reaction to environmental concerns is the concept of a *RF utility*, which refers to a base site that can couple many different kinds of RF services (cellular, paging, and PMRs) onto the air interface from different providers of mobile radio services for a fee. This concept takes advantage of one of the most valuable assets a RF service provider may own in a crowded RF landscape: a good radio site.

11.4   Baseband signal processing

In this and the following seven sections of this chapter we take a closer look at how a GSM *baseband implementation* can be realized. The term *baseband* refers to all signal processing functions related to audio (voice) signals, user rate data, and the radio bits that are transmitted to and received from the radio/IF section. It also refers to data converters and signaling control functions. What are the issues? What is behind such functions as speech codecs, equalizers, Viterbi decoders, converters, and security features? What are typical performance figures and what resources do they require? We reserve space in Sections 11.5 and 11.6 for a closer look at two uniquely digital functions in GSM phones: speech coding and equalization. Although our focus is on a GSM terminal, some of the subjects also apply to the remainder of a GSM network: the base station (signal processing) and the MSC (speech coding and the handling of protocol functions). Among the key technologies that enable and secure the success of wireless services and products, we can easily identify two important ones: semiconductor components and digital signal processing. Together they allow products to shrink in size, cost, and power consumption and still take on new features with no compromises in performance. Silicon technology and other relevant hardware technologies (filters and amplifiers) continue to go through cycles of innovation that allow even lower cost and smaller products. New and powerful algorithms and a growing skill at implementing them efficiently on target hardware, such as *digital signal processors* (DSPs), provide additional design freedom and product enhancements. The wireless communications industry will continue to drive new technology and clever innovations of older technologies.

Figure 11.6 shows the functional partitioning and terms associated with a typical baseband implementation in a GSM phone. A DSP performs layer 1 functions, speech coding, modem, and channel coding and decoding tasks. The so-called mixed signal functions are A/D and D/A conversions. They bring the analog and digital domains together for voice and baseband signals as well as for some control functions. A microprocessor handles GSM protocol functions, general control functions, and the user interface, which is called the *man/machine interface* (MMI), and which includes the keypad, display, and the user menus. TDMA timing needs to be maintained and supported by a dedicated timer unit. Different kinds of memory hold the microprocessor program and its data (flash ROM), non-volatile programmable data (EEPROM for equipment identity, phone book, and other more permanent data), active program and data RAM,

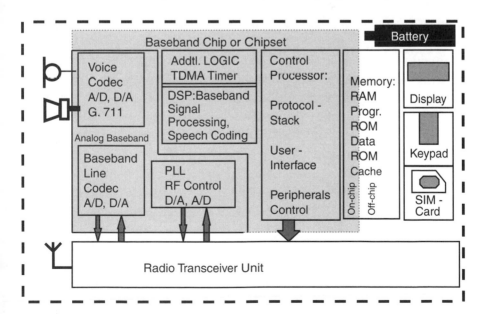

Figure 11.6 Functional block diagram of the digital part of a GSM mobile station.

and, finally, the DSP program and its own data, which are typically located on the same chip with the DSP itself for fast access.

In the block diagram shown in Figure 11.1, we saw that many functions were needed to provide the speech and signal processing in both the transmitting and receiving directions. Flexibility, gate count, and power consumption let today's implementation of most of these functions appear in software running on DSPs. All the functions share a single processor, which has to be dimensioned according to constraints on processing power and the available memory. The software modules that demand most of the processor power in GSM are (1) the modem (equalizer), (2) encryption, (3) speech encoding and synthesis, and (4) channel encoding and decoding. The data handling procedures for GSM data and fax services are performed in the place of voice coding functions, and each has its own channel coding mechanism. The processor load for some of the functions can be reduced with the use of hardware accelerators that carry out nonstandard, operation-intensive functions or functions that have to be repeated many times for certain algorithms. Examples are accelerators used for the Viterbi algorithm (equalizer and decoder) and for the ciphering algorithm (encryption engine). Input data are handed over to an accelerator. When the computations are completed, the results (engine output) are available in memories or registers.

All the signal processing functions in a full-rate GSM phone (modem, channel coding and decoding, speech processing, etc.) require a certain number of instructions per second. In the processor world the unit of measure for these is expressed in units of *millions of instructions per second* (mips), and the mips figure for the functions listed earlier are different for each type of DSP, the DSP's instruction set, and the use of accelerators. The range for a full-rate GSM implementation—using typical hardware accelerators—is around 13 to 20 mips. If we assume one instruction per clock cycle, then this means that a DSP clock speed of 26 MHz is required. This is twice the 13-MHz reference clock frequency from which all GSM timing is derived (see Section 11.10).

Furthermore, the software in object code will require a certain amount of program and data memory space as well as someRAM. Direct memory access with no-wait states is secured by reserving a certain amount of program and data memory "on-chip" with the DSP. This ensures that the crucial real-time operations are carried out very quickly.

11.5 Speech coding and speech quality in GSM

The purpose of speech coding in mobile radio is to reduce the data rate required to transmit speech information from the transmitter to the receiver with minimum adverse impact on the perceived voice quality. Lower data rates can be achieved at the expense of voice quality. Efficient speech coding without sacrificing quality, similar to radio design, is an art of its own. It holds an important place in the world of wireless digital communications. An abundance of speech codecs is available for "wireless" use, each of which features different methods, algorithms, complexities, and performance. The interested reader is referred to [14–17] for additional information on the subject.

11.5.1 Speech coding tutorial

There are three ways to encode speech: waveform coding, parametric coding, and hybrid coding [18,19].

Parametric coders are strikingly different from waveform coders in both performance and construction. Whereas waveform coders can be viewed as elaborate variations on simple A/D converters, parametric coders analyze the input speech signal within strict bounds of, and with full regard for, the processes that created the sounds in the first place. Parametric coders are extremely efficient at removing redundancy from human speech, and offer good voice quality at very low data rates.

As a generic introduction into the subject, let us think of ways to transmit acoustic information over an imaginary transmission line, say, a pair of simple telephones connected by a twisted pair of copper wires. One simple way is a very direct one; let us look at an example. The discovery of some new and interesting chords may motivate a guitar player to call up his friend and play the tune to her over the a landline phone, or—why not?—over his brand new PCS phone. In this way the musician's friend on the other side of the connection is able to listen to the chords instantaneously, but with some restriction in sound quality due to the limited bandwidth in the phone system. Now, let's consider a more indirect way. Our guitar player might ask his friend to use her own guitar as he instructs her exactly where to place her fingers and otherwise manipulate the strings to duplicate the sounds that he discovered for himself earlier in the day. The result of this more indirect approach is that the original sounds are reproduced, after some delay, with better fidelity than if the sounds were transmitted directly over the phone lines. A similar indirect approach is followed in digital phones. The instructions for recreating the sounds in the near end are transmitted to the far end where some apparatus is placed that can transform the instructions back into the original sounds. Confining our attention to human speech, low-frequency sound (3.4 kHz), such as the low-pass filtered human voice, can be linearly digitized (time and amplitude quantization) at 8 Ksps with 8-bit resolution (A-law or μ-law). This technique is often referred to as *pulse code modulation* (PCM), which yields a 64-Kbps bit stream, which is the rate used on time-multiplexed digital wireline transmissions such as ISDN. It achieves the expected, and quite adequate, high-quality speech that serves as a reference today. For wireless applications in which the bandwidth for such a high data rate is not available, compromises must be made as additional signal analysis is applied to reduce the redundancy inherent in PCM data. The three main categories of speech processing—(1) *waveform coders*, (2) *vocoders* or *parametric codecs*, and (3) *hybrid codecs*—are discussed below [15].

11.5.1.1   Waveform coders

Waveform coders [18,19] work in the time or frequency domain and exploit properties of the source signal *waveform* (spectral envelope/harmonics/pitch) through methods such as short-term correlation, for example, by the *linear prediction* (LP) method, which attempt to predict waveform samples from the values of previous samples. There are many kinds of waveform coders [18,20]. Some derive their codes by examining the input speech in the frequency domain, and others do so in the time domain. Waveform coders tend to be relatively simple, take minimum advantage of the redundancy in human speech, and are not particularly efficient at reducing the data rate on the radio channel. Though their data rates tend to be high, and they are not very efficient users of radio channels, waveform coders exhibit excellent voice quality and are tolerant of background noise. Many can even encode music and other sounds that are not of human origin. They work independently from how the signal was generated and attempt a close reconstruction of the original signal. Different varieties work in various ways in both the time and frequency domains. In the frequency domain, for example, sub-band coding schemes split the speech signal into sub-bands that are coded independently. The coding expenditure in terms of bits per second is related to, and in some way proportional to, the energy in the particular band.

Waveform codecs have low complexity and, in general, use relatively high bit rates (above 16 Kbps). Linear PCM (64 Kbps) represents the simplest form of waveform coding. ADPCM (at 32 Kbps, CCITT/ITU G.721 standard), which is used in some wireless access systems (DECT and PHS), is a more complex algorithm that still carries a relatively high rate waveform. There is also a variable rate ADPCM, which operates at 16, 24, 32, and 40 Kbps (CCITT/ITU G.726/727).

11.5.1.2 Vocoders

Vocoders (voice coders) are parametric coders that model the voice signal generation in the human vocal tract through a time-varying filter, which gets "excited" in order to produce sounds. This modulating filter is characterized by a mathematical polynomial description through poles and coefficients. The approach is to model the human vocal tract (acoustic filter), which gets excited through a source of energy (vocal cords). This source can be described through two main characteristics of speech: loudness (amplitude), and pitch (frequency). Human speech consists of two types of sounds: (1) voiced sounds (vowels) with a more regular and periodic structure (low frequency pitch, higher energy, and longer duration) and (2) unvoiced sounds (consonants) with a less predictable noise-like characteristic of lower energy that is spread over the whole spectrum.

Voiced sounds come from the vocal cords, which form an oscillator with lots of harmonic distortion. A human speaker can change the rather simple oscillations from the vocal cords from one kind of voiced sound to another (different vowel sounds) by changing the shape of a complicated filter, which is the vocal tract. The vocal tract includes the throat, the soft pallet, the hard pallet, the nasal cavity, the oral cavity (including the tongue), and the mouth itself. A human speaker can form dozens of speech sounds with his vocal tract "filter," but all human vocal tracts are slightly different from each other; each vocal tract has its own presets, which is why we can recognize the identity of a speaker by merely listening to him on the phone. Some talented people can make themselves sound like other people by carefully manipulating their vocal tract dimensions to mimic the vocal tract filter settings of those whose voices they choose to imitate.

Unvoiced sounds are the hissing noises that make the consonant sounds, such as the "t" in "top," the "s" in "sit," and the "sh" in "shall." The noise is created by blowing air over or through a constriction to the air flow, for example, between the teeth or between the tip of the tongue and the hard pallet. These unvoiced sounds can be altered by the same vocal tract filter to which the voiced sounds are subjected, which is why we can whisper whole sentences to each other without using voiced sounds, and we can even recognize who is whispering to us in the dark.

In the speech coding process, vocoders analyze the original speech signal regarding its contents (voiced or unvoiced, gain, pitch and filter coefficients). The coder assembles a certain set of information for transmission:

▶ Filter coefficients;

▶ Indication of voiced or unvoiced speech;

▶ Gain/loudness values or parameter;

▶ Pitch information for voiced speech.

Reproduction of such sounds in the receiving speech synthesizer, as well as in the loop filter in the encoder, has to be different for different sets of voiced and unvoiced speech samples. Voiced speech will be reproduced by a filter excitation through periodic or regular (glottal) pulses. The pitch of the sounds can be represented by the distance (in the time domain) between the pulses. For unvoiced sounds, the excitation of the filter is random (white) noise. See Figure 11.7 for a depiction of speech synthesis.

These kinds of coders are excellent at moving intelligible voice over narrow channels at astonishingly low bit rates. The disadvantages are that parametric coders are much more complex than even the most elaborate waveform coders, and they can only handle human speech sounds. Vocoders operate at low bit rates (down to 2.4 Kbps) and, though their voice reproductions are fully intelligible, they tend to sound rather synthetic.

The problem with this simple vocoder approach is that it resolves only very short intervals between voiced and unvoiced sounds. So, in the gray area between different sounds, the determination of the characteristics of the speech may not be adequate. In addition, interactions between the excitation and the filter (sound source and vocal tract) are disregarded.

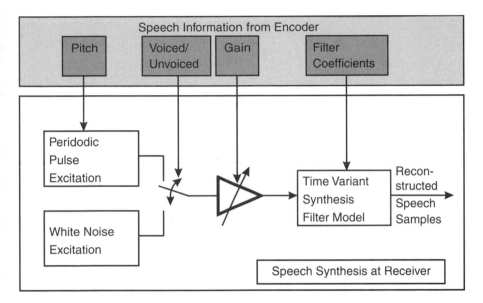

Figure 11.7   Speech synthesis with vocoders.

Thus, the use of vocoders is not recommended for normal telephony, but they find wide use in professional radio systems such as in public safety applications.

11.5.1.3   Hybrid coders

Hybrid coders, which are generally known as *analysis by synthesis* (AbS) *codecs*, introduce a more distinct analysis of the excitation source. They transmit either a residual excitation model, or an excitation pulse shaping, sizing, and spacing model, or a codebook-oriented selection of excitation pulses (*code excitation linear prediction* [CELP]). They fill the gap between the two methods described earlier. They make use of linear prediction (waveform coding), and they replace the vocoder method, which simply distinguishes voiced and unvoiced sounds, with a better excitation signal processing scheme. Hybrid coders try to minimize the error between the input speech waveform and the reconstructed speech waveform by finding the ideal excitation signal.

Figure 11.8 illustrates a simple but imaginary hybrid vocoder. Though the encoding and decoding processes in speech coders are usually accomplished within the same physical device, the illustration in Figure 11.8 has

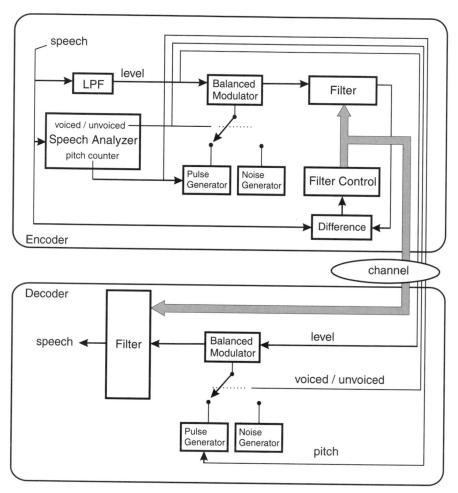

Figure 11.8 Simple hybrid speech coding example.

the two processes separated for clarity. The encoding function is in the top half of the figure, and the decoding function is in the bottom half of the figure. The two halves are separated by the radio channel as indicated in the oval. The imaginary speech coder regards human voice, in short frames of typically 20 ms, as either voiced vowel sounds, or unvoiced hissing sounds.

The speech analyzer in Figure 11.8 separates voiced from unvoiced sounds, determines the level of the voice, estimates the fundamental pitch frequency in voiced sounds and sends data representing these three simple voice characteristics over the channel. The voiced or unvoiced parameter can be represented by a single bit. A filter in the encoder gets some adjustment instructions from a block that compares the filter's output with the actual speech input, thus closing a *control loop* that maintains the filter's output as a faithful copy of the original speech. If the control loop's information is also sent over the channel to an identical filter (in the receiver), then the filter settings will alter the properly adjusted voiced or unvoiced sources such that a copy of the original speech leaves the filter. Consider, again, our guitar player teaching his friend some new chords over the telephone. He can either play the chords into the telephone's mouthpiece and hope the connection is clear enough that the music can be discerned in the student's ear (waveform coding), or he can tell his friend which chords to play and how to play them, (parametric coding). The reader may note that all the functional blocks in the lower receive portion of Figure 11.8 also exit in the upper portion of the drawing. This serves to illustrate how functional blocks can be used for two purposes, which is not particularly difficult if all the functions are actually data manipulations in a DSP.

As we explained before in our example of a hybrid coder, the error minimization is already done in the *coder*, which passes the excitation recursively through the synthesis filter that already has some presets (AbS). The speech signal gets chopped up into short frames of 20 ms, over which the analysis is carried out. Once the ideal excitation and filter parameters are found, they are transmitted to the synthesizer. In CELP codecs, a codeword within a codebook is indexed by a so-called vector (a pointer). Such vectors (the only information that is actually transmitted) point to excitation values stored in the codebook in the receiver (decoder). This procedure reduces the number of bits needed to transmit the excitation information. However, codebook searches in CELP codecs are very complex tasks that require relatively high processing power. Codebooks also use up quite some nonvolatile memory (ROM tables), which needs to reside with the signal processor. The whole transmitted speech information frame consists of synthesis filter parameters (eventually including long-term prediction and pitch information) and the appropriate codebook excitation vectors for the speech synthesizer at the other end. At the speech synthesizer, an excitation of the synthesizer is typically performed with nonzero pulses at a rate of 4 to 10 pulses every 5 ms. This implies that for a 20-ms speech frame, up to 40 excitation pulses (positions and amplitudes) must be found by the analyzer and generated in the synthesizer. Hybrid codecs, which also appear in many variations and combinations, typically work at net bit rates that are somewhere between the ones known for vocoders and waveform codecs, which puts them between 4 and 16 Kbps.

Some speech coders, such as the Qualcomm Q5414 variable rate vocoder (which is a hybrid codec [14]), can actually reduce their output data rates in accordance with the lack of need for speech coding work.

11.5.2 Speech quality

The line quality of a digital communication link is, to a substantial degree, determined by the speech codec. Other effects including delays, echoes, interference, and interruptions, which do not fall within the speech codec's responsibility, may have an impact on the perceived voice quality. The speech coder's stability, or its *susceptibility*, remains a determining factor that will be tempered by the channel coding regime and modem performance. Disturbing effects may come from the network (echo cancellation, round-trip delays), environmental influences (background noise), and signal reception conditions (interference and fading). The perceived speech quality is also affected by the terminal's acoustic design. *Voice activity detection* (VAD), which is used for speech-dependent transmission (DTX), may introduce annoying clipping effects.

One network-related factor influencing end-to-end speech quality is *tandeming*. Conversion of GSM speech to PCM, or even further to analog signals, and back to GSM speech hurts speech quality, because distortions are introduced that increase with each conversion. Tandem-free operation avoids transcoding as much as possible. In mobile-to-mobile calls, the encoded speech can be transferred directly at 13 Kbps without any conversions to PCM within the PLMN. Another annoying effect on perceived speech quality is the delays and echoes that cannot be compensated for in dedicated echo canceling units in a GSM MSC.

On the matter of quality, how do we measure a speech codec's performance? *Acoustic* measurements on the terminal side are performed with an artificial mouth and ear [21]. The relevant parameters are (1) loss, (2) *sending* and *receive loudness ratios* (SLR and RLR), (3) frequency response, (4) acoustic SNR, (5) sidetone, (6) out-of-band signals, and (7) crosstalk. For a GSM network, TS 03.50 specifies the transmission performance requirements [22].

Then, categorizing speech codecs is, in general, achieved with tests performed by trained listeners. So-called *mean opinion score* (MOS) values are obtained for different speech coding algorithms by subjective auditory tests performed by trained human listeners under controlled conditions. The conditions, or *error profiles* (EPs), include the static *carrier-to-interference ratio* (C/I) at the radio receiver, for example, at 13, 10, 7, and 4 dB, or 0%, 5%, 8%, and 13% for average gross bit error rates and dynamic radio path conditions (fading, Doppler, and frequency hopping). MOS figures can rank between 0 and 5. An MOS of 4.0 already indicates relatively high perceived wireline quality (as achieved through PCM), which is often referred to as toll quality (the transmission quality maintained between toll centers).

There is a tight relationship between speech codec rate, achievable quality, and the complexity of speech codecs. The better the perceived quality and/or the lower the net bit rate, the more complex (mips and memory requirements) a speech codec implementation will be. Other important quality aspects relate to robustness (resistance to bit errors and burst errors) and to the efforts that must be included in channel coding and decoding. The better the protection through channel coding, the better the speech quality.

11.5.3   DTMF and signaling tones

Dual-tone multifrequency (DTMF) is commonly used in telecommunications systems to transmit user data or application-specific information (digits 0–9 and the characters A, B, C, D, *, and #) with audible tones generated from two audio sources selected from a four-by-four matrix. Applications include the submission of a PIN, a credit card number, or other access and control codes. DTMF is used for accessing and controlling answering machines and voice mail services, as well as automatic teller services. In GSM, DTMF is handled in one of two ways that depend on the origin of the tones. The GSM speech codec does not handle DTMF tones (monotonous sine waves) too well. It introduces a non-negligible amount of distortion. So, in the uplink direction (MS to network), the DTMF digits are delivered through signaling. A DTMF *message* is sent over a signaling channel (FACCH) for each digit. This is a normal application case. In the downlink direction (network to MS), the tones can be actually carried in the speech channel, thus treating them exactly like voice signals, even though the speech codec would introduce distortion. However, so far the application of a GSM MS having to decode DTMF tones has not appeared and also was not anticipated. Still, a GSM speech codec has to be transparent, with a minimum of audible distortion added to DTMF so that an analog DTMF decoder still could detect the tones. The same requirement for such transparency is true for other acoustic *signaling tones* from a network. Some of these other signals are the familiar ringing sound (ring-back indicator), the busy line indicator, and the warning signal.

11.5.4   GSM full-rate speech coding

The GSM full-rate speech coding function is described in the 06 series of the GSM specifications for the *digital cellular telecommunications system* [23]. The specifications are the only reference that can be given for engineering purposes. More complete treatments of the subject can be found in [2,24]. GSM full-rate speech coding makes use of a RPE/LTP combination of processes. This combines *linear predictive coding* (LPC), which exploits short-term correlation in the speech signal, with long-term correlation (through LTP). The residual signal, which excites the vocal tract model in the receiver, consists of a set of regular pulses. The RPE-LTP contender received the highest average MOS score among the six final proponents for the GSM full-rate speech codec.

All the codecs were tested with seven languages under various transmission conditions (three different input levels and bit error rates, one and two transcodings, and two environmental noise forms) [25]. It is interesting to note that the actual language spoken has an effect on the speech codec efficiency. This phenomenon is true for many low-complexity predictive vocoders, and it has to be taken into account when qualifying improved codecs for future use in systems deployed in many countries. There is also a distinction between female and male voices, which some codecs handle one or the other better.

11.5.5   GSM half-rate speech coding

The GSM half-rate speech processing function is described in the 06.20 series of the GSM specifications (ETS 300 581 series) [4]. It makes use of a

5.6-Kbps CELP/VSELP combination of speech codecs. The net 5.6-Kbps data are channel coded (GSM 05.03) with a different coding scheme than that used for GSM full-rate channels. Because each bit is more important in a low-rate process, a higher complexity convolutional coder with more memory is used. That is, instead of a constraint length of five ($k = 5$), the half-rate data are protected through a constraint length of seven ($k = 7$). The convolutional coding, which is applied to the most important class 1 bits, is combined with a puncturing mechanism. If we include the cyclic code parity bits and the tail bits, then a gross rate of 228 bits per 20-ms frame (11.4 Kbps) is used for block diagonal interleaving over four bursts. This represents exactly half the gross rate of that used for full-rate traffic channels (22.8 Kbps).

The motivation for this speech codec is that it allows for the use of one time slot in *every other* TDMA frame in one dedicated traffic channel, which doubles the physical channel capacity. This is regarded as an important feature by GSM network operators. Because cellular spectrum is a rare resource, and some densely populated areas already have capacity problems during peak hours, half-rate would relieve this situation. However, as nothing comes to us for free, there are some drawbacks. First, only a significant number (30% to 50%) of half-rate users in a network can generate noticeable relief in capacity inside bottleneck areas. This is difficult to achieve with an existing user base because the exchange/replacement period for terminals is between 18 and 36 months. Second, half-rate terminals need to be able to operate in full-rate environments. This is because a network may not be completely covered by half-rate capability, or the user may want to roam within other full-rate networks. Such dual-voice-codec terminals would be slightly more complex and slightly more expensive. Third and finally, the speech quality offered by half-rate vocoders is regarded as lower than full-rate in MOS trials. The MOS figures fall in the range of 3.0 and a comfortable operating region in terms of error profiles was more limited. The presence of background noise and/or tandeming conditions can make half-rate coded voice almost unintelligible.

11.5.6 GSM enhanced full-rate speech coding

The GSM EFR speech processing function is described in the 06.50/06.60/06.80 series of the GSM specifications (ETS 300 723–300 730) [5]. The coding scheme was adopted by the American ANSI T1P1 committee as a higher quality speech codec for PCS 1900 systems based on the GSM platform. It was also adopted within ETSI through GSM Phase 2, thus becoming a *European Telecommunications Standard* (ETS).

The GSM EFR is a so-called *algebraic code(-book) excited linear prediction coder* (ACELP). This indicates the use of an algebraic codebook, in which an algebraic code is used to arrange or populate the excitation vectors within the codebook. This makes the search algorithm more efficient. The EFR generates speech compression information at the same bit rate as the full-rate process does (13 Kbps), and the coding and interleaving schemes are identical. The EFR was finally accepted after some debate on whether channel coding should be adapted to better fit the EFR codec and to further improve performance in low C/I conditions. This alternative was eventually rejected because some manufacturers had already committed to hardware implementations.

The perceived voice quality achieved through the EFR is much better than the full-rate scheme. It has been tested thoroughly, and the MOS figures are above 4.0 in office and street noise environments. It is said by some to even outperform the ADPCM in the presence of background noise. The use of EFR codecs in established GSM markets, where there is a large population of phones already deployed, only makes sense in combination with the GSM full-rate standard as a fallback in both the networks and the terminals. In newly established networks, such as in PCS 1900 networks in North America, the EFR is seen as the main standard.

It turns out that all the world's digital cellular systems have recently specified and deployed improved codecs of similar quality to GSM's EFR. For example, the ACELP has replaced the VSELP in some of the IS-136 systems, and the 13.3-Kbps coder and *enhanced variable rate codec* (EVRC) have replaced the 8-Kbps process in the IS-95 systems. Time brings steady improvements to speech coding technology; better quality speech can be realized with the same data rates of only a few years ago. Further improvements are sure to be proposed in the future: the *advanced multirate* (AMR) proposal may eventually compete with some wideband (50-Hz to 7-kHz) voice coders for attention in the GSM community. The former is proposed as a way to get half-rate service deployed, and the latter are proposed to allow GSM to compete directly with wireline services. Because GSM systems exist next to and compete with IS-136 and IS-95 systems in North America, the Americans have had to push the EFR specification through their own T1P1 standards body rather than wait for the slower ETSI process.

11.5.7 Complexity comparison FR-HR-EFR

We have to consider the required resources before implementing a speech codec in a GSM mobile phone or in a network component. A certain amount of signal processing power must be reserved to make sure that all the real-time coding and decoding functions are carried out in time. Program and data memory resources must be provided with adequate access time, and temporarily accessible memory (RAM) has to be available.

Table 11.1 compares optimized full-rate, half-rate, and enhanced full-rate speech codec implementations. These are generic figures that take into account assembler programmed, standard 16-bit, fixed-point DSPs that are used in wireless applications. Different DSPs perform differently for some operations and in some implementations. There is also a human factor in that all the figures are also programmer dependent. For half-rate codecs, note that due to the higher constraint length in the convolutional coder ($k = 7$), the peak decoder-complexity increases. This is, however, partially compensated by the longer receiving intervals (double) through the half-rate operation.

11.5.8 The future for GSM speech coding

Most GSM network operators find themselves in very competitive environments with other operators that use digital and analog, and fixed-line

Table 11.1
Comparison of Voice Coding Complexities

Speech Codec	Maximum mips	Program and Data ROM	RAM
GSM FR	2.5–4.5	4–6 kWords	1–2 kWords
GSM HR	17.5–22	16–20 kWords	~ 5 kWords
GSM EFR	17–22	15–20 kWords	~ 5 kWords

or wireless, technology. Because voice telephony remains the most widely used service, audio and speech transmission quality can be made a major differentiator in the market for both networks and terminals. With traditional full-rate technology, GSM offers more or less adequate quality at a relatively high transmission rate at the radio interface. But other options are being explored. As speech processing technology matures and improves, lower bit rates may provide even better quality. This reduction in net bit rates can be used to increase robustness through channel coding, or to increase capacity through high-quality half-rate vocoders. The gap between full rate and half rate may eventually fade away: variable rate vocoders can fill the gap and allow on-demand scaling of speech data rates and transmission rates. When available, when necessary (bad radio conditions), or when subscribing to a premium service, the user gets a high-quality, full-rate channel with a robust wireline-like transmission quality. When radio conditions are good and/or more capacity is needed, the rate falls back to a lower bandwidth that still provides adequate quality. ETSI has responded to industry requirements for open standards on the next-generation speech codecs. The specifications within SMG 11 for an AMR codec are being elaborated. The GSM specifications describe the entire PLMN and interworking protocols, so there is much more than just a particular coding algorithm to be worked out and agreed on. There are also a number of signaling and control issues, channel coding, and testing requirements to be considered. The introduction of the AMR codec was requested by the network operator community to be ready shortly after the turn of the millennium, and the standards bodies are obliged to respond.

Tandem-free operation (TFO) refers to mechanisms that avoid multiple speech coding and decoding steps such as when, for example, a mobile phone asks for a voice channel with another mobile phone within the same PLMN. The lack of TFO in this case would mean that the encoded speech is converted into the wireline system's PCM scheme at one BSC, and the PCM code is converted back into the wireless scheme through a codec in another BSC; each conversion diminishes the speech quality. TFO is being tackled by a working group within ETSI SMG 11. Definitions and specifications have to be elaborated in order to provide for the appropriate control mechanisms. This is necessary in order to engage TFO whenever possible in calls within a GSM PLMN as well as among GSM PLMNs.

11.5.9 Speech coding and...

There are some other matters that need to be considered when processing speech and audio/acoustics. Two important ones are hands-free operation and voice recognition.

11.5.9.1 Hands-free operation

Though the practice by automobile drivers is increasingly seen as lacking in social responsibility, mobile telephones are used while on the move. The use of a hand-portable phone in a car is supported with a connection to a hands-free kit. This kit extends four interfaces within the terminal: the microphone, the speaker, the battery, and the antenna. The microphone used in a car kit is often installed close to the speaker's mouth, perhaps just beneath the inside rearview mirror. It is small and has a directional reception characteristic. A loudspeaker is placed wherever there is room to spare. A connection to the car's battery saves the terminal's own battery capacity, and even allows for the possibility of charging the phone's battery when the kit is used. The outside antenna included with the kits provides greatly improved radio signal reception and transmission.

As far as the phone is concerned, a car is a Faraday cage that isolates a radio transceiver from the host cell sites. With only the internal (built-in) antenna of a hand-portable unit, the reception is impaired because the power control mechanism drives the radio's transmitting power level to the highest power levels. Thus, the external antenna affords improved signal reception, longer battery life, and better voice quality, and reduces radio interference with the car's electronic systems and other potential influences inside the car. There is growing evidence that a driver using a hands-free kit to engage himself in a phone conversation is sufficiently distracted from the driving task to pose a safety concern. Some countries do not allow drivers to talk on the phone without the hands-free feature, and the prudent automobile operator may want to consider restricting the use of the hands-free feature to passengers.

Hands-free operation takes place within a complex acoustic system consisting of a loudspeaker, a human speaker, a "room" or chamber, and a microphone. Figure 11.9 shows the acoustic system at the near end. Without any further treatment, the incoming audio signal from the far end would emerge from the loudspeaker and be received by the

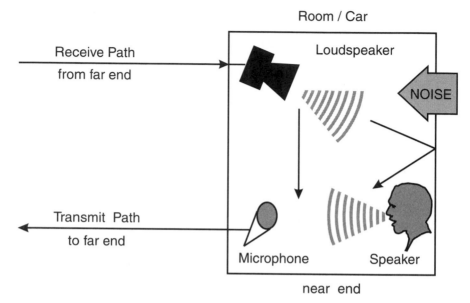

Figure 11.9   Hands-free environment.

microphone through a direct path as well as through some additional indirect reflections that depend on the "room." Because there is not much attenuation in the acoustic link, an echo would result (perhaps with some feedback) at far end. Under the best circumstances, the far end speaker would hear herself with an echo. Under the worst circumstances, the feedback path would quickly build up an annoying howling sound. Some tricks are employed to prevent these annoying effects. Two of the countermeasures are *echo suppression* and *echo cancellation*.

Echo suppression   Echo suppression means that the inactive audio path gets attenuated while the active path is left open as in normal non-hands-free operation. Echo suppression introduces a *half-duplex* flavor to the conversation mode; only one path is active at a time. Echo suppression requires some kind of VAD at the near end for the signals from both ends. Voice activity detection can be quite complex, especially in the presence of background noise so common in moving vehicles. Many VAD schemes exhibit some kind of annoying clipping effect that is generated when the voice is clipped off when a certain energy threshold is achieved. Because the speech coders in GSM terminals need to support VAD and DTX anyway, the information is present for use in hands-free logic: (1) conflicts arise when both ends speak at the same time, (2) an idle status is resumed when there is no speech from either end, and (3) hangover periods are introduced to cover the time when a speaker resumes talking after a short pause. All this is handled in the hands-free control logic, which performs its tasks through relatively simple digital signal processing. Attenuation and volume control can also be carried out digitally just after, or immediately before, the PCM conversion in the handset.

These kinds of tasks consume something in the range of 1 mips or less in a standard DSP. The memory requirements for program, data storage, and RAM space are relatively modest so that echo suppression is an attractive solution for low-cost implementations. It is not a very elegant solution. At the time of this writing, echo suppression was the only technique widely employed by mobile terminal and accessory manufacturers. The hands-free control logic can add lots of perceived value to otherwise similar products; some are much better than others. In general, echo suppression provides adequate quality, especially when some speaking discipline is applied at both ends. Figure 11.10 shows the concept of echo suppression for hands-free operation.

Echo cancellation   A more elegant, but much more complicated and costly solution is echo cancellation. Rather than abruptly clipping off voice signals and ruthlessly attenuating audio paths, echo cancellation maintains a full-duplex link between both ends. How can this be achieved without the annoying echo effects? Complex dynamic or adaptive echo

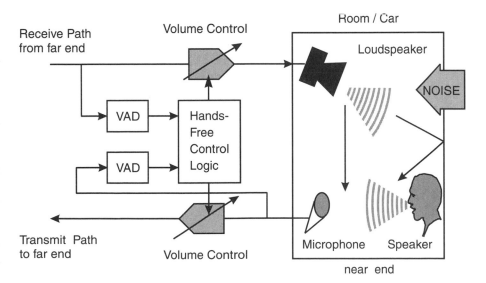

Figure 11.10   Echo suppression for hands-free operation.

cancellers work with a system model that describes the closed room system. The concept is similar to the more static echo cancellers used in fixed networks and switches. In particular, all of the acoustic signal paths between the loudspeaker and the microphone form an *acoustic system*. Such a system can be modeled through a complex time-variant filter function, which is used to calculate and then subtract out the echo signal that would have occurred if the acoustic system were left alone. This echo signal resembles the original far end signal that was modeled or shaped by the acoustic system on its way to the far end. The concept of adaptive echo cancellation is shown in Figure 11.11.

For simplification we will use the terms of the z-transformation for time-discrete signals. The z-transformation does not apply to the analog time-continuous audio signals of the "room" system. In Figure 11.9, the incoming signal, X, is shaped through the "room" filter, $H(p)$. The system response estimator calculates and maintains this digital filter function, $S'(z)$, which models $H(p)$. The calculated echo signal, $X'(z)$, is subtracted from the return signal, $Y(z)$.

The problem with the implementation in adaptive echo cancellers for hands-free operation is that the ideal processing solution has not been found yet. Trade-offs have to be made in complexity, performance, and adaptivity (rate). Especially when the echo mechanism is not a static one, but rather undergoes changes (additional noise, movements, and when

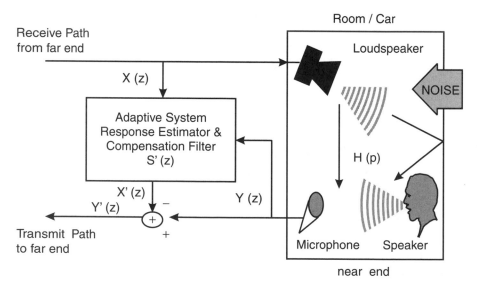

Figure 11.11   Concept of echo cancellation.

both parties are speaking from both ends), the ideal path can get lost very quickly. Digital filters tend to have a dependency on their complexity and their training period, which can cause trouble in some cases. Even though there are already some good solutions, which require us to add another DSP into the system, research in this area is ongoing and will certainly yield better solutions in the future. The solutions need to have adequate performance with reasonable cost and processing effort. Technical implementations require something in the range of 20 mips and above, and are realized in a DSP located within the handset cradle (which is connected to the hands-free speaker and microphone) or in the handset itself.

11.5.9.2 Voice recognition

Voice recognition technology, similar to echo cancellation, has received a lot of attention in the communications industry. For the personal communications world, this means that terminal functions become accessible through simply "telling" the device what to do. Simply say, "Call Mom," and the device dials your home number so that you can check what's for dinner. Voice recognition works either with a trained set of personalized commands, which is a less complex implementation that requires some upfront tuning and storage work, for example, for the user's 16 favorite wishes, or with a generic and much more complex approach which reads *all* characters and words. A similar approach is evident in character and handwriting recognition on touch-sensitive screens where a computer recognizes handwriting which it converts into machine text and characters.

11.6 Equalizers

Under ideal situations a transmitted symbol should arrive at the receiver, greatly attenuated, but undistorted and occupying only its intended time interval. Unhappily, such is seldom the case in mobile radio. When one symbol is so distorted that it occupies the time reserved for other symbols, we have *intersymbol interference* (ISI). The situation is made even worse in GSM as the GSM transmitter itself contributes is own ISI. There are two contributions to ISI in the GSM system: (1) uncontrolled ISI from the radio channel and (2) controlled and deliberate ISI from the transmitter's partial response modulator.

The reader is cautioned that a complete treatment of the causes of, and the solutions to, ISI are not particularly intuitive. First, the mathematics involves vector quantities rather than scalar ones. Add the scalar quantities, 5W and 10W, to get an intuitive 15W. One dollar plus another dollar is two dollars. Vector numbers (also called complex numbers) have both an *amount* and a *direction*. One kilometer to the east (90°) plus another kilometer to the east equals 2 km to the east; but 1 km to the east (90°) plus 1 km to the north (0°) equals 1.414 km to the northeast (45°), a not particularly intuitive result without some personal reflection. Engineers work with complex numbers so often that they eventually gain considerable intuitive sense of their manipulations. Further, the mobile radio channel's impairments, fading, multipath, and Doppler spread, are tempered by the modulation scheme in subtle ways that require sophisticated simulations on computers in order to deduce the BER performance or the probability of an outage. Finally, the effects of ISI are influenced by the effectiveness of signaling techniques. We enlist the relatively simple linear examples in phone lines and our experience with analog FM radio to explain what an equalizer is and bring some appreciation for how this important device deals with the impairments of the radio channel. The engineer who must effectively deal with radio channels and equalizers can turn to the books suggested in Section 11.12 and to his computer.

An equalizer can repair some of the damage caused by ISI in a radio channel or in a phone line, and Figure 11.12 shows that the equalizer is the first function in the receiving path after the demodulator.

All digital receivers have a decision circuit that constantly passes judgment on the state (e.g., "1" or "0") of each and every noisy and distorted symbol emerging from the demodulator. The equalizer is inserted between the demodulator and the decision circuit, thus removing some of the ambiguity from the distorted symbol stream.

Equalizers are neither free nor do they draw negligible power; they are only used if their cost and burden on the available power can be justified. Most implementations of GSM receivers employ a type of adaptive, nonlinear, *maximum likelihood sequence estimation* (MLSE), soft output equalizer known as a *Viterbi equalizer* [26,27]. Many different types of Viterbi equalizers have appeared each differing from the others in details, complexity, and performance [28]. Their details are usually shrouded in secrecy as the different equipment makers seek competitive advantage in the market by improving the performance of their receivers with

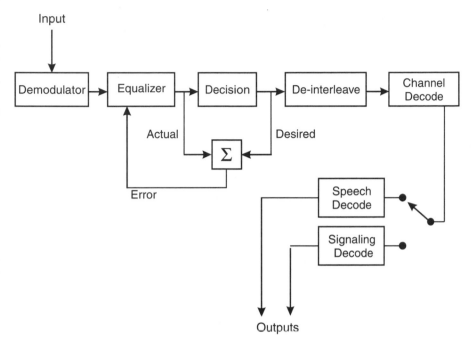

Figure 11.12   A digital receiver.

equalizer improvements. The secrecy in the market is further compounded by the highly specialized nature of present-day baseband techniques. Because a rigorous and complete treatment of Viterbi equalizers is too specialized for this chapter, we endeavor to ford the river that separates the specialist from the general reader as we describe these devices intuitively with pictures and simple examples.

11.6.1   The problem—ISI

The primary cause of ISI is multipath fading. It is not likely that a radio signal will propagate unimpeded from a transmitter to a receiver through free-space along only one path; the transmitter's signal typically follows many paths to the receiver. In addition to the direct path, different types of alternative paths are (1) *reflected* off a structure that is large compared to the wavelength of the signal, (2) *scattered* off an object that is small compared to the wavelength of the signal, and (3) *refracted* around the corners of objects. Another cause of ISI is Doppler spread, which is due to the relative motion between a receiver and a transmitter; each path the signal's wave takes from the transmitter to the receiver experiences an apparent shift in frequency. The signal that finally arrives at the receiver is the sum of many signals with random amplitude (the *amount*) and phase (the *direction*). The amplitude of each individual signal changes with the loss on its path and the phase of each individual signal changes with the length of its path. In some instances the sum of all the signals will add constructively, improving the received signal, and at other instances the signals will add destructively degrading the received signal. When a large number of such signals from a transmitter is present at the receiver, each a function of two random variables, amplitude and phase, we have a Rayleigh distribution. This means that we know the probability of encountering a fade of some specific amplitude in a multipath environment. Let's set up an experiment in a city, or any typical place GSM handsets are likely to be used. If we measure the received signal strength, at a test receiver located anywhere in our city, of the emissions from a small transmitter carried around in a circle of, say, 1-km radius, from the test receiver's antenna, plot the recorded signal strengths on the vertical, logarithmic axis of a piece of graph paper, then we will get a curve that looks something like that shown in Figure 11.13. The curve shows occasional deep dips, or fades, in the received signal strength. Now, let's add a *sorter* to Figure 11.13. A sorter is a set of small *boxes* arranged on the vertical axis. Each box represents a small range of received signal strengths. As

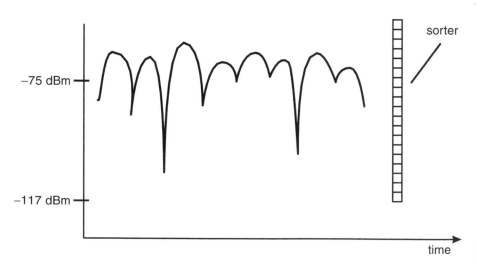

Figure 11.13 Rayleigh fading.

we move steadily along the horizontal time axis, we will add a count to whatever box represents the observed signal strength at every instant we stop to record them. When we finally arrive at the end of our time axis we will see that the distribution of counts in the boxes is not symmetric. In the case depicted in Figure 11.13, the box that includes the –75-dBm received signal strength will hold the largest count, the counts in the boxes that represent slightly stronger received signal strengths will be almost as high (decreasing quickly as we examine the contents of boxes representing even stronger received signal strengths), and the boxes representing very weak received signal strengths may have accumulated counts. The counts in the boxes that represent signal strengths below –75 dBm decrease more slowly than counts in the boxes above –75 dBm; the distribution of the counts around the –75-dBm box is not symmetric.

The *normal* (Gaussian) distribution has a symmetric distribution of sorter box counts around some most likely value. This would occur, for example, if we observe the weights of cookies emerging from a machine designed to manufacture cookies of 150g in weight. Most will weigh very nearly 150g, but it is just as likely for one to weigh 140g as another is to weigh 160g. Both processes, the possible received signal strength from a transmitter and the weights of cookies, are random ones, but they are described in different ways and have different distributions of possibilities around some dominant value. Yet another distribution is the Rician distribution that applies when the direct path from the transmitter to the receiver dominates all the other paths.

If we move our test transmitter in a straight line away from the test receiver instead of moving it in a circle around the receiver, then the "bumps" in Figure 11.13 will fall as the test transmitter moves further and further away from the receiver.

Most of us have heard the effects of multipath fading in analog FM mobile radio. As we carry a portable radio in a large circle around a test receiver, we note that the signal strength varies in a manner depicted in Figure 11.13. We can sometimes move the radio around (over a distance of ½ *wavelength, λ/2,* which in the typical 900-MHz cellular band is about 15 cm) until the resulting input to the receiver is weak enough—destructive interference—to allow noise to dominate in the speaker; we have thus positioned the radio in a deep fade. If we drive around the test receiver fast enough, our receiver will pass through the fades quickly enough to cause pops in the speaker. The pops are sometimes described as "popcorn" or "cooking bacon" sounds, and are caused by abrupt phase rotations in the carrier-plus-noise phasor that yield unwanted phase modulation. The phase modulation manifests itself as clicks and pops from the FM detector. The physics that causes these amusing events in analog FM mobile radio conspire to cause ISI in digital radio by spreading or smearing the symbols that arrive at a receiver.

One bit interval in GSM—in which the payload data symbols are transmitted using a kind of phase modulation (GMSK)—is a mere $3.69\,\mu s$, and the typical delay spread (the time interval over which signals following different paths from a transmitter arrive at a receiver) in GSM systems is about $5\,\mu s$, which is longer than one bit time [2,27,29]. Because there is only one bit in a symbol, we can equate the bit time with the symbol time. The short symbol time means there are plenty of opportunities for ISI, and the relatively short burst duration and reoccurrence time does not allot us much time to repair the symbol damage in the receiver before the next burst arrives. Figure 11.14 depicts a simple case of ISI, where the time interval, t, between points 4 and 5 could be the interval the original transmitted symbol actually occupies. The short original pulse (symbol) arrives at the receiver spread over eight symbol times.

There are many causes of ISI. Multipath fading is one. Shoving pulses (representing symbols for the payload data) through a filter too narrow to

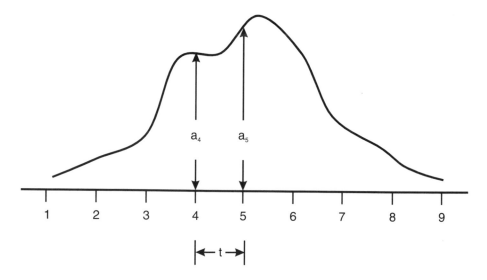

Figure 11.14   Intersymbol interference (ISI).

accommodate them is another cause. Indeed, the limited bandwidth of the mobile radio channel is the chief constraint on symbol rates in digital radio systems. Narrow filters spread the modulation pulses over time as they round off the edges of the pulses and the mobile radio channel has its own time-varying and largely unknown filter-like influences: Doppler spread and multipath fading. They make the radio channel appear to be a filter the characteristics of which change with time. Because the destructive influence of the mobile channel changes with time, the compensation techniques must be *adaptive*. A static channel, such as we may find in a long cable, exhibits a fixed influence on the pulses. We can use much simpler nonadaptive and linear techniques to clear ISI in such well-behaved channels. Equalization is only one of the techniques used to clear the effect of ISI. Two others are *diversity* and *channel coding*. All three techniques are used in parallel to counter the effects of ISI.

Diversity works on the causes of fading channel impairments; it attacks at least some of the causes of the problem by reducing the depth of fades. Diversity techniques are well known. The use of multiple receiver antennas, or diversity reception, is an example of *space diversity*, and it works by exploiting the highly unlikely possibility that deep fades, which occur at half wavelength intervals, will be experienced in two antennas, separated by some distance, at the same time. There is 1 chance in 6 that a rolled die will fall with two dots face up, but there is only 1 chance in 36 that two dice will land with two dots face up on both cubes. Frequency hopping is an example of *frequency diversity* [30]. Diversity measures are considered successful if the duration and depth of fades are reduced to a manageable level.

Channel coding is an inoculation technique against ISI. The channel coder adds carefully contrived and clever redundancy to the transmitted data which the receiver can use to detect and, within certain limits, repair the underlying information data [31,32]. The receiver can perform its own repairs or it can ask for repeated transmissions. Though they are not perfect in their ability to repair destroyed data, channel coding techniques have become very powerful in recent years [32–35]. Just as is the case with the voice codec, both the receiver and the transmitter must agree on a common channel coding scheme for the techniques to work.

Equalization works on the symptoms rather than the causes of ISI. In one sense, the more simple forms of equalization are compensation techniques; whatever the modulator and the mobile radio channel do to the original transmitted data to cause ISI, equalization does the opposite to compensate for the influences [36]. Another form of equalization is the MLSE that is used in GSM. This form tries to figure out what the transmitter's original information (input) bits were from the received (distorted) symbols (output) together with an estimate of the channel that caused the distortions. Equalizers "undo" the observed ISI in the receiver in widely different ways, and the most advanced techniques use signal processing techniques within the baseband portion of the receiver to perform their work.

The difference between the three ISI countermeasures, diversity, channel coding, and equalization, are not always clear. The CDMA system mentioned in Chapter 3 uses an elegant combination of all three methods. The three methods come together at one point in the RAKE receiver [37,38]. The RAKE receiver itself can be viewed as an elaborate equalizer that takes advantage of *time diversity* to make the system work. Time diversity is a form of multipath fading in which the paths differ so much in time that appropriate techniques can be employed to actually improve overall performance.

A salient characteristic of the type of GMSK employed in GSM is the relatively low filter bandwidth, B, and bit duration, T, product, which is the BT product: $BT = 0.3$. The low BT product implies a gradual change of the transmitted carrier from one symbol to the next in such a way as to spread each symbol change over several symbol periods. A bit change is spread over three bit times when $BT = 0.3$. Were the BT infinite, a modulation bit's influence would be confined to only one bit time, but the resulting signal would be simple MSK, which responds to modulating symbol changes through abrupt phase shifts in the transmitter. Figure 11.15 illustrates the difference between MSK and GMSK. The data applied to the modulator and the resulting phase change created by the modulator are plotted over time for both the MSK case (the fine line with sharp corners) and the GMSK case (the dark line with rounded corners).

Figure 11.15 shows that GMSK is a form of MSK. MSK exhibits a constant phase change during a symbol time, but the abrupt phase changes at symbol transitions cause MSK to occupy more spectrum than the equivalent rate GMSK. GMSK rounds off the phase transitions between symbols thus more tightly confining the spectral power to the allotted bandwidth. The smaller the BT product, the more power is confined to a portion of the spectrum by spreading the symbol changes over even more symbol

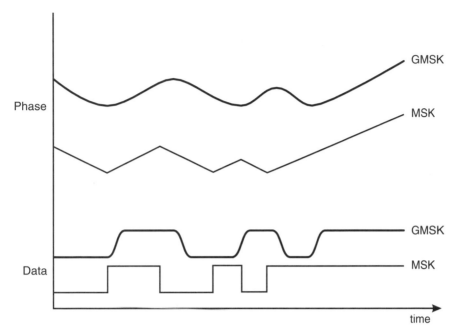

Figure 11.15 MSK and GMSK.

periods. GMSK is said to have greater spectral compactness and efficiency than MSK, because GMSK confines the spectral emissions to a relatively narrow channel. The bandwidth-limited emissions of GMSK means we can use somewhat less power in the transmitter than we would if we used MSK. But the spreading of each information bit over three bit times, in the GSM case, adds ISI to the modulation even before it leaves the transmitter. So, the price we pay for the savings in spectrum and the power consumed in the transmitter by using GMSK, is picked up by the receiver, in which we are obliged to add even more complexity with an equalizer.

11.6.2 General equalizers

An equalizer is any device that compensates for ISI in a receiver. Equalizers have been around for a long time. Simple impedance matching networks that perform echo canceling functions are examples of early devices still used today [39,40]. Figure 11.16 depicts a general form of a linear equalizer. Whenever we see an equalizer for the first time we should look for three parts: (1) the filter, (2) the control function, and (3) the feedback path.

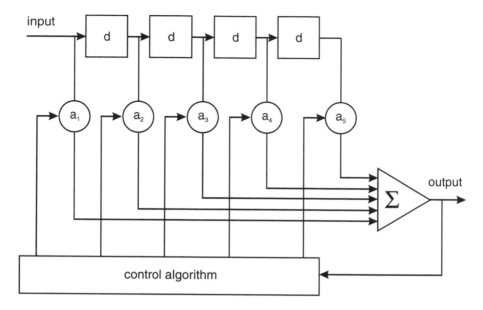

Figure 11.16 A linear equalizer.

The filter part in Figure 11.16 is a transmission line with taps separated from each other by a delay, d, of one symbol period, t, or $d = t$. As the distorted input from the receiver's demodulator (see Figure 11.14) propagates down the filter, samples of the signal appear, at the filter's sampling period, into the weighting functions, a. The outputs of the weighting functions are summed to form the equalizer output. The output will be a faithful reproduction of the original transmitted pulse if the weighting functions are adjusted properly. A look at the a_n levels in Figure 11.14 are an indication of what the weighting values should be.

The weighting functions are adjusted by the control part in Figure 11.16 in such a way as to yield a flat frequency response with linear phase response. It can do this by amplifying the attenuated frequencies and attenuating the peaked ones. Such a filter is called a *matched filter*. The control part is usually some kind of algorithm contrived from a detailed knowledge of the channel it has to compensate for. If the channel characteristics do not change, such as in wire applications, then it is enough to hard wire a fixed control process. This makes the equalizer a fixed or static one. We can increase the resolution with which we define and compensate for the communications channel by oversampling the input, $d < t$.

The feedback path in Figure 11.16 is a simple analog connection, which works fine for fixed and relatively well-behaved channels such as are found on phone lines. In mobile communications it is more common to use nonlinear feedback as is the case in the general adaptive device depicted in Figure 11.12. There is always some kind of decision device in digital systems between the demodulator and the decoder, which makes hard decisions on the state of symbols recovered in the demodulator. The decision device can be as simple as a comparator. Nonlinear feedback uses the decision device, which allows us to change the control algorithm in the equalizer with changing circumstances such as we would see in mobile radio channels. Such an adaptive equalizer is called a *nonlinear feedback equalizer*. Conceptually, the simplest forms of this kind of equalizer decode each symbol, or a group of symbols, adjust the equalizer's actual output for the best agreement with the desired output, and then use the latest adjustments for the next round of symbols.

There are several variations on the nonlinear technique. One of them is called the *decision feedback equalizer* (DFE), which inserts another matched filter, just like the forward one, in the feedback path coming back from the output of the decision circuit. The second filter in the feedback path is called the *backward* filter, and the gain outputs from its taps are summed with the forward filter's gain outputs in the control block in Figure 11.16. This doubles the number of influences on the control algorithm; half of the inputs, the backward ones, are derived from the experience repairing the delay spread damage to the to previous time slot.

Another nonlinear technique is called MLSE which is an elaborate and powerful technique that is covered in detail in Section 11.6.3. The MLSE technique can only work properly if it has some knowledge of what the desired data output from the decision device *should* be. It gets this information from a known symbol sequence that the transmitter sends to the receiver at regular intervals. The symbol sequences are called *training sequences* or *sounding sequences*. Though the appropriate symbol sequences do not look particularly simple to humans, they yield so-called *ideal responses* that only an equalizer likes and appreciates; they create signals that are relatively easy for the equalizer to examine, even through a poor channel, in order to figure out what the original, undistorted signal may have been. A bright green laser light is easier for us to distinguish through a dense fog among a riot of other lights and neon signs of various colors than a small white light would be. So, the symbol sequences are carefully constructed, through experiments or computer searches, to yield a simple output at the equalizer's own filter. If the ideal response, in the absence of ISI, is simple enough, then the equalizer's control algorithm can work backwards on a training sequence, which was received in the presence of ISI, to determine what the channel must have done to the sequence to account for the mess that was actually received.

11.6.3 Viterbi equalizer

A Viterbi equalizer includes a local signal generator and something called an *increment calculator* [27–29]. Viterbi equalization performs four daunting tasks. It constantly:

1. Generates its own versions of all possible data sequences that could come from the transmitter;

2. Calculates the receiver inputs that correspond to each and every transmitted possibility;

3. Compares the actual receiver inputs with the calculated ones;

4. Selects the locally generated data sequences that have the greatest probability of being the ones that were actually transmitted.

The comparison task in step 3 is performed with *metric* calculations, which are the same as those used in convolutional decoders that use the Viterbi process. Because channel coding depends so much on convolutional codes these days, we will take the opportunity to expose the processes to the light of day in Sections 11.6.3.1 and 11.6.3.2. A generalized Viterbi equalizer is shown in Figure 11.17. At this point the reader may feel that the equalizer's tasks dwarf all others in the GSM radio, or that, perhaps, something has been lost in the text. For now, we assure the reader that the tasks this kind of equalizer must perform would be, indeed, impossible were it not for the *Viterbi algorithm* (VA), which is the VA block in the figure, and which will be explained in due course. The Viterbi equalizer performs its work with vector quantities (complex numbers) and soft outputs that try to add some additional resolution to the digital symbol stream from the receiver's demodulator. We will avoid obfuscation in our discussion by ignoring complex numbers and soft outputs as we confine ourselves to binary examples.

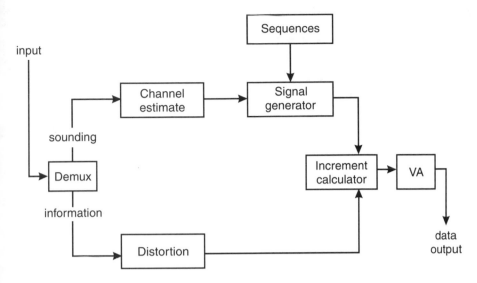

Figure 11.17 Viterbi equalizer.

There are two paths in the Viterbi equalizer that terminate in the increment calculator just before the VA. The symbol stream from the demodulator enters a demultiplexer that separates the training sequence found in all the time slots and bursts (except in the FCCH) in the GSM air interface. The training sequence follows the upper path in Figure 11.17. The *normal burst* (NB) reserves 26 of its centermost symbols for one of eight time slot–specific training sequences, hence, the term *midamble*. We confine our attention in this discussion to the NB after we recall that the synchronization burst has 64 training bits and the access burst has 41 training bits. Remember, a symbol period is the same as a bit period in GSM. The bottom path in Figure 11.17 holds the remaining data symbols.

An estimate of the channel characteristics is derived in the upper path. Take the simple case where a training sequence is extracted from a burst that arrived at the receiver over an ideal channel. If the training sequence is applied to a matched filter (Section 11.6.2), then a narrow impulse emerges from the filter. This is because the eight training sequences in the GSM time slots are carefully chosen to have this effect even when a great deal of delay spread occurs in the channel [27,29]. If the channel is not ideal, then delay spread will distort the pulse from the matched filter. Because the equalizer knows what the matched filter's output should be for all eight of the training sequences, it can estimate the radio channel's effects on the received signals by examining the output of the filter. This examination is also called a *channel estimate*. Such a channel estimate also renders a frequency offset value that can be used to synchronize (fine-tune) the receiver's reference oscillator, and thus adjust it to the more stable frequency reference of the downlink or forward channel transmitted by the base station. Then, it adjusts the matched filter to get the proper response back again, and makes the same adjustments in the lower path of Figure 11.17 for the data that represent actual information.

The signal generator in Figure 11.17 gets all possible transmitted data sequences from the sequences block above it. The local signal generator duplicates the distant transmitter's modulator as it allows its output to be appropriately distorted by the current channel estimate. The bottom path has a filter that distorts the received data symbols the same way the locally created symbols are distorted in the modulator. The actual received data are compared with all the locally generated possibilities in an incremental calculator that yields confidence values, *metrics*. The VA uses the metrics to select the most likely data sequence that could have originated at the transmitter. The Viterbi equalizer differs from the other equalizers we looked at; it does not recover the received symbols, or even blocks of symbols, as linear techniques and the DFE does. Instead, the VA block in Figure 11.17 creates the most likely received symbol sequences itself and, hence, the data that probably came from the transmitter. Except for the input and output of the VA block in Figure 11.17, all the paths (the darker paths in the figure) carry I and Q values that describe the modulation from whatever source: (1) the recovered training sequences from the received time slots, (2) the received data, and (3) the output of the local signal generator. Moreover, in a practice reminiscent of the $d < t$ oversampling technique described in Section 11.6.2, the symbol sequences described here are actually sequences of phase changes, which can be equated to symbol sequences. Finally, all the processes described here are performed numerically in a DSP after the demodulator's baseband output is converted into appropriate digital values.

11.6.3.1 Convolutional encoder

Figure 11.18 depicts a small machine, which we will call a convolutional encoder, that illustrates the typical channel encoding process in digital transmitters [29,30,32]. The actual encoders used in GSM phones and

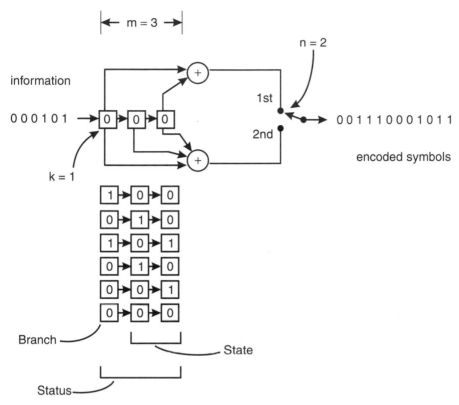

Figure 11.18   Convolutional encoder.

base stations are much more complex than the simple machine depicted here; they have five stages (cells) in the shift register and are too complicated to serve as reasonable illustrations. The general principles are the same. The raw information input is on the left, and the channel coded output, sent out in pairs read left to right, is on the right. We will start with all zeros loaded into the input shift register on the left, and then shift the information $k = 1$ bit at a time into the shift register, which is $m = 3$ stages (cells) long. The *status* of the shift register is shown in the bottom of Figure 11.18 for all six input instances. The *state* of the encoder is the 2 bits already in the shift register, which will change with each new bit that appears in the *branch* cell, the darker cell on the far left of the shift register. The 2 bits stuck in the shift register (state) delay or hold the effect of the inputs on the left; we say that the encoder has a delay of 2. The encoder generates $n = 2$ bits output on the right for each single bit shifted into the left. The encoding rate, r, is written as $r = k/n = \frac{1}{2}$; the output bit rate is twice the input rate. With each input instance, the entire shift register is regarded through two modulo-2 adders (which each are characterized by $0 + 0 = 0$ and $1 + 1 = 0$ and $0 + 1 = 1$ and $1 + 0 = 1$); the two-position switch on the output of the encoder passes through both of its positions. There is nothing particularly significant about the wiring between the modulo-2 adders, and the stages (cells) in the shift register in Figure 11.18; the machine serves only to illustrate several important baseband processes. *Convolutional coders* (CCs) and their codes are described thus: $CC(n,k,m)$, which in this case is a $CC(2,1,3)$ code.

Figure 11.18 shows all 12 encoded outputs for the six inputs. Every 2 bits generated by the encoder is influenced by 3 consecutive bits on the input. The effect of the "smearing" of the encoder's input over many output bits gives a properly equipped receiver the ability to correct errors in the received signal [41,42]. If $n = 3$, which would be a $CC(3,1,3)$ code, then the receiver can make even more dramatic repairs to erroneous inputs, but the bit rate in the channel has to be increased to do so. The following paragraphs and figures illustrate how a receiver, with a detailed knowledge of how the transmitter's encoder works, can correct errors in received data, and how these same techniques are used correct for ISI.

11.6.3.2 Convolutional decoder

A graphic way to describe undoing the convolutional code in Figure 11.18 is depicted in Figure 11.19, in which we accept the encoder's undistorted and error-free output to see how a trellis diagram gets our original information back again [43]. The graphic rules we use to decode the bit stream is shown in the top of the figure. The four possible states of the encoder (in the two rightmost cells) appear on both sides of the small *rule trellis* in the top of Figure 11.19, and a state is represented by a dot. All eight possible transitions from one state to another state are indicated by a line over which an oval is drawn and the resulting two output symbols that are the result of the state change are written into the oval. The transition line can be dotted or solid depending on the *branch bit* as shown in the legend to the right of the rule trellis.

Moving to the larger trellis in the bottom of Figure 11.19, the original information is represented by half the number of bits the encoder generated. The trellis diagram, so called because it resembles the framework of crossing strips of wood used to support climbing plants, is a tabulation of

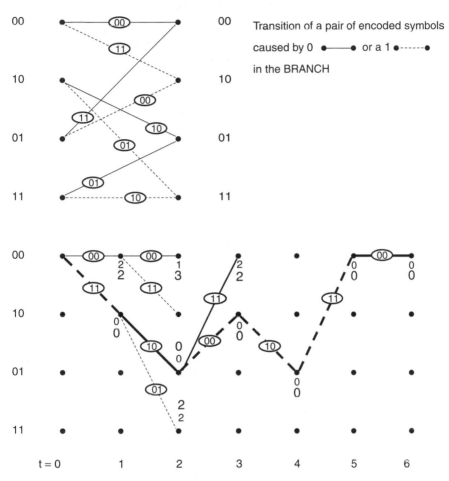

Figure 11.19 Decoding rules and example.

all possible encoder state transitions at every symbol instant exactly as in the rule trellis [30,32]. We consider six transition instances starting with $t = 1$. As the receiver's decoder recovers pairs of symbols, we plot all possible corresponding encoder states between the matrix of dots in Figure 11.19. Because there are four possible encoder states, there are four dots for each of the decoding instances, t. Because the encoder starts with all zeros, the decoder will begin the same way in the upper left corner of the figure at $t = 0$. As each pair of symbols is recovered in the receiver, at $t = 1$, for example, we draw a line from the stating dot ($t = 0$) to the next encoder state, which records the influence of the first information bit entering the machine on the left of Figure 11.18.

There are $2^k = 2$ (with $k = 1$) possible branches from one state to the next, which is also the number of branches that could terminate in a dot (encoder state). A transition of encoder states caused by a "1" shifted into the encoder's input is represented by a dotted line, and a solid line represents a state transition caused by a "0" shifted into the encoder's input. Each pair of encoded output symbols from the encoder, which are recovered in the receiver, is shown in a small oval in the diagram in accordance with the rules in the top of the figure.

At $t = 1$, we recover "11" and then calculate the branch metric, which is the confidence we have that the received pair of bits is, in fact, what was generated in the transmitter's channel encoder. A transition from state "00" to "10" would mean the output symbols were "11," which is no different from what we observed.

$$00 \text{ to } 10 \rightarrow 00 \text{ to } 11 \Rightarrow 0$$

The symbolic shorthand contrived here is read: *Encoder state "00" changed to the next encoder state "10" yields* ($\rightarrow$) *encoder output pair "00" changed to subsequent encoder output pair "11."* The *number of encoder output bits that would need to be inverted to give the observed output* is ($\Rightarrow$) 0, which is the branch metric.

The only other possibility is that the input to the encoder did not change ("00" to "00"), which would mean that the transmitted pair of symbols was "00." This, however, would mean that we received both symbols in error (2 errors), which is less likely than the previous case.

$$00 \text{ to } 00 \rightarrow 00 \text{ to } 00 \Rightarrow 2$$

We write the branch metric as a small number next to both of the possible states in the $t = 1$ column. Next, we write the current cumulative path metric as a larger number next to both states. The next pair of decoded symbols is "10," which, if we also consider the second less likely case from the previous pair of decoded symbols, gives us four possibilities. We write the branch metrics next to each of the four possible states (dots) at $t = 2$.

$$00 \text{ to } 00 \rightarrow 10 \text{ to } 00 \Rightarrow 1$$

00 to 10 → 10 to 11 ⇒ 1 (which we exclude from the figure to avoid clutter)

$$10 \text{ to } 01 \rightarrow 10 \text{ to } 10 \Rightarrow 0$$

$$10 \text{ to } 11 \rightarrow 10 \text{ to } 01 \Rightarrow 2$$

But, at $t = 2$, one of the pairs of symbols already has a high path metric (less likely possible occurrence relative to the "00" state). If we calculate the four possible path metrics we get the following results, where the bold numbers indicate path metrics:

$$00 \text{ to } 00 \text{ to } 00 \Rightarrow \mathbf{2} + 1 = \mathbf{3}$$

$$00 \text{ to } 00 \text{ to } 10 \Rightarrow \mathbf{2} + 1 = \mathbf{3}$$

$$00 \text{ to } 10 \text{ to } 01 \Rightarrow \mathbf{0} + 0 = \mathbf{0} \text{ (survivor)}$$

$$00 \text{ to } 10 \text{ to } 11 \Rightarrow \mathbf{0} + 2 = \mathbf{2}$$

We write the path metrics in larger numbers next to all four possible dots at $t = 2$, and declare the path with the lowest path metric to be the *survivor*. We perform all of these tasks again for the next pair of symbols, "00" recovered at $t = 3$ as shown in Figure 11.19. Because trellis diagrams can quickly become too cluttered to be useful, we mark only a few of the branch and path metrics on the figure for $t = 3$ through $t = 6$.

The convolutional decoder illustrated with the trellis diagram uses its input sequence to estimate the most likely stream of bits that caused the state transitions in the transmitter's encoder. Because the example in Figure 11.19 has no errors in the recovered symbol stream, we choose the path with the lowest metric (0), which is the path with the highest confidence. Figure 11.19 distinguishes the most likely path from the less likely ones with bold lines. To determine the original information bits that entered the transmitter's encoder, we simply note if a line between state transitions is dotted or solid. In Figure 11.19 the results are as follows:

 ▶ dotted line = 1;

 ▶ solid line = 0;

⏵ dotted line = 1;

⏵ solid line = 0;

⏵ solid line = 0.

This "10100..." sequence is exactly what we saw enter the channel encoder in Figure 11.18.

The trellis diagram in Figure 11.20 illustrates how the same decoder can determine the original information sent into the transmitter's convolutional encoder even when errors are present in the symbol pairs recovered in the receiver. To avoid calculating tedium and to keep the process clear, we change our transmitted data to a constant stream of zeros. In our example, the fourth bit recovered in the receiver is wrong: the "1" should have been a "0," which means the second recovered pair of symbols is "10" instead of "00." The *process* device between the encoder's input information bits and the output symbol pairs in the top of Figure 11.20 represents whatever distorts the output stream, which is everything between

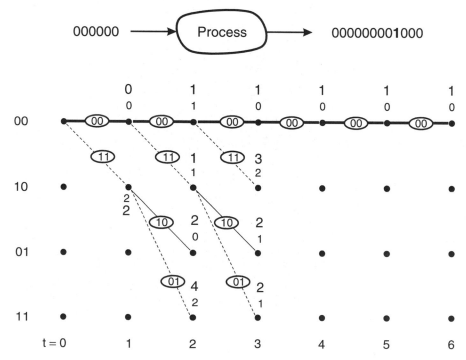

Figure 11.20 Correct decoding in the presence of errors.

the transmitter's modulator and the receiver's demodulator; the radio channel is the prime cause of errors.

At $t = 1$, at which instant we received "00," we calculate our branch metrics as before:

$$00 \text{ to } 00 \rightarrow 00 \text{ to } 00 \Rightarrow 0$$

$$00 \text{ to } 10 \rightarrow 00 \text{ to } 11 \Rightarrow 2$$

At $t = 2$ we recover the symbols "10" and continue with our branch and path metrics as if nothing were wrong. As before, the bold numbers are path metrics.

$$00000 \text{ to } 00 \text{ to } 00 \Rightarrow \mathbf{0} + 1 = \mathbf{1}$$

$$00 \text{ to } 00 \text{ to } 10 \Rightarrow \mathbf{0} + 1 = \mathbf{1}$$

$$00 \text{ to } 10 \text{ to } 01 \Rightarrow \mathbf{2} + 0 = \mathbf{2}$$

$$00 \text{ to } 10 \text{ to } 11 \Rightarrow \mathbf{2} + 2 = \mathbf{4}$$

We note that at $t = 2$, we cannot declare a surviving path, because we have two paths with identical path metrics, namely "1." We will need to decode two more symbols when $t = 3$ before we can declare a survivor. We can also note that at $t = 2$, even as we have a branch metric of zero (at state "01"), it follows an unlikely event with a high branch metric of 2. At $t = 3$ we finally have a survivor (branch metric = 1) and all remains well through $t = 6$.

Figure 11.21 illustrates what happens when the receiver is overwhelmed with symbol errors. In this example, we enlist the same stream of zeros into the transmitter's channel encoder, but invert three out of the first four symbols at the receiver's input. This makes the first pair of output bits "11" instead of "00" and the second pair "01" instead of "00."

Even if we look all the way out to $t = 7$, we see that the decoder interprets the encoder's input incorrectly (1010101...) even if all the received symbols after $t = 2$ are received without error. One way to prevent this kind of decoding catastrophe is to add a little analog character to the decoder, which relieves the receiver's decision function from making

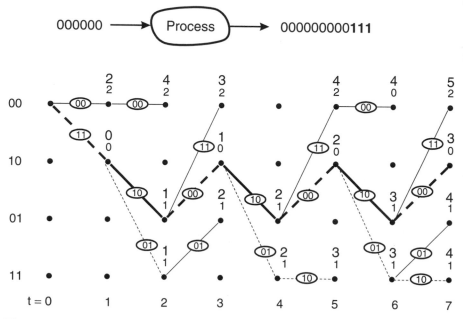

Figure 11.21   Incorrect decoding in the presence of too many errors.

hard "yes" and "no" ("1" and "0") decisions. The technique is called *soft decision* decoding.

Soft decision decoding adds some additional resolution to the judgments that the receiver's decision circuits make as symbols are recovered [29]. Figure 11.22 shows a simple case where a received symbol can be classified into one of eight categories, which could be something like a detected voltage in our simple example. It is not clear what the third symbol should be in the figure.

A hard decision circuit may have declared the third symbol a zero ("0"), because there is an instant of high negative voltage at the beginning of the symbol time, or it may declare the third symbol a one ("1"), because the symbol, perhaps, spends more time in the "yes" or "1" territory relative to the "no" or "0" territory.

If we take the same catastrophic situation depicted in Figure 11.21, but look closer and make the received symbol decisions soft ones, we can determine the correct transmitted information. In this case the surviving path metrics will have the greatest value rather than the value closest to zero. We will use the same scale in Figure 11.22 to resolve the received symbols as depicted in Figure 11.23. At $t = 1$ the branch metrics are

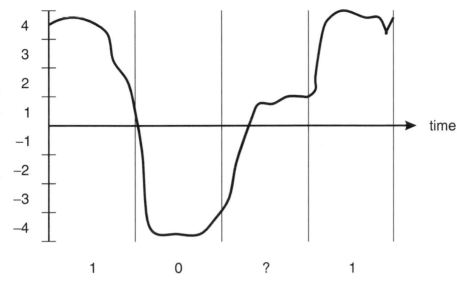

Figure 11.22 Soft decisions.

$$00 \text{ to } 00 \rightarrow 00 \text{ to } 00 \Rightarrow (-2) + (-1) = (-3)$$

which reads: to go from 00 to 00 state, the signal has to be *reduced* ("–") by a 2 and 1 metric, at their respective instances, to be in the negative domain. The (–3) reflects the probability, or our confidence, that the actually sampled values represent two 0s.

$$00 \text{ to } 10 \rightarrow 00 \text{ to } 11 \Rightarrow 2 + 1 = 3$$

which reads: to go from 00 to 10 state (to render 11 at the imaginary encoder output), the signal has 2 and 1 "overshoot" or surplus metrics at their respective instances. There is a higher probability (+3), that the actual samples represent two 1's.

At $t = 2$ the branch metrics are

$$00 \text{ to } 00 \rightarrow 00 \text{ to } 00 \Rightarrow 1 + 3 = 4$$

which means: to go from 00 to 00 state (to render 00 at the imaginary encoder output again), the signal overshoots, into the negative domain, by 1 and 3 metrics at their respective instances. The probability that the samples represent 0s is high.

$$10 \text{ to } 01 \rightarrow 11 \text{ to } 10 \Rightarrow (-1) + 3 = 2$$

similarly, to go from state 10 to 01, we have 1 negative metric too much for the first sample to go to 0 (–1) and we lack 3 metrics for a 1.

$$10 \text{ to } 11 \rightarrow 11 \text{ to } 01 \Rightarrow (-1) + (-3) = (-4)$$

Next, here we lack 1 metric to get a 1 and 3 metrics to get another one. Consecutively, at $t = 3$ the branch metrics are

$$00 \text{ to } 00 \rightarrow 00 \text{ to } 00 \Rightarrow 4 + 4 = 8$$

$$01 \text{ to } 00 \rightarrow 10 \text{ to } 11 \Rightarrow (-4) + (-4) = (-8)$$

$$01 \text{ to } 10 \rightarrow 10 \text{ to } 00 \Rightarrow 4 + 4 = 8$$

At $t = 4$ and beyond, the received samples seem to have settled down in their representations of "0" symbols (–4) and the path metric climbs to 32 at $t = 6$ as the branch metrics settle to 8 on the surviving path.

The absolute value of the level at any sampling instant shown at the top of Figure 11.23 is an indication of our confidence in the received symbol at that instant. The first sample is +2 and the second sample is +1. This means that we are a bit *more* sure that the symbol at the first sampling instant represents a "1" than we are of the sample at the second instant. Similarly, we are much more confident that the value sampled at the fourth instant represents a "0" than we are of the value sampled at the third instant. The first two recovered symbols, 2 and 1, respectively, are compared with both possible "clean" symbol representations, which are –4 and 4 on our high-resolution scale. We have almost no confidence the "2,1" pair of values declared during the first two sample instants actually represents 0,0 (–4,–4), so we assign a –3 branch metric:

$$00 \text{ to } 00 \rightarrow 00 \text{ to } 00 \Rightarrow (-2) + -1) = -3$$

We have more confidence that the same two symbols, 2 and 1, actually represent 1,1. We, therefore, assign a higher metric:

$$00 \text{ to } 10 \rightarrow 00 \text{ to } 11 \Rightarrow 2 + 1 = 3$$

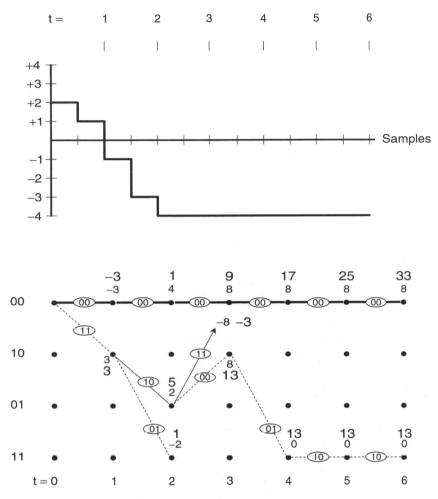

Figure 11.23 Soft decision decoding.

If we calculate the branch metrics in this way out to $t = 6$, and then calculate the path metrics, we see that the path with the greatest metric gives the original transmitted data back to us.

So, even as this discussion clarifies the channel coding and decoding blocks in the GSM transceiver, what does all of this have to do with equalizing the channel and working around the ISI problem? If we consider the GSM case with five stages (cells) in the encoder's shift register, four stages of which constitute the status of the encoder, we turn back to Figure 11.17. At each sampling instant the received bit pattern, and one of 16 possible locally generated bit streams, is presented to the increment calculator as 32 inputs: 16 inputs for the received bit stream and another 16 for one of the local trials. The 16 local streams are constant until another time slot, with its training sequence, arrives to change the local signal generator's output to reflect the latest channel conditions. The increment calculator issues 16 metrics to the VA block in Figure 11.17, one for each pair of inputs. In the manner demonstrated in Figure 11.23, the VA block generates the most likely data stream that was sent from the distant transmitter. Now, the symbols that are actually examined in a GSM receiver are not the simple soft values on a single scale seen in Figure 11.22. Instead, the instantaneous I and Q phase states are examined, and the dark lines of Figure 11.17 point out the signal paths that are actually complex numbers or I and Q values represented by complex numbers. The received signal's phase state at any instant can be resolved into units of radians, or portions of a radian, in proportion to how closely we want to examine a recovered signal in order to make soft decisions on what the actual transmitted symbol probably was. But the use of complex numbers, and additional resolution in examining them, adds more complexity to the equalization process than Figure 11.23 may indicate.

Some GSM-specific criteria for channel equalization can be found in [29] in which the reader will discover how easy it is to become overwhelmed in the details without computer simulations. Unlike the simple telephone line equalizers from which we have drawn liberal references in the chapter, the Viterbi equalizer hasn't the faintest clue what the actual transmitted data may be; it merely knows how it was coded and the true nature of all the possible training sequences. It generates the most likely sequence of source data given the messy results from the receiver's demodulator and its knowledge of the transmitter's construction. The occasional training sequences help determine what is happening to the coded data on its way to the receiver.

11.6.3.3 Implementation

Typical implementations for the Viterbi equalizer, and the subsequent decoding functions, are done in a DSP. Some computations are suitably done by using so-called Viterbi accelerators, which are accessed by the DSP when carrying out the algorithms. For instance, the selection (by comparison) and update of soft path metrics, and the respective sequences, can be done in hardware whereas other functions like matched filter related calculations are realized in software. A GSM equalizer would use up to 16 states to recover the initially transmitted information. Less channel delay spread implies fewer states (perhaps only four states), which would demand less computational effort. Typical figures for equalizer implementations, including the help of accelerators, are in the range of 5 mips for the modem/equalizer, and one additional minute for the actual channel decoding (supported by a Viterbi accelerator). Other functions such as CRC calculations and burst assembly demand relatively little processing power.

11.7 Encryption and security in GSM

GSM incorporates certain security features that prevent (1) unauthorized service access by maintaining the confidentiality of a subscriber's identity and authentication on the SIM, and (2) over-the-air eavesdropping thanks to the ciphering of user and signaling data. The ciphering feature is optional. Subscriber anonymity is further protected by a number of features including the *temporary mobile subscriber identity* (TMSI), which, because it is a much shorter number than the IMSI, also frees up some signaling space on the air. References [44–50] provide relevant GSM specifications and other discussions on security and encryption.

11.7.1 Algorithms and keys

However the operators choose to use the security features included in GSM, the details and procedures behind them are thoroughly specified. Security mechanisms are realized with *algorithms* (A), *keys* (K), *random numbers for challenges* (RAND), and *signed responses* (SRES) to the challenges. Sensitive information, such as keys and algorithms, are never transmitted over the air. The mechanisms are transparent to the operators for interworking, and only sets of challenges and responses need to be exchanged between operators, or more precisely, between their networks on PCM links.

Authentication keys and algorithms are stored in the SIM (see Chapter 8) on the mobile's side of the channel, and within the authentication center on the network side. The SIM processes the 128-bit RAND challenge (a random number) from the network together with the 128-bit *individual subscriber authentication key*, K_i, as inputs to the *authentication algorithm*, A3. The 32-bit result is called the *signed responses*, SRES, which the MS sends back to the network. The network, which has a table of current K_i assignments listed by subscriber identity (in the VLR and/or HLR), compares the mobile's SRES with its own calculations to determine if the subscriber is, in fact, who he says he is before granting access to the system. Refer to Chapter 8 for graphical explanations.

A similar approach is taken with ciphering. The ciphering algorithms are implemented in the *mobile equipment* (ME) and in the network (BTS). Because they need to run in real time during a phone call, their implementation needs to exhibit good performance. Typically, an encryption engine is built in silicon hardware rather than in DSP software, and consists of only a few thousand transistors [20]. One parameter, the cipher key, K_c, is used in the encryption engine, A5, to cipher the coded transmitter data and perform the corresponding decoding process in the receiver. The K_c value is computed in a manner almost identical to the way the SRES value is computed during authentication. The SIM, which holds the A8 cipher key algorithm, accepts the same RAND challenge the network passed to the mobile (usually on a SDCCH) during authentication, and combines it with K_i within A8 to deliver K_c.

Additional security is evident by noting that the SIM does not reveal K_i or the A3 and A8 algorithms, which can be specific to an operator. Key K_c is derived from K_i and RAND, which means that K_c can change during a conversation, and will certainly be different with each call setup.

11.7.2   Ciphering in GSM

GSM ciphering algorithms are applied to protect user and signaling data immediately before they are sent over the air. The protection is accomplished by manipulating the 114 radio bits that fill each normal burst in a TDMA frame. A reverse manipulation is performed in the receiver immediately after the data stream is recovered in the receiver.

A *ciphering sequence* and a *deciphering sequence* have to be generated for *each* TDMA frame by a stream cipher algorithm. A *stream cipher* algorithm with so-called *linear feedback shift registers* (LFSRs) is used in GSM. The A5 is first fed with the 64-bit cipher key K_c (for initialization) and the current 22-bit TDMA frame number. Note that K_c may actually contain less than 64 bits of significance; 64 significant bits implies maximum security. A short (not-so-secure) K_c is accompanied by enough zeros in order to fill all 64 bits of the A5 register. Because the output is also a function of the current TDMA frame number, the output of the clocked A5 shift registers, the ciphering sequences, is different for each TDMA frame and has two different uplink and downlink *cipher sequences* of 114 bits each. Finally, the payload data (received bursts and transmitted bursts), each of which holds 114 bits, is XORed with the current ciphering sequence. To recover the transmitted payload data, the receiver needs to use the same key for decryption (XOR with received radio bits) that was used by the transmitter for encryption.

11.7.3   Regulations

The A5 encryption algorithm, developed in Britain, was originally intended to be used in the Pan-European cellular system for which the GSM specifications were initially defined. Its use was confined to CEPT countries (Western Europe). Since the A5 algorithm represents a sophisticated ciphering technology, there were valid concerns from Western governments about protecting it and preventing its misuse. Tight export regulations (for export from Great Britain) are applied with nondisclosure obligations that must be observed by the manufacturing industry, which receives the *technical details* of the A5 algorithm. To address the need to export GSM technology to countries outside Western Europe, a "weaker" A5 algorithm, dubbed the A5/2 algorithm, was drafted and adopted. The original A5 was subsequently renamed the A5/1 algorithm. This leaves GSM networks with three options for payload data protection: A5/1, A5/2, or no ciphering at all. One has to look where (in which country) a network operates, and then look further at the network operators' choices to sort out the ciphering possibilities. Export restrictions on A5/1 and A5/2 implemented in silicon within mobile terminal equipment are not particularly strict. Today's GSM terminals are usually fitted with both algorithms. The GSM specifications allow the use of up to seven different A5s. At the time of this writing, no further A5 algorithms have been specified beyond the A5/2. The administration of the GSM encryption secrets around the A5/1 and A5/2 algorithms are left with the GSM MoU's security group.

11.7.4   Security vs. fraud

Together with the SIM, a GSM terminal can provide a very high level of security for both the equipment and the subscriber. We have all heard the reports on how revealing authentication keys and cracking ciphering keys impairs GSM security, but the technical effort for doing so renders the tasks fruitless. We have also heard the reports about simple FM scanners affording easy and low-cost eavesdropping on some analog cellular and cordless systems. With the introduction of digital technology, the technical effort for over-the-air snooping has increased dramatically. The pure digital nature of the information transmitted over the radio channel, and the fact that digital speech coding is used together with channel coding, further raises the price of unauthorized listening sessions. Moreover, the TDMA effect in GSM and IS-136 systems, the signal spreading used in CDMA systems, and frequency hopping and digital encryption techniques move the cost of casual radio interceptions beyond what most people who engage in such activity can afford. Protecting digital information is relatively simple and inexpensive, and it can be very effective. In particular, the GSM ciphering algorithms, and the related radio techniques, can be regarded as extremely safe barriers against determined interceptors. Some safety considerations, crime prevention efforts, and criminal investigations ask for particular police and law enforcement agencies to be able to listen into phone conversations. This can be done quite easily through appropriate technology in the fixed backbone network in which we find access points where encryption is not applied to the signaling and user data, and the difficulties of dealing with the TDMA effect are absent.

Finally, we are all aware of the techniques employed in *cloning* cellular phones through, for example, copying *electronic serial numbers* (ESNs), which in some analog cellular systems is sufficient to place calls at a legitimate subscriber's expense. Again, digital technology on the air interface and the introduction of sophisticated authentication mechanisms help to prevent this kind of fraud. Progress has been made in network management software, which alerts the operator and bars the service for a particular subscription when, for instance, calls are simultaneously made from what seems to be the same terminal in different locations, or when too short a time passes for a legitimate terminal to travel between two locations.

11.8   Mixed signals

Digital telecommunications technology cannot work without access to the analog domain. Human voice originates and terminates with analog signals. Today, even though we can reach multigigabit per second transmission speeds, the physical transmission medium remains an analog one. Signals need to be converted from the familiar analog domain into the digital domain in which computers, processors, and control logic perform their work so well, and then back to the analog domain again. As we saw in our block diagrams of the different signal processing paths in a GSM mobile station, there are three two-way interfaces: (1) the voice codec, (2) the baseband codec, and (3) the radio control interface.

Silicon implementations that combine all three functions are common in analog *complementary metal-oxide semiconductor* (CMOS) processes, which are not suited for adding too many digital functions due to size, cost, yield, and power consumption. Today we can even see mixed signal technology realized with "digital" CMOS processes. This implies the integration of "analog" functions (converters) and cells with all the additional digital baseband logic on one piece of silicon. Typical approaches primarily use the "digital" *sigma-delta* ($\Sigma\Delta$) concept for conversions in interfaces 1 and 2 just mentioned. This means that analog signals are modulated with a high sampling rate that yields ones ("1") for increasing analog signals (the voltage of the new sample is higher than the previous sample) and zeros ("0") for decreasing signals (the voltage of the new sample is lower than previous sample). A digital decimation filter delivers PCM words at the required resolution and rate. For a 10-bit A/D converter in the GSM RX path (at 270,833 Ksps), using an 8× oversampling rate, the minimum sampling/modulation rate is $10 \times 270,833 \times 10^3 \times 8 = 21.666$ MHz. This is almost the double rate of the 13-MHz GSM reference clock that is used for this purpose (26 MHz), thus leaving some overhead for serialization and deserialization.

Advantages of the $\Sigma\Delta$ concept are the suppression of quantization noise (through shifting the noise spectrum above half the sampling frequency), and the fact that further filtering can be performed digitally. Resolution and accuracy depend on the order and sampling rate of the $\Sigma\Delta$ modulator and the low-pass, postfilter characteristics. Depending on which technology is available, analog filtering (for anti-aliasing or reconstruction) can be supported on-chip with capacitors and resistors. By this, interfacing serialized data with other components, which has its own power consumption penalties and spurious signal generation tendencies, is not necessary. A reduced component and pin count on the printed circuit board allows for greater and more cost-effective integration and manufacturing. Some external passive and active components are required when on-chip technology is too big or inaccurate. A $\Sigma\Delta$ modulator can be used in the receiving path of the voice codec for D/A conversion that reconstructs the analog voice signal from a digital bit stream received from the speech codec through an interpolation filter.

D/A conversion and filtering, for example, in transmit path 2, has to deliver a band-limited analog signal from a digital data stream. The signal shaping (modulation), which is GMSK with $B \times T = 0.3$, is typically done through a look-up table that holds symbol transition curves. If, for example, a particular GSM realization selected an 8× oversampling rate, the modulated data stream would have a rate of 8 × 270,833 Ksps, which is represented by an input signal to the D/A converter (typically of the switch capacitor type in CMOS implementations) with a 2.1667-MHz (13/6-MHz) rate.

Other RF functions (interface 3) include D/A converters for:

▶ *Automatic frequency control* (AFC), which is the tuning reference of a VCO;

▶ *Automatic gain control* (AGC), which is used in LNA stages;

▶ *Power amplifier control* (PA), which performs the ramping and control of RF transmitter output power levels.

The requirements for speed and resolution are typically in the range of several microseconds and a resolution of 10 bits is typical.

Additional A/D converters are called on for:

▶ Received signal strength measurements (for RF carrier acquisition and monitoring);

▶ Battery voltage and supervision;

▶ Temperature supervision in PA stages.

Again, the required conversion time is slow (a few microseconds) and a resolution of 6 to 10 bits is adequate to serve the purposes. A single A/D converter can be shared among different functions, because not all of them need to be done simultaneously and constantly. This implies the use of a signal path multiplexer.

Other mixed signal functions include a system PLL synthesizer, which generates all of the required system clocks off the reference oscillator frequency, which is 13 MHz in GSM.

11.9 Microprocessor control

As we can see when we look back at Figure 11.6, the heart of the GSM mobile is a control processor (near the center of the figure). The mobile equipment in the GSM networks was assigned many control and management functions. The mobile terminal is a miniature counterpart to the whole GSM network. It has to support the network's connection management, mobility management, and radio resource management tasks, sundry user communication protocols (data and voice), and a more or less sophisticated user interface. The higher level intelligence above the complex real-time digital signal processing layers are built into the GSM mobile station as software running on the microprocessor. The microprocessor is quite often a *reduced instruction set computer* (RISC) with a 16-bit instruction set and external (off-chip) memories for programs and data (flash ROM and RAM). Usually a *real-time operating system* (RTOS) supports preemptive and interrupt-based software design. A TDMA timer (see Section 11.10) furnishes interrupts to the microprocessor. High-priority control tasks are executed before lower priority protocol and user interface tasks. In general, the microprocessor treats lower layer functions implemented in the DSP, interfaces, and other proprietary logic as black boxes. Information is polled as it is needed, and instructions and parameters are exchanged through commands in shared memory or registers.

The following high-level tasks are typically covered by the microprocessor:

▶ User interface and application software (menus, keypad/display interactions, and access to services above layer 3);

▶ Protocol (GSM layer 1; logical channel handling; frame multiplexing for hyperframes, superframes, and the multiframes; time and frequency tracking; and the layer 2 and layer 3 functions);

▶ Interface to, and communications with, the SIM;

▶ Interface to the digital signal processor;

▶ Interfaces to hardware (e.g., the radio for control functions);

▶ Interface to the serial data port for fax and data services.

A typical GSM implementation requires relatively low processing power, compared to the DSP, and is generally in the range of several million instructions per second and several hundreds of kilobytes of program ROM. The *user interface* (UI) software occupies a huge chunk of the program and data memories. The more complex, sophisticated, elaborate, and the more services that are supported, the larger the UI software effort and code size. UI design is an art of its own and is a very important one. The user interface, its ease of use, and its general look and feel are some of the key differentiating factors among manufacturers apart from size, price, weight, and standby time. It also represents the aspect of current GSM phones that needs the most work; ask any owner of a GSM phone how to activate call divert, how to retrieve his short messages, or how to retrieve a stored phone number from her phone, and you are likely to wait a long time for the correct reply.

11.10   GSM timing

All GSM terminal timing references are derived from a 13-MHz clock. The oscillator does not have very stringent free running stability requirements (1 to 3 ppm), because exact synchronization (0.1 ppm) is derived from the downlink signal from the GSM base station (see lbelow). The baseband bit rate is 270,833 kHz/Kbps, which is 13 MHz divided by 48, thus one symbol lasts 3.69 μs. One TDMA frame is 4.615 ms, which is eight time slots. Every time a 13-MHz counter runs up to 60,000, a new TDMA frame begins. Thus, a 26 multiframe (see Figure 11.24)—consisting of 24 frames for TCH, 1 for SACCH (S) , and 1 left idle (I)—lasts 120 ms [2,51]. Two traffic channel multiframes consist of 52 basic TDMA frames. This is one TDMA frame more than makes up the 51 multiframe, which is used in all kinds of signaling channel combinations including, for example, the *base station control channel*, which is a combination of FCCH, SCCH, BCCH, and CCCH resources. This 51–26 relation allows a mobile station

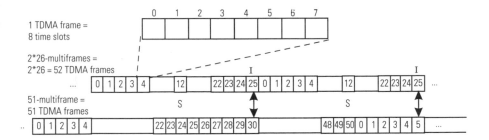

Figure 11.24 The 26- and 51-multiframe structures.

engaged in a traffic channel to monitor the base channel as the idle frames of the 26 structure slowly *slide* over the 51 structure of the base channel. In Figure 11.24 the TCH idle frame (I) occurs at signaling frames 30 and 5 (51-multiframe structure). The next two TCH idle frames would fall on signaling frames 31 and 6, respectively. So the MS can sequentially grasp all the necessary information (synchronization, cell/system information) of the serving base station as well as that from adjacent cells.

The lower level logical timing in GSM base stations and mobile stations is typically and practically performed on a per-TDMA-frame basis. The events include setting radio parameters for receiving, transmitting, or monitoring functions (channel number/frequency, channel parameters for PLL settings, power, low noise amplifiers, etc.) as well as logical information. Do we need to generate a traffic channel burst or a SACCH, or do we need to send a SDCCH? The logical channels need to be scheduled. The use of interrupt controlled preemptive operating systems, for both the controller and the DSP, assists in the flexible scheduling of different tasks according to predefined priorities. Still, because timing in GSM systems may shift or drift (through frequency offsets or a time delay when moving), and correction mechanisms are in place (timing advance commands), a higher resolution than the single-bit period (e.g., 1/4 or 1/8 bit periods) is called on for precise adjustments.

A GSM base station derives its highly accurate absolute timing reference off the digital Abis network interface clock. This clock is locked to a rubidium reference somewhere in the nodes connected with PCM links. A GSM mobile station locks onto its master, which is the BTS. Coarse timing and frequency correction is achieved through initial coarse synchronization with the FCCH and then with the SCH. Once locked onto the base channel, a mobile station maintains fine synchronization through its tracking of the paging channel as the equalizer provides sufficient frequency error information off the downlink training sequences for reference clock adjustments. Further coarse timing correction is carried out through timing advance commands, which are calculated and triggered by the BTS [2].

11.11 Components and technology

Advances in digital technology let us put more and more baseband processing into chips, thus reducing component count, increasing reliability, diminishing the size of handsets, and lowing power requirements. Intense processing functions, such as voice coding, microprocessor control (protocol stack, user interface), signal processing (modem), RF/IF stages, and mixed signal functions used to be integrated on separate components. This was due to the lack of highly integrated components (RISC microprocessor and DSP). CMOS silicon technology (above 0.6 μm) did not allow integration of multiple entities onto one or two chips. Today's technology and transistor density, which is typically 0.35- and 0.25-μm technology, allows comprehensive integration of GSM processing functions that incorporate all of the digital baseband entities:

▶ RISC microprocessor;

▶ *Digital signal processor* (DSP);

▶ Memory (for DSP and microprocessor cache), except flash memory that holds microprocessor code and data, as well as related microprocessor RAM;

▶ Supporting logic (timer, registers, accelerators, encryption engines, etc.).

When silicon process technology allows the integration of mixed signal blocks with sufficient suppression of substrate noise and other prerequisites, then all the required baseband functions can be integrated onto one piece of silicon. This is the trend for GSM baseband chips, ASICs, ASSPs, and chip sets. Integrating mixed signal logic with the digital logic does away with the need to drive serialized data signals between components over printed circuit boards, thus saving power.

Lots of progress has been made in the RF sections even though the evolution of the technology has not come as fast as it has in digital CMOS technology. Initially, the RF circuits were built from standard discrete components. With the market drive behind digital cellular technology, RF and IF *integrated circuits* (ICs), bipolar silicon technology, BiCMOS, and *gallium arsenide* (GaAS) *monolithic microwave integrated circuits* (MMICs) have appeared that allow small size and low-power RF front-ends and IF implementations [52].

Shrinkage of silicon technology and related silicon process enhancements allow a reduction of the operation voltages. Earlier solutions depending on a 5V supply for both baseband and IF/RF circuits can be substituted for newer technologies that run off a nominal 3V supply (±10%). Power requirements for CMOS baseband ICs are moving from 2V down to 1V. The RF power amplifiers still require higher voltages. Trade-offs occur between the desired RF power output (in GSM handheld applications this can be up to 2W), supply and breakdown voltages, and impedance matching [53,54].

Typical figures for average GSM power consumption (off a 3V supply) are in the range of a few milliamps, where 5 mA is a typical example for standby operation (listening to a paging channel and monitoring adjacent cells) and several hundreds of milliamps (150 mA is typical) for the conversation mode. These figures depend on many parameters and only typical values can be given. For example, the periodic appearance of the paging channel for a particular serving cell is variable, and can triple the best case figure for the worst case for battery life between charges.

The organization of paging messages is intermittent with up to a few seconds between phone-specific pages. Phones need to monitor periodically the serving cell and adjacent cell signals, and this affects the *duty cycles* during which a mobile phone needs to activate some sections of its circuitry for receiving and processing the information. Transmission in standby mode is normally not required. There can be intermittent short periods when a mobile phone needs to leave the idle status and communicate (transmit) with the network: location updating/registration with the network when roaming. A GSM terminal may have a (baseband) duty cycle for standby activity (reading the paging channel and full monitoring) of as little as 0.5%!

Designs can exploit *sleep modes*. The processor, baseband functions, and the radio blocks can be turned off during the idle periods. Even the whole reference oscillator section (13 MHz) can be switched off during such periods; in such case it can be substituted by a low-frequency, low-power oscillator operating at 32 kHz.

Conversation or talk time largely depends on the output RF power control levels, and the use of voice activity detection and discontinuous transmission. The RF power amplifier in conversation mode uses up most of the energy, whereas in standby mode the baseband section has at least an equal impact on the available energy compared to the whole radio. This means that a battery with a typical 500 mA-hour (mAh) capacity will support about 100 hours of standby and several hours of talk time.

11.12 Guide to the literature

The following paragraphs organize some of the references in this chapter so that they can be pursued in an order anyone new to digital radio will find helpful. They are listed in the order in which they should be read. Some of the books are old and may be hard to find, but the effort will be rewarded. The references listed here should not be considered the result of an exhaustive search of everything available; they are merely what the authors have found particularly helpful in their own work.

11.12.1 General radio design

Bowick's work [8] is a small workbook that is an excellent introduction to RF circuits. Vizmuller [9] provides a more modern treatment of the same subject, and his book is an excellent introduction to the design process. It deals with the passive components and circuit details very well, and is helpful in sorting through all the compromises in RF design. Larson [10] gives many excellent examples of digital techniques, and offers clear and thorough explanations of modulation and demodulation.

11.12.2 Coding and its mathematics

Pierce [42] provides an introduction to information theory, and Tarasov [41] introduces probability; both are aimed at the general reader. Wiggert

[43] followed immediately by Peterson [35] take the mystery out of error codes. The Berlekamp publication [34] is a collection of important papers, some rather old, on coding theory written by key figures in the field.

11.12.3 Digital radio

Aside from such classics as Taub and Schilling's *Principles of Communications Systems* (McGraw-Hill), and Proakis [33], which should be in every engineer's private library; the following are worth looking for. Start with [39], which, though Bellamy deals mostly with wireline systems, is a simple and clear introduction to digital communications. Follow this with Rappaport's book [28], which is a relatively short but complete survey of wireless practice. Feher [32] concentrates on digital radio, and his book has beautifully clear drawings and illustrations. Finish off with Steele [26], who provides a collection of separate works from several authors that deal exclusively with digital cellular radio. There is a separate chapter at the end that uses GSM as a review of everything in the book.

Once the reader is familiar with digital radio, four more books will fill in the empty spots in GSM. Start with [36] as an excellent survey of cellular radio. Follow this with the Redl et al. book [2], which is, as its title says, an easy introduction to GSM. Finish up with a new book by Mehrota [27], which is an amplification and elaboration of the material in [2], but without the material on testing.

References

[1]   GSM TS 06.10 (ETS 300 580-x), "Digital Cellular Telecommunications System, GSM Full Rate Speech Transcoding," ETSI, Sophia Antipolis.

[2]   Redl, S. M., M. K. Weber, and M. W. Oliphant, *An Introduction to GSM*, Norwood, MA: Artech House, 1995, Chap. 5.

[3]   GSM TS 05.03 (ETS 300 575), "Digital Cellular Telecommunications System, GSM Channel Coding," ETSI, Sophia Antipolis.

[4]   GSM TS 06.20 and related documents (ETS 300 581/1–8), "Digital Cellular Telecommunications System, Half Rate Speech Transcoding," ETSI, Sophia Antipolis.

[5]   GSM TS 06.51 and related documents (ETS 300 723–730), "Digital Cellular Telecommunications System, Enhanced Full Rate Speech Transcoding," ETSI, Sophia Antipolis.

[6] GSM TS 05.04 (ETS 300 576), "Digital Cellular Telecommunications System, Modulation," ETSI, Sophia Antipolis.

[7] Redl, S. M., M. K. Weber, and M. W. Oliphant, *An Introduction to GSM*, Norwood, MA: Artech House, 1995, Chap. 10.

[8] Bowick, C., *RF Circuit Design*, Indianapolis, IN: Howard W. Sams & Company, 1982, Chap. 7.

[9] Vizmuller, P., *RF Design Guide Systems, Circuits, and Equations*, Norwood, MA: Artech House, 1995, Chap. 1.

[10] Larson, L. E. (ed.), *RF and Microwave Circuit Design for Wireless Communications*, Norwood, MA: Artech House, 1996, Chap. 4.

[11] Redl, S. M., M. K. Weber, and M. W. Oliphant, *An Introduction to GSM*, Norwood, MA: Artech House, 1995, Chap. 10.

[12] McMahan, Khatzibadeh, and Shah, *Wireless Systems and Technology Overview*, Dallas, TX: Texas Instruments, (www.ti.com), 1996.

[13] Maoz, B., and A. Adar, "GaAs IC Receivers for Wireless Communications," *Microwave Journal*, Vol. 39, No. 1, January 1996.

[14] Winter, E., "Speech Coding," *Communications Systems Design Magazine*, May 1996.

[15] Xydeas, C. "Speech Coding," in *Personal and Mobile Radio Systems*, R. C. V. Macario (ed.), IEE Telecommunications Series 25, London: Peter Peregrinus, 1991.

[16] http://www.speech.su.oz.au/comp.speech/ *and* http://www-mobile.ecs.soton.ac.uk/speech_codecs *and* http://wwwdsp.ucd.ie/speech/tutorial/speech_coding/.

[17] Steele, R., "Speech Codecs for Personal Communications," *IEEE Communications Magazine*, November 1993.

[18] Rappaport, T. S., *Wireless Communications Principles & Practice*, Upper Saddle River, NJ: Prentice Hall PTR, 1996, Chap. 7.

[19] Qualcomm Data Sheet for the Q5414 Variable Rate Vocoder, San Diego, CA: Qualcomm ASIC Products.

[20] Steele, R., et al., *Mobile Radio Communications*, London: Pentech Press, 1992, Chap. 3.

[21] GSM TS 11.10 (ETS 300 607), "Digital Cellular Telecommunications System, GSM Mobile Station Conformity (Test) Specifications," ETSI, Sophia Antipolis.

[22] GSM TS 03.50, (ETS 300 540), "Digital Cellular Telecommunications System, Transmission Planning Aspects of the Speech Service in a GSM PLMN," ETSI, Sophia Antipolis.

[23] GSM TS 06.10 and related documents (ETS 300 580 x), "Digital Cellular Telecommunications System, Full Rate Speech Transcoding," ETSI, Sophia Antipolis.

[24] Spanias, A., "Speech Coding: A Tutorial Overview," *IEEE Proceedings*, Vol. 82, No. 10, October 1994.

[25] Vary, P., "GSM Speech Codec," *Digital Cellular Radio Conf. Proc.*, Hagen, Germany: Deutsche Bundespost, France Télécom, Fern Universitaet, 1988.

[26] Steele, R., et al., *Mobile Radio Communications*, London: Pentech Press, 1992, Chap. 4.

[27] Mehrotra, A., *GSM System Engineering*, Norwood, MA: Artech House, 1997, Chap. 6.

[28] Rappaport, T. S., *Wireless Communications Principles & Practice*, Upper Saddle River, NJ: Prentice Hall PTR, 1996, Chap. 6.

[29] Steele, R., et al., *Mobile Radio Communications*, London: Pentech Press, 1992, Chap. 8.

[30] Yacoub, M. D., *Foundations of Mobile Radio Engineering*, Boca Raton, FL: CRC Press, 1993, App. 6A.

[31] Gibson, J. D. (ed.), *The Mobile Communications Handbook*, Boca Raton, FL: CRC Press, 1996, Chap. 5.

[32] Feher, K., *Wireless Digital Communications Modulation & Spread Spectrum Applications*, Upper Saddle River, NJ: Prentice Hall PTR, 1995, Chap. 5.

[33] Proakis, J. G., *Digital Communications*, 2nd ed., New York: McGraw-Hill, 1989, Chap. 2.

[34] Berlekamp, E. R. (ed.), *Key Papers in the Development of Coding Theory*, New York: IEEE Press, 1974.

[35] Peterson, W. W., and E. J. Weldon, Jr., *Error Correcting Codes*, 2nd ed., Cambridge, MA: The MIT Press, 1990.

[36] Balston, D. M., et al., *Cellular Radio Systems*, Norwood, MA: Artech House, 1993, Chap. 7.

[37] Viterbi, A. J., *CDMA Principles of Spread Spectrum Communication*, Reading, MA: Addison-Wesley, 1995, Chap. 4.

[38] Redl, S. M., M. K.Weber, and M. W. Oliphant, *An Introduction to GSM*, Norwood, MA: Artech House, 1995, Chap. 13.

[39] Bellamy, J., *Digital Telephony*, 2nd ed., New York: John Wiley & Sons, 1991, Chap. 1.

[40] Ericsson home page, *Echo in Telephony Systems*,

http://www.ericsson.com/echo/edu/we.htm.

[41] Tarasov, L., *The World Is Built on Probability*, Moscow: Mir Publishers, 1988, Chap. 3.

[42] Pierce, J. R., *An Introduction to Information Theory*, New York: Dover Publications, 1961, 1980, Chap. 8.

[43] Wiggert, D., *Codes for Error Control and Synchronization*, Norwood, MA: Artech House, 1988, Chap. 6.

[44] Redl, S. M., M. K. Weber, and M.W. Oliphant, *An Introduction to GSM*, Norwood, MA: Artech House, 1995, Chap. 3.

[45] GSM TS 02.09 (ETS 300 506), "Digital Cellular Telecommunications System, Security Aspects," ETSI, Sophia Antipolis.

[46] GSM TS 02.17 (ETS 300 509), "Digital Cellular Telecommunications System, Subscriber Identity Modules," ETSI, Sophia Antipolis.

[47] GSM TS 03.20 (ETS 300 534), "Digital Cellular Telecommunications System, Security Related Network Functions," ETSI, Sophia Antipolis.

[48] Vedder, K., "GSM Security, Services and the SIM—Developments in Technology in the Fight Against Fraud," *GSM '96 Congress*, Cannes, France, Feb. 20–22, 1996. London: IBC Technical Services.

[49] Brookson, C., "Has GSM's Security Been Compromised?" GSM '96 Congress, Cannes, France, Feb. 20–22, 1996. London: IBC Technical Services.

[50] Schneier, B., *Applied Cryptography*, New York: John Wiley & Sons, 1994.

[51] GSM TS 05.02 (ETS 300 574), "Digital Cellular Telecommunications System, Multiplexing and Multiple Access on the Radio Path," ETSI, Sophia Antipolis.

[52] Abidi, A. A., "Low Power Radio Frequency ICs for Portable Communications," Chap. 3 in *RF and Microwave Circuit Design for Wireless Communications*, L. E. Larson (ed.), Norwood, MA: Artech House, 1996.

[53] Baringer, C., and C. Hull, "Amplifiers for Wireless Communications," Chap. 79in *RF and Microwave Circuit Design for Wireless Communications*, L. E. Larson (ed.), Norwood, MA: Artech House, 1996.

[54] Vizmuller, P., *RF Design Guide*, Norwood, MA: Artech House, 1995.

Appendix

Coding of the default GSM alphabet

Each character is represented by a pattern of 7 bits (b1 ... b7). These characters are packed into the user field of a short message according to Sections 6.1.2 and 6.1.3.

					b7 0	b6 0	b5 0					

b4	b3	b2	b1		0	1	2	3	4	5	6	7
			b7		0	0	0	0	1	1	1	1
			b6		0	0	1	1	0	0	1	1
			b5		0	1	0	1	0	1	0	1
0	0	0	0	0	SP	0	¡	P	¿	p		
0	0	0	1	1	£	1)	!	1	A	Q	a	q
0	0	1	0	2	$	F	"	2	B	R	b	r
0	0	1	1	3	¥	G	#	3	C	S	c	s
0	1	0	0	4	è	L	¤	4	D	T	d	t
0	1	0	1	5	é	W	%	5	E	U	e	u
0	1	1	0	6	ù	P	&	6	F	V	f	v
0	1	1	1	7	ì	Y	'	7	G	W	g	w
1	0	0	0	8	ò	S	(	8	H	X	h	x
1	0	0	1	9	Ç	Q	)	9	I	Y	i	y
1	0	1	0	10	LF	X	*	:	J	Z	j	z
1	0	1	1	11	Ø	1)	+	;	K	Ä	k	ä
1	1	0	0	12	ø	Æ	,	cell L	Ö	l	ö	
1	1	0	1	13	CR	æ	-	=	M	Ñ	m	ñ
1	1	1	0	14	Å	ß	.		N	Ü	n	ü
1	1	1	1	15	å	É	/	?	O	§	o	

Glossary

μs   Microsecond, 1/1,000,000 of a second

ΣΔ   Sigma-delta

3GIG   Third-generation Interest Group, a technology group within the GSM MoU looking after third-generation systems

A/D   Analog-to-digital conversion

AAeM   Automatic answer of eMLPP service, supplementary service

ABM   Asynchronous balanced mode, state of the radio link protocol

AbS   Analysis by synthesis

ACC   Access control class, elementary file on the SIM

ACELP   Algebraic code(-book) excited linear predictive

ACM   Accumulated call meter, elementary file on the SIM

ACMmax   ACM maximum value, elementary file on the SIM

ACTE   Advisory Committee for Terminal Equipment (type approval committee of EEC)

ACTS   Advanced Communications Technologies and Services

AD   Administrative data, elementary file on the SIM

ADC   Analog-to-digital converter

ADM   Asynchronous disconnected mode, state of the radio link protocol

ADN   Abbreviated dialing number, elementary file on the SIM containing the phone book

ADPCM   Adaptive differential pulse code modulation

AF   Audiofrequency

AFC   Automatic frequency control

AGC   Automatic gain control

AGCH   Access grant channel

AIG   Arab Interest Group, an interest group within the GSM MoU

AIReach   Hughes Network Systems' version of PACS

ALS   Alternate line service, PCN-defined service that allocates two phone numbers (lines) to one subscriber

AM   Amplitude modulation

AMPS   Advanced mobile telephone system

AMR   Adaptive multirate (speech codec)

ANSI   American National Standards Institute

AoC   Advice of charge, supplementary service (SS)

AoCC   Advice of charge charging, subgroup of the AoC SS

AoCI   Advice of charge information, subgroup of the AoC SS

APCO   Association of Public-Safety Communications Officials International

APIG   Asia Pacific Interest Group, an interest group within the GSM MoU

ARCH   Access response channel (IS-136)

ARQ   Automatic repeat request

ASIC   Application-specific integrated circuit

ASSP   Application-specific standard poduct

BABT   British Approvals Board for Telecommunications (UK)

BAIC   Barring of all incoming calls, subgroup of call barring SS

BAOIC   Barring of outgoing international calls, subgroup of call barring SS

BAOIC-excHC   Barring of outgoing international calls except those directed to the home PLMN country, subgroup of call barring SS

BARG   Billing and Accounting Rapporteur Group, a working group within the GSM MoU looking after billing and accounting issues

BB   Baseband

BCCH   Broadcast control channel, channel on which the network supplies the mobile station with information on parameters currently used in the network

BCF   Base control function

BDN   Barred dialing numbers, elementary file on the SIM card that is organized similar to ADN and contains numbers that are barred

BER   Bit error ratio

BiCMOS   Bipolar complementary metal-oxide semiconductor

BICroam   Barring of incoming calls when roaming outside the home PLMN, subgroup of call barring SS

BOAC   Barring of outgoing calls, subgroup of call barring SS

BOC   Bell Operating Company

BOIC   Barring of outgoing international calls, subgroup of call barring SS

BS   Base station

BSC   Base station controller, control entity for the base station

BSS   Base station subsystem, made up of BSC and BTS

BT   Bandwidth (B) multiplied by bit period (T)

BTA   Basic Trading Area

BTS   Base transceiver station

BZT   Bundesamt für Zulassung in der Telekommunikation, German approval body for telecommunications

C/I   Carrier-to-interference ratio

CAI   (1) Charge advice information, set of parameters used by the mobile station to calculate the charging information; (2) common air interface

CAMEL   Customized applications for mobile network enhanced logic, intelligent network and tool for network operators to make their features available to subscribers roaming abroad

CB   Call barring

CB   Cell broadcast, allows transmission of short information from the network to the mobile station within a geographical area

CBC   Cell broadcast center, schedules the transmission of cell broadcast messages within a network

CBCH   Cell broadcast channel, physical channel used by cell broadcast service

CB-DRX   Cell broadcast discontinuous reception, new Phase 2+ feature allowing the scheduling of the reception of new information coming from the network

CBE   Cell broadcast entity, originates cell broadcast messages

CBMI   Cell broadcast message identifier selection, elementary file on SIM used in conjunction with cell broadcast

CBMID   Cell broadcast message identifier for data download, elementary file on SIM used in conjunction with SIM application toolkit

CBMIR   Cell broadcast message identifier range selection, elementary file on SIM used along with cell broadcast

CBS Cell broadcast short message service, allows the transmission of short information units from the network to the mobile station within a geographical area

CC Convolutional code

CCBS Completion of call to busy subscriber, call completion supplementary service

CCCH Common control channel

CCIR International Radio Consultive Committee

CCITT International Telegraph and Telephone Consultative Committee

CCM Current call meter, elementary file on the SIM

CCNRc Completion to call when subscriber not reachable, call completion supplementary service

CCNRy Completion to call when no reply, call completion supplementary service

CCP Capability configuration parameters, elementary file on the SIM

CCS Call century seconds

CDCS Continuous dynamic channel selection

CDG CDMA Development Group

CDL Coded DCCH locator (IS-136)

CDMA Code division multiple access

CDVCC Control digital verification color code (IS-136)

CELP Code(-book) excited linear predictive (speech codec)

CEPT European Conference of Posts and Telecommunications Administrations (Conférence Européenne des Administration des Postes et des Télécommunication)

CF Call forwarding, call offering supplementary service

CFB   Call forwarding on mobile busy subscriber, call offering supplementary service

CFNRc   Call forwarding on not reachable, call offering supplementary service

CFNRy   Call forwarding on no reply, call offering supplementary service

CFU   Call forwarding unconditional, call offering supplementary service

CH   Call hold

CHV1   Card holder verification 1, parameter on SIM usually referred to as the PIN1

CHV2   Card holder verification 2, parameter on SIM usually referred to as PIN2

CIS   Commonwealth of Independent States

CLIP   Calling line identification presentation, line identification supplementary service

CLIR   Calling line identification restriction, line identification supplementary service

CMOS   Complementary metal-oxide semiconductor

CNL   Cooperative network list, elementary file on the SIM

CODEC   Coder-decoder

COLP   Connected line identification presentation, line identification supplementary service

COLR   Connected line identification restriction, line identification supplementary service

CPU   Central processing unit

CRC   Cyclic redundancy code/check

CS   Coding scheme, for example, used for different channel coding for GPRS to achieve different (higher) data rates

CSAIG   Central/Southern Africa Interest Group, an interest group within the GSM MoU

CSE   CAMEL service environment, logical entity supporting CAMEL

CSFP   Coded superframe phase (IS-136)

CSG   Communication strategy group, a working group within the GSM MoU

CSI   CAMEL subscription information, subscriber-related information stored in the network

CSIC   Customer-specific integrated circuit

CSPDN   Circuit-switched public data network

CT   (1) Call transfer, call completion supplementary service; (2) cordless telephone

CTIA   Cellular Telecommunications Industry Association (USA)

CTM   Cordless telephone mobility

CUG   Closed user group, supplementary service

CW   Call waiting, call completion supplementary service

D/A   Digital-to-analog conversion

DA   Destination address, within a short message, usually a phone number of the addressee

DAC   Digital-to-analog converter

dBm   Decibels relative to 1 mW

DCA   Dynamic channel allocation

DCCH   Digital control channel (IS-136)

DCK   Depersonalization control keys, elementary file on SIM

DCS   Data coding scheme, of the short message content; (2) digital cellular system

DD   Differential decoder (PACS)

DECT Digital Enhanced Cordless Telecommunications

DF Dedicated file, directory on the SIM

DFE Decision feedback equalizer

DGPT Directorate General of Post and Telecommunication (France)

DISC Disconnect, command for the radio link protocol

DLC Data link control (Layer)

DM Disconnect mode, a state of the radio link protocol

DQPSK Differential quadrature phase shift keying

DRX Discontinuous receive

DSA Direct subscriber access, supplementary service

DSAR Direct subscriber access restriction, supplementary service

DS-CDMA Direct sequence code division multiple access

DSP Digital signal processor

DSSS Direct sequence spread spectrum

DTC Digital traffic channel (IS-136)

DTE Data terminal equipment, equipment that is normally connected to a MS for the exchange of data

DTI Department of Trade and Industry, United Kingdom

DTMF Dual-tone multifrequency

DTX Discontinuous transmission

EBCCH Extended broadcast control channel (IS-136)

EBRC EMC Bio-Effects Review Committee Group, a working group within GSM MoU

ECAIG East Central Asia Interest Group, an interest group within the GSM MoU

ECC Emergency call codes, elementary file on SIM

ECT Explicit call transfer, supplementary service

EEC European Economic Community

EEPROM Electrical erasable and programmable read-only memory

EF Elementary file, data record on SIM

EFR Enhanced full-rate speech codec, improved speech codec used in GSM

EIA Electronics Industry Association (USA)

EIG European Interest Group, an interest group within the GSM MoU

EMC Electromagnetic compatibility

eMLPP Enhanced multilevel precedence and preemption service, supplementary service

EMR Electromagnetic radiation

EP Error profile

ESN Electronic serial number

ETS European Telecommunications Standard

ETSI European Telecommunications Standards Institute

EVRC Enhanced variable rate (speech) codec

FBCCH Fast broadcast control channel (IS-136)

FC Fast channel (PACS)

FCC Federal Communications Commission (USA)

FCCH Frequency correction channel

FCS Frame check sequence, portion of a frame used by the radio link protocol to detect errors

FDD Frequency division duplex

FDMA Frequency division multiple access

FDN   Fixed dialing numbers, elementary file on the SIM card that is organized similar to ADN and contains numbers to which the phone's use can be restricted

FEC   Forward error correction

FF   Fraud Forum, a working group within the GSM MoU

FM   Frequency modulation

FOCC   Forward control channel (IS-136/AMPS)

FP   Fixed part (base station)

FPLMN   Forbidden PLMN, filed on the SIM

FPLMTS   Future Public Land Mobile Telecommunications System, a proposal for third-generation cellular system, see IMT-2000

FR   Full-rate speech codec, standard speech codec used in GSM

FTA   Full type approval

FVC   Forward voice channel (IS-136/AMPS)

GaAs   Gallium arsenide

GAP   Generic access profile

GCR   Group call register, register in the network which is used with voice group call service

GFSK   Gaussian-filtered frequency shift keying

GGSN   Gateway GPRS support node, gateway entity for GPRS support

GID1   Group identifier level 1, elementary file on the SIM used for personalisation

GID2   Group identifier level 2, elementary file on the SIM used for personalization

GMSC   Gateway mobile switching center

GMSK   Gaussian-filtered minimum shift keying

GPRS   General packet radio service, packet data service defined for GSM

GPS Global positioning system, a satellite system used for accurate positioning information

GSM Global system for mobile communications

gsmSCF GSM service control function, function in the network related to CAMEL

gsmSSF GSM service switching function, function in the network related to CAMEL

HLR Home location register, network register for home subscribers, always associated with a MSC

HOLD Call hold, supplementary service

HPLMN HPLMN search period, elementary file on SIM

HR Half-rate speech codec, optional speech codec used in GSM, allowing for increased system capacity

HSCSD High-speed circuit-switched data, data service defined for GSM allowing the combining of up to 8 time slots for transmission or reception

IC Integrated circuit

ICCID ICC identification, elementary file on SIM

ICM In-call modification

IF Intermediate frequency

IIG Indian Interest Group, an interest group within the GSM MoU

IM Intermodulation

IMSI International mobile subscriber identity, elementary file on SIM holding the unique number identifying the subscriber within a GSM network

IMT-2000 International Mobile Telecommunications 2000 System, replacement for the FPLMTS designation, see FPLMTS

IN Intelligent networks, CAMEL being an example for IN as used in GSM

I/O   Input/output

IP   Internet protocol

IP3   Intercept point of the third order

IP-M   IP multicast, transmission of packet data (GPRS) from one subscriber to multiple subscribers via the Internet protocol

IPR   Intellectual property right

IREG   International Roaming Expert Group, a working group within the GSM MoU looking after international roaming issues

IS   Interim standard (US)

ISDN   Integrated services digital network

ISI   Intersymbol interference

ITA   Interim type approval

ITU   International Telecommunication Union

IWF   Interworking function, interface between cellular network and outside PSTN or data network

JTC   Joint Technical Committee (TIA body)

Kbps   Kilobits per second (1000 bits per second)

KB   Kilobyte

K_c Ciphering key

K_i Internal key

Ksps   Kilosamples per second (1,000 samples per second)

L&RG   Legal and Regulatory Group, a working group within the GSM MoU

LAI   Location area information

LAN   Local-area network

LAPM   Link access procedure for modems, protocol for data exchange between modems

LEC   Local exchange center, local switch in a PSTN or ISDN

LEO   Low-earth orbit

LFSR   Linear feedback shift register

LNA   Low-noise amplifier

LND   Last numbers dialed, elementary file on the SIM

LO   Local oscillator

LOCI   Location information, elementary file on the SIM

LP   Language preference, elementary file on the SIM

LP   Linear prediction/predictive (speech coding)

LPC   Linear predictive (speech) coding

mA   Milliampere

MAC   Medium access control (layer)

MAH   Mobile access hunting, supplementary service

mAh   Milliampere-hour

MCC   Mobile country code

MCID   Malicious call identification, supplementary service

ME   Mobile equipment (GSM terminal without SIM)

MEO   Medium-earth orbit

MF   Master file, main directory on SIM

mips   Millions of instructions per second

MLPP   Multilevel precedence and preemption service, supplementary service in ISDN

MLSE   Maximum likelihood sequence estimation

MMI   Man/machine interface, also called user interface

MMIC   Monolithic microwave integrated circuits

MMS More-messages-to-send, status flag within the short message to indicate that more messages are scheduled to be sent

MNC Mobile network code

MNP Microcom networking protocol, data protocol to secure data transmission and compression of data between two modems

MO Mobile originated

MOS (1) Metal-oxide semiconductor; (2) mean opinion score

MoU Memorandum of Understanding, within a group of GSM operators

MPT Ministry of Posts and Telecommunications (Japan)

MPTY Multiparty communication, supplementary services

MR Message reference, mobile station internal referencing of messages to which the network also makes reference

ms Millisecond, 1/1,000 of a second

MS Mobile station (in GSM, terminal with SIM)

MSC Mobile services switching center, switch within the GSM network

MSIN Mobile subscriber identification number, portion of the IMSI within a dedicated network

MSISDN Mobile station international ISDN number, international phone number of a GSM subscriber

MSK Minimum shift keying

MSP Multiple subscriber profile, supplementary service

MT Mobile terminal/termination

MTA Major Trading Area

MTI Message type indicator, indicating for short messages the purpose of the sent message

MTSO Mobile telephone switching office

MUX Multiplexer

NAIG North American Interest Group, an interest group within the GSM MoU

NB Narrowband

NCH Notification channel, logical channel used for voice group call service

NDUB Network-determined user busy

NITZ Network identity and time zone, service defined for GSM networks to transmit identity and time zone regularly through the network

NMT Nordic mobile telephone

ns Nanosecond, 1/1,000,000,000 of a second

NT Nontransparent services, data service that uses the radio link protocol between the MS and the MSC

NTA National Telecom Agency (Denmark)

NTP Nominal transmitted power

NTT Nippon Telephone and Telegraph Company

NULL Null information, frame used for the radio link protocol

OA Originating address, details, for example, the phone number of the sender

ODB Operator-determined barring

OQPSK Offset quadrature phase shift keying

OSI Open systems interconnection

OSS Operator-specific services

OVHM Overhead message (IS-136)

P25 Project 25 (formerly APCO Project 25)

PA Power amplifier

PACS Personal Access Communications System

PAD   Packet assembler/disassembler, access to a packet data network coming from a circuit-switched network, such as GSM

PAMR   Public access mobile radio

PBX   Private branch exchange

PCC   Power control channel (PACS)

PCH   Paging channel (IS-136)

PCM   Pulse code modulation

PCN   Personal communications network, also referred to as DCS 1800

PCNIG   Personal Communications Networks Interest Group, a technology group within the GSM MoU

PCP   Power control pulse (IS-661)

PCS   Personal communications system, also referred to as PCS 1900 or GSM NA

PDA   Personal digital assistant (palmtop organizer/computer)

PDC   Personal digital cellular

PDN   Packet data network

PDS   Packet data on signaling channels, implementation of packet data services for GSM

PDSS2-SN   PDSS2-support node, support node for second service of PDS

PFC   Paging frame class (IS-136)

PH   Packet handler, access to a packet data network coming from a circuit-switched network such as GSM

PHP   Personal handy phone

PHS   Personal Handy Phone System

PID   Protocol identifier, identifies to which protocol a short message should be transferred

PIN   Personal identification number

PLL   Phase-locked loop

PLMN   Public land mobile network, cellular network

PLMNsel   PLMN selector, elementary file on SIM

PMR   (1) Private mobile radio; (2) professional mobile radio

POTS   Plain old telephony service

PP   Portable part (mobile station)

ppm   Parts per million

PRBS   Pseudorandom bit sequence

pr-ETS   Preliminary European telecommunications standard

PRS   Premium rate services

PSID   Private system identifier (IS-136)

PSPDN   Packet-switched public data network

PSTN   Public-switched telephone network, standard wireline network

PTM   Point-to-multipoint, transmission between one single point subscriber or base station to several subscribers; this applies for packet data services or short message service

PTM-G   PTM group call, point-to-multipoint service for general packet radio service allowing the transmission of data to a predefined group

PTM-M   PTM multicast, point-to-multipoint service for general packet radio service allowing the transmission of data to multiple subscribers within an area

PTP   Point-to-point, transmission between two dedicated subscribers; this can apply to packet data services or short message service

PTT   Push-to-talk

PUCT   Price per unit and currency table, elementary file on SIM

PUK   Personal unblocking key (for GSM SIM)

PWT   Private wireless telecommunications

QCELP   Qualcomm codebook excited linear predictive

QCPM   Quadrature continuous-phase modulation

QPSK   Quadrature phase shift keying

RA   Rate adaptation, data transmission speed adaptation between different interfaces

RACE   Research and Development of Advanced Communications Technologies and Services (in Europe)

RACH   Random access channel

RAM   Random access memory

RAND   Random value, used for authentication and calculation of the ciphering algorithm

RDI   Restricted digital information, direct digital data exchange between a GSM and a digital network

RDS   Radio data system, enables the display of characters on the radio's display

RECC   Reverse control channel (IS-136/AMPS)

REJ   Reject, layer 2 frame/information

RF   Radio-frequency

RISC   Reduced instruction set computer

RLL   Radio local loop

RLP   Radio link protocol, link protocol to reduce the error rate via the air interface

RLR   Receive loudness ratio

RNR   Receive not ready, layer 2 frame/information

ROM   Read-only memory

RP   (1) Reply path, feature within SMS that allows the sending of replies to messages free of charge for the sender; charge will be picked up by the recipient; (2) radio port (PACS)

RPCU   Radio port controller unit (PACS)

RPE-LTP   Regular pulse-excited–long-term prediction

RR   Receive ready, layer 2 frame/information

RSID   Residential system identifier (IS-136)

RTOS   Real-time operating system

RVC   Reverse voice channel (IS-136/AMPS)

RX   Receiver

SABM   Set asynchronous balanced mode, a state of the layer 2 or the radio link protocol

SACCH   Slow associated control channel, a logical channel type containing regular control messages between the network and the mobile station

SAPI   Service access point identifier, identifier for different logical access points

SAW   Surface acoustic wave (ceramic-based filter)

SC   (1) Service center for short message services; (2) service code, used for supplementary services; (3) slow channel (PACS)

SCAG   Smart card application group, a group within the GSM MoU

SCF   Shared channel feedback (IS-136)

SCH   Synchronization channel

SCT   Single-step call transfer

SCTP   Service center time stamp, time stamp for a short message that arrives at the service center

SDCCH   Stand-alone dedicated control channel, signaling channel

SDN   Surface dial numbers

SEQAM   Spectrally efficient quadrature amplitude modulation (IS-661)

SERG   Services Expert Rapporteur Group, GSM MOU subgroup

SG   Security Group, a working group within the GSM MoU looking after security issues

SGSN   Serving GPRS support node, entity needed to support general packet radio service

SI   Supplementary information, used for supplementary services

SIM   Subscriber identity module

SLR   Sending loudness ratio

SME   Short message entity, entity originating a short message, for example, a mobile station

SMG   Special mobile group

SM-MT   Short message–mobile terminated

SMR   Specialized mobile radio

SMS   Short message service, service that allows the transmission of messages up to 160 characters

SMSCH   SMS message channel (IS-136)

SMS-GMSC   Short message service–gateway mobile services switching center, logical entity connected to the MSC that is used to coordinate the delivery of SMS

SMS-IWMSC   Short message service–interworking mobil services switching center, local entity connected to the MSC which is used to coordinate originating SMS

SMSP   Short message service parameter, elementary file on SIM

SMSS   SMS status, feature of SMS

SNR   Signal-to-noise ratio

SOR   Support of optimal routing

SPACH   SMS/PCH/ARCH channel (IS-136)

SPN   Service provider name, elementary file on SIM

SPNP   Support of private numbering plan, supplementary service

sqkm   Square kilometer, area of 1 km × 1 km

SREJ   Selective reject, layer 2 frame/information

SRES   Authentication result

SRI   Status report indication, indication that a status report will be issued for this message

SRR   Status report request, indication that a status report is requested for this message

SS   Supplementary services

SSB   Single sideband

SS No. 7   Signaling System Number 7

SST   SIM service table, elementary file on SIM describing the services supported by a SIM

SU   Subscriber unit (PACS)

T   Transparent, data service that does not use the radio link protocol between the MS and the MSC

T1P1   Network Interfaces Committee of ANSI (USA)

TA   Type approval; terminal adapter

TADIG   Transferred Account Data Interchange Group, a working group within GSM MoU

TAF   Terminal adaptation function, dedicated function between the mobile station and the data terminal for data/fax transmission

TBR   Technical basis for regulation

TDD   Time division duplex

TDMA   Time division multiple access

TE   Terminal equipment

TETRA   Terrestrial trunked radio (formerly trans-European trunked radio)

TFO   Tandem-free operation

TIA   Telecommunications Industry Association (USA)

TMSI   Temporary mobile subscriber identity

TRAU   Transcoder rate adaptation unit

TWG   Terminal Working Group, a working group within the GSM MoU

TX   Transmitter

UA   Unnumbered acknowledge, layer 2 frame/information

UCI   Universal computer interface

UCS2   Universal Coding Scheme 2, 16-bit based coding scheme defined for SMS to support, for example, Arabic and Chinese

UD   User data, actual message that is to be transmitted with the SMS

UDI   Unrestricted digital information, direct digital data exchange between a GSM and ISDN network

UDL   User data length, the length of the user data to be transmitted with the SMS

UDUB   User-determined user busy

UI   (1) Unnumbered information, layer 2 frame/information; (2) user interface

UIC   Union Internationale de Chemin de Fer, an organization of European Railways

UMTS   Universal Mobile Telecommunication System, ETSI standardization proposal for third-generation cellular system

UPT   Universal personal telecommunications

USSD   Unstructured supplementary services data, supplementary service

UUI   User-to-user information, data transmitted by user to user signaling

UUS   User-to-user signaling, supplementary service

UWCC   Universal Wireless Communications Consortium (IS-136/IS-41)

VA   Viterbi algorithm

VAD   Voice activity detection

VBS   Voice broadcast service, supplementary service and elementary file on SIM

VBSS   Voice broadcast service status, elementary file on SIM

VCO   Voltage-controlled oscillator

VGA   Variable gain amplifier

VGCS   Voice group call service, supplementary service and elementary file on SIM

VGCSS   Voice group call service status, elementary file on SIM

VGS   Voice group service

VLR   Visitor location register, network register for visiting subscribers, always associated with a MSC

VP   Validity period of a sent short message

VPF   Validity period format, the format of the validity period, either absolute or relative

VPN   Virtual private network

VSELP   Vector sum excited linear predictive (IS-136)

WACS   Wireless Access Communications System

WAFU   Wireless access fixed unit (PACS)

WB   Wideband

WCDMA   Wideband CDMA

WIN   Wireless intelligent network (usually refers to IS-41/IS-136)

WLAN   Wireless local-area network

WLL   Wireless in the local loop

WTR   Wireless Technology Research

XID   Exchange identification, layer 2 frame used by the RLP

About the authors

Siegmund M. Redl graduated from the Technical University of Munich in 1989 with a degree in communications engineering. He joined Schlumberger Technologies, Communications Test Division (known as Wavetek since 1994), Munich, in early 1990 as a support engineer for GSM Test Systems. He continued to hold positions within Schlumberger in marketing and product management for digital wireless communication test systems as he participated in the ETSI GSM standardization efforts. In 1995 he joined LSI LOGIC, a leading supplier of integrated circuits, where he is director of the Wireless Communications Business Unit.

Matthias K. Weber graduated from the Technical University of Munich in 1989 with a degree in communications engineering. He joined Schlumberger Technologies, Communications Test Division (known as Wavetek since 1994), Munich, in October 1989. After a period in R&D on Schlumberger's GSM test equipment, he turned to providing technical product support for GSM radio test systems before he assumed responsibility for product management of GSM/DCS-1800 test equipment. In 1995 he joined Sony Personal Communication Europe as a technical product manager. He now is a manager in the Product Planning group for GSM-related products for Sony PCE.

Malcolm W. Oliphant graduated from Hawaii Loa College, Kaneohe, Hawaii, with a degree in biology. During the next 20 years he worked in a variety of technical fields, including avionics, metrology, and mobile radio systems engineering. He joined Schlumberger Technologies, Communications Test Division (known as Wavetek since 1994) in early 1990 where he worked in marketing and applications engineering. In late 1996 he joined IFR Systems, Wichita, Kansas, where he manages the marketing, applications engineering, and training departments. He is active in North American GSM and public safety radio standards committees.

Index

F

R

W

Z

The Artech House Mobile Communications Series

John Walker, Series Editor

For further information on these and other Artech House titles, including previously considered out-of-print books now available through our In-Print-Forever™ (IPF™) program, contact:

Artech House
685 Canton Street
Norwood, MA 02062
781-769-9750
Fax: 781-769-6334
Telex: 951-659
e-mail: artech@artech-house.com

Artech House
Portland House, Stag Place
London SW1E 5XA England
+44 (0) 171-973-8077
Fax: +44 (0) 171-630-0166
Telex: 951-659
e-mail: artech-uk@artech.house.com

Find us on the World Wide Web at: www.artech-house.com